COGNITION, COMPUTING, AND COOPERATION

edited by

Scott P. Robertson

Wayne Zachary

John B. Black

ABLEX PUBLISHING CORPORATION
NORWOOD, NEW JERSEY

Printed in the United States of America.

Library of Congress Cataloging-in-Publication Data

Cognition, computing, and cooperation / [edited by] Scott P. Robertson, Wayne Zachary, John B. Black.
p. cm.
Some papers were first presented at symposia during the 150th annual meeting of the American Association for the Advancement of Science in New York, 5/84, sponsored by the AAAS Psychology Section and various other sections of AAAS.
Bibliography: p.
Includes index.
ISBN 0-89381-536-X
1. Interactive computer systems—Congresses. 2. Human-computer interaction—Congresses. I. Robertson, Scott P. II. Zachary, Wayne. III. Black, John B. (John Benjamin), 1947- IV. American Association for the Advancement of Science. Section J—Psychology.
QA76.9.I58C62 1989 89-14895
004'.33—dc20 CIP

Ablex Publishing Corporation
355 Chestnut Street
Norwood, New Jersey 07648

Table of Contents

Preface

The chapters in this volume represent work from a broad sample of disciplines including anthropology, computer science, information science, and psychology. In this sense we hope that the book is an exercise in *cognitive science*, yielding something more interesting than a collection of papers from within any one of these disciplines alone. Because many disciplines are involved, different methodologies are strikingly apparent. Unfortunately, cooperation is not taken for granted in interdisciplinary endeavors, but we have tried to craft the section introductions and concluding remarks so that common features of the papers are highlighted. It is a working hypothesis that important parts of a theory of cooperation will emerge from comparison of these different approaches.

This book grew out of a day-long symposium, under the general title *Cognition, Computing, and Interaction*, presented at the 150th annual meeting of the American Association for the Advancement of Science that took place in New York in May, 1984. Scott Robertson and John Black organized a half-day program on *Human-Computer Systems*, and Wayne Zachary organized a half-day program on *Models of Cooperation*. The original contributors and participants in that symposium, John Anderson, Richard Frankel, Arthur Graesser, John Goodson, Carl Hewitt, Jerry Hobbs, Robin Jeffries, Thomas Landauer, Victor Lesser, Robert Mack, Kelly Murray, Denis Newman, Mary Beth Rosson, Charles Schmidt, Lucy Suchman, and Bonnie Lynn Webber, provided us with an exciting, interdisciplinary array of research on the structure and behavior of multi-agent systems. Seven of the chapters are based on work that was presented at that symposium, now much elaborated and extended. Other chapters represent work that has been done since or that we have added to round out the contributions and present a broader view of cooperative systems.

We are grateful to the sponsors of the original AAAS meeting, specifically the AAAS Psychology section and AAAS Information, Computing, and Communication section as main sponsors; and the AAAS Mathematics section, AAAS Education section, and AAAS Office of Science and Technology Education as cosponsors. Arthur Herschman, head of the Meetings and Publications Center for AAAS at the time, was responsible for putting together our two, initially separate symposia.

We owe a tremendous debt of gratitude to the authors who stayed with us throughout the editing process. We also owe considerable thanks to the authors who joined us later. Their contributions together provide an exciting perspective on cognition, computation, and cooperation.

1

Introduction to Cognition, Computation, and Cooperation

Wayne W. Zachary

CHI Systems Incorporated
Spring House, PA

Scott P. Robertson

Department of Psychology
Rutgers University
New Brunswick, NJ

INTRODUCTION

New computer technologies have, in rapid succession, opened new vistas for interactive computing, graphical interaction, and personal computing. A major consequence of these technological solutions is a new kind of technological problem—the need to engineer intelligent interactions between people and computers, sometimes even between computers and other computers. The simple fact that so many human users of computers express displeasure with their human–computer interfaces makes it clear that this is indeed a problem. Until the very recent past, intelligent interaction was studied only as a human phenomenon. It was observed, analyzed, and theorized about by behavioral scientists, but always from the perspective of explanation, not that of design. Now, computer interface designers are looking to the behavioral sciences for theories, models, and tools to understand intelligent interaction, just as the behavioral sciences are beginning to see human–computer interaction as an important context in which to collect data and test theories. The opportunity and need to design interaction thus seems to be leading both the design sciences and the human sciences in new directions. This volume is an attempt to sort out some of these new directions.

In this book we have brought together several studies that are concerned with intelligent interaction, regardless of whether this interaction occurs between humans, machines, or some combination of the two. This selection principle has resulted in papers that cut across some conventional distinctions. Some are studies of interaction among intelligent machines, originating from the work in distributed artificial intelligence. Other papers focus only on intelligent interaction among humans. These address concerns that have deep roots in the behavioral sciences but are important applications to interface design and engineering. And still other papers deal directly with human factors engineering of human–computer interaction, a hybrid field that is at the same time both behavioral science and engineering.

All the papers, however, do have three points in common. The first point has to do with *cognition*. Interaction has often been approached from a strictly behavioral perspective, particularly in the social sciences and some branches of psychology. The last 30 years of research into human problem solving, however, overwhelmingly suggests that a cognitive perspective which postulates unobservable internal information processing is needed to adequately model intelligent activity. This perspective seems particularly appropriate for intelligent interactions, which require the interacting agents to understand each others' goals and motives (or at least behave as if they do) and figure out how to integrate these goals with their own. Such a cognitive perspective focuses on the information processing that gives rise to the interactive behavior—the problem solving, the knowledge about the relationships of actions to goals and subgoals to superordinate goals, and the knowledge about the behavior of distributed systems. We understand that humans, as *cognitive* systems, exhibit such capabilities, making the study of human cognitive abilities a primary path to the understanding and successful designing of cooperative systems. One goal of this book is to present studies of human cognition in situations that involve cooperation, especially situations involving human–computer interaction. Another related goal is to present research on how models of human cooperation can be applied to designing cooperative computer systems.

The second point of commonality among the papers in the book is that they all have to do with *computing*. There are two ways in which computing is relevant to the studies of interaction presented here—as a model for interaction and as a context for interaction. Abstract computation has, beginning with Chomsky (1957), become the primary metaphor for studying human cognition (Pylyshyn, 1984, recounts this process elegantly). In this framework, theories and models of cognitive phenomena are expressed in abstract computational terms (i.e., Turing computable frameworks), and often simulated via computer programs. Journals, research communities, and whole professional societies have formed around this computational metaphor of cognition. Although the overwhelming majority of these computation and cognition studies have considered individual behavior, we believe that this computational metaphor may be equally valid in the context of cooperation. Several papers included here frame models of interaction in explicitly computational terms.

Just as abstract computation can provide a language for modeling interaction, instrumental use of computing forms a concrete and crucial example of interactive systems. Computer systems are coming to be used more like consultants or even collaborators, and less like tools. In these contexts, people come to expect interaction with computer systems to have the same properties as interaction with other human beings. At a minimum, users expect computers to make available certain options at certain times and suppress others, just as a person would. In order to do this, a system must embody a task description of some kind and be able to employ it when making decisions. In more sophisticated systems, the computer is expected to take on parts of the task. An air traffic control computer, for example, should do more than monitor the objects within an airspace; it should also warn controllers about potentially dangerous situations. This requires information, not only about the task, but also knowledge about the consequences of actions in the task domain. Several contributions here deal with the design of cooperative computer systems. These papers argue that *cooperative* computer systems require an understanding of the problem-solving strategies employed by the person in the application domain. Additionally, these papers suggest, a computer system that is cooperative must make ongoing decisions about the plans of its user(s), other agents, or other systems that affect the domain in which it is acting.

Perhaps the most important contribution that such studies can make is an empirical one. When building an intelligent interactive system, there is ultimately feedback on the efficacy of the design—the system either works or it doesn't. Actually, the results are usually somewhere in between the two. The program exhibits some expected behaviors, fails to exhibit others, and may even produce some that are totally unexpected. Because these successes and failures can be traced back to elements of the original design, the relationship between theory and behavior can be examined empirically. This is never possible in purely human studies because the original design for human intelligence seems to have been lost somewhere along the way.

Finally, all of the papers here are concerned in one way or another with the concept of *cooperation*. As we use it, *cooperation* is a special form of interaction in which the cooperating agents share goals and act in concert with them over an extended period of time. It differs from other computational studies of interaction, such as most of those in economics, or evolution (e.g., Axelrod, 1981), which examine competitive rather than cooperative behavior. Every chapter in this book deals in one way or another with cooperation situations and cooperative interaction. Each contribution provides a different view of cooperation and a different requirement of participants in cooperative situations. These views are based on the kind of cooperative system studied, as well as on the disciplinary vantage point of the authors. Our primary goal in this volume is to find a common thread or concept that underlies cooperative behavior in the varied contexts in which it is examined—in human interaction, through human–computer interfaces, or in the design and analysis of abstract cooperative systems. In conclusion, we attempt to draw out this common thread as a preliminary theory of cooperation and cooperative systems.

BACKGROUND TO THE STUDY OF COOPERATIVE SYSTEMS

The chapters in this volume represent scholars from anthropology, computer science, information science, linguistics, human factors, and psychology. Because divergent disciplines are involved, different methodologies, assumptions, and forms of data are apparent. We have deliberately chosen to not offer any discussion of the history of research in this area, or any detailed review of the literature, as is normally the case in the introduction to an edited volume. Cross-disciplinary research is something that is always discussed and encouraged (and to that extent is nothing new), yet only rarely makes a lasting impact on major, established fields of study. Bibliometric research on cross-disciplinary citation (Chubin, Rossinni, Porter, & Connelly, 1986; Neeley, 1981; Rigney & Barnes, 1980; Small & Crane; 1979) has shown that mainstream fields are associated with well-defined research journals, and that publications in those primary journals show low rates of cross-disciplinary citation that do not vary much through time. While citation is not the only (or the best) measure of the *interdisciplinary-ness* of a given piece of research, it is nonetheless a reasonable one, and one which leads to the conclusion that most of the time, most fields of study tend to look inward toward resolving well-defined problems that have (local) theoretical significance. Even the study of cognitive science, with its recent origins as a multi-disciplinary field, may have already begun to grow introspective and focus on problems that are well-defined and important only in terms of cognitive science.

We therefore want to view this volume as a new endeavor. The study of cognition, computation, and cooperative interaction does not appear to be anything remotely resembling a well-defined field. Researchers from many different disciplines are getting involved, often with an original intent to study something else. This has resulted in some cases of the wheel being re-invented. It has also led to other cases where researchers are clearly unaware of important and relevant results in related fields. For example, Suchman's chapter argues that most cognitive scientists working on interaction are woefully unaware of the highly relevant social science work in ethnomethodology. The point is that we are not arguing that the convergence of interests contained in this book is any sort of historical inevitability or theory-driven breakthrough. Instead, we see it as an interesting accident that arose from the need to design intelligent interactions. We hope that this accident might produce results of interest to many.

A PRELIMINARY DEFINITION OF COOPERATION

What is meant by cooperation? In this book many types of cooperative situations will be discussed. Before plunging into the various contributions it is reasonable to ask just what constitutes cooperative behavior. One approach to the problem of cooperation is to compare example scenarios and discuss whether they should be categorized as demonstrating *cooperative* behavior. Here we will present several scenarios and follow the tradition from linguistics by marking an asterisk by those which we feel are not members of the class of cooperative behaviors. By comparing cooperative and non-cooperative acts, distinguishing features should begin to emerge.

The position that cooperation is a special form of goal-directed behavior among several agents will be set forth in more formal analyses of the examples. This position is shared by all of the contributers to this book. In fact, it could be argued that only by analyzing the interaction of multiple agents' goals can we understand cooperation—or any other coordinated action undertaken by several rational actors. There are many ways that the goals of multiple agents can interact, but when they are combined or distributed so that the agents work as a goal-directed system we have examples of *cooperative behavior*. Mechanisms for achieving and controlling this task are the topics of this book. First, though, some examples.

Cooperating With Shared Goals

First, consider several example scenarios in which two agents share a goal:

1. *Two children want a piece of pie. They fight over it and one gets it.
2. Two children want a piece of pie. They talk about it and cut it in half.
3. Two children want a piece of pie. They talk about it and one of them decides to skip the pie today if he or she can have some tomorrow.
4. *Two children want a piece of pie. They fight over it and it breaks in half. Each grabs a half and eats it.
5. *Two children want a piece of pie. Their mother brings each of them a piece.

In cases 1 and 3, one agent achieves the shared goal and the other agent fails. In cases 2 and 4, both agents achieve their mutual goal (at least partially), and in case 5 both agents fully realize their common goal. Cases 1, 4, and 5 are judged to be not cooperative while cases 2 and 3 are judged to be cooperative.

Examples 1–5 suggest that it is not necessary that agents achieve their goals to be considered cooperative, and it is not necessarily the case that agents have cooperated just because they achieve mutual goals. Case 5 reinforces the latter point by showing that agents can share goals and realize them completely, but still not be considered to have cooperated. The distinguishing feature of the two cooperative cases (2 and 3) is that the outcome is the result of coordinated behavior of the agents. In both cases, the original goal is renegotiated in such a way that the new goal formulation can be achieved. In case 2, for example, each agent decides to have less pie so that both can have some.

While the outcome in case 4 is the same as in case 2, it is not achieved in the same way and is judged to be noncooperative because the goal reformulation is not arrived at by mutual problem solving but is the result of a change in the state of the world. Similarly, in case 5, the mutual goal is arrived at by the action of another agent and is not accompanied by mutual problem solving. In case 5, in fact, the agents need not have been in communication at all for the scenario to have developed. Thus, it seems that the nature of problem solving, more than the outcome, is critical to cooperative behavior.

Figure 1 presents an analysis of the situation depicted in examples 2 and 3 in terms of the goal configurations of the agents. The goals of the two agents are shown in different columns. The superordinate goal of both agents has the form (Eat *agent* Pie). In both cases, we assume that the agents have generated a subgoal of the form (Have *agent* Pie) in order to satisfy a precondition for eating the pie. (This analysis is consistent with any number of planning or goal decomposition mechanisms in the problem-solving literature.)

At the second level of decomposition, where both agents have generated the subgoal to have pie, we cannot say that they have a mutual goal. In fact, if pie is a limited resource, then the agents have a goal conflict. All of the situations depicted in examples 1–5 are consistent with the goal analysis in Figure 1 up to the second level of decomposition. A precondition for the cooperative acts in examples 2 and 3 is that the agents formulate

Figure 1.

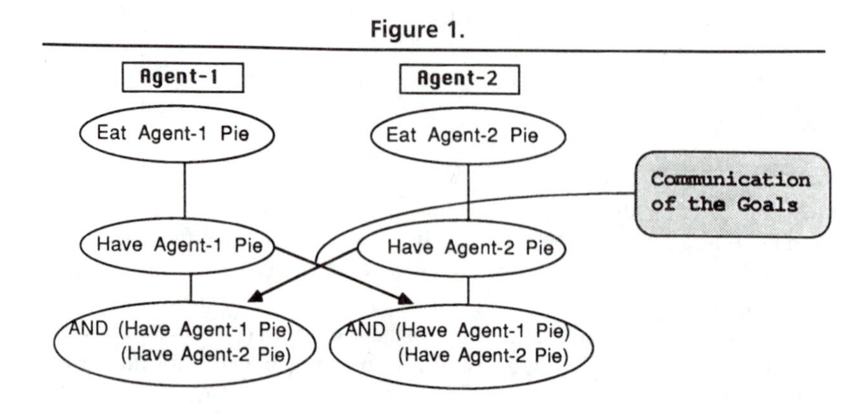

a truly mutual goal, a goal with the same form and content. In order to do this, each agent must incorporate the goal of the other. This is indicated at the third level of decomposition by a conjunctive goal which includes both agents' goals to have pie. Also indicated is the fact that in order to formulate a goal with another agent as the subject, communication about this goal must take place. In examples 2 and 3 this is achieved by conversation about the mutual goal.

The point of working toward the conjunctive goal in Figure 1 is that planning processes will be qualitatively different under the conjunctive goal than under its parent goal. For example, under the parent goal (Have *agent* Pie), grabbing the pie and running off is a satisfactory plan, but it is not satisfactory under the conjunctive goal. Also, because the conjunctive goal contains its parent, satisfying the conjunctive goal means satisfying the parent goal. Thus the conjunctive goal is really a reformulation of the parent goal which allows planning toward satisfaction of both the parent goal and another agent's goal.

Figures 2 and 3 show further goal decompositions of situations 2 and 3

Figure 2.

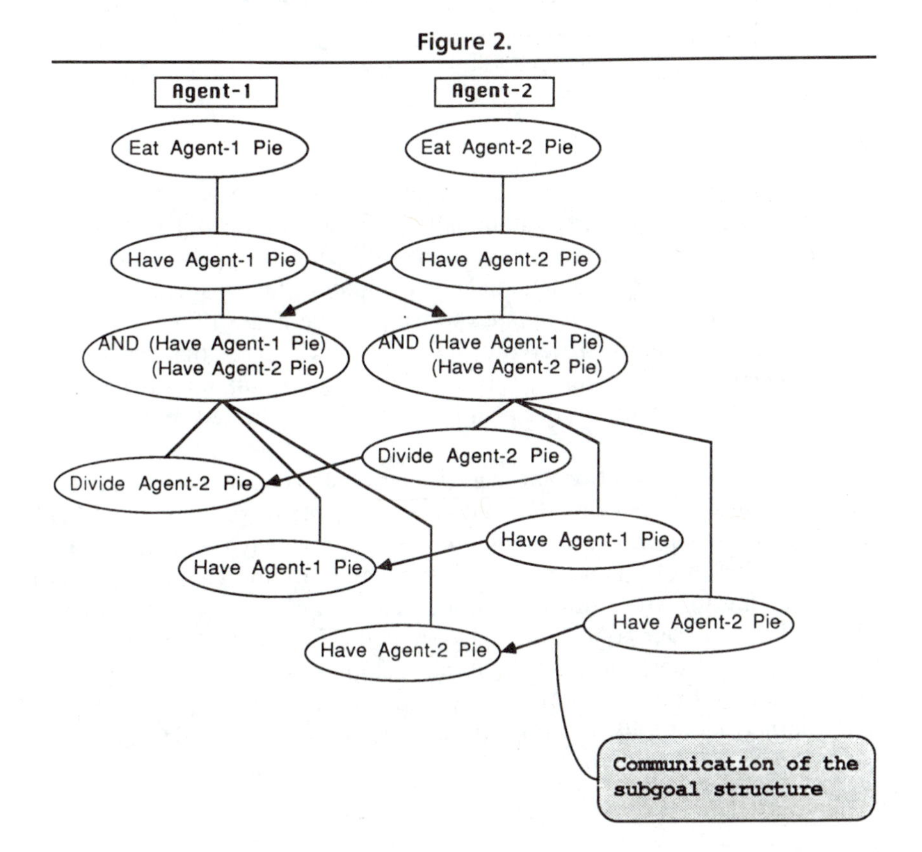

Figure 3.

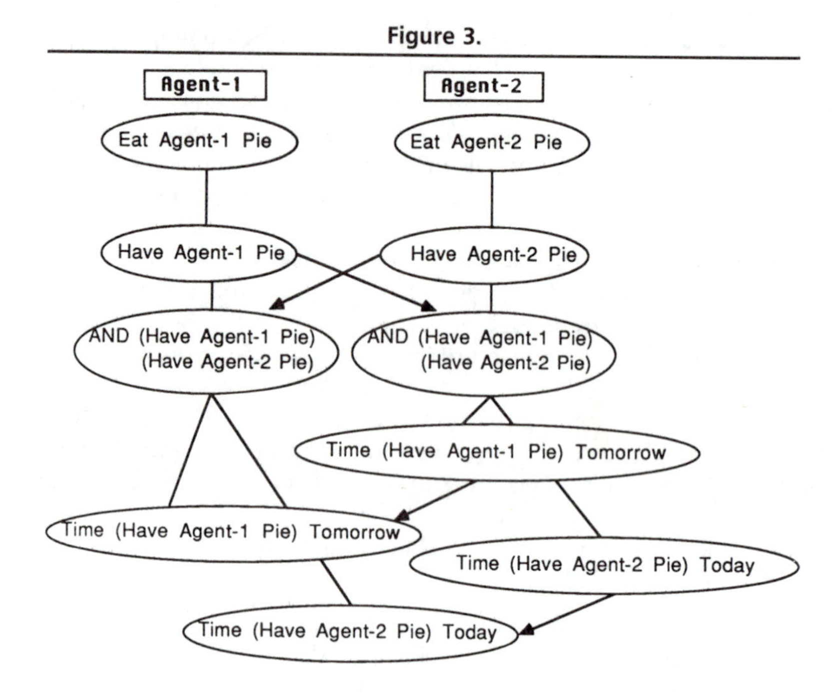

respectively. In Figure 2, agent-2 decides that the pie object can be redefined such that having and eating half of a piece of pie will be sufficient to satisfy the goal. Thus a plan to divide and share the pie resolves the goal conflict. Note that this plan must be communicated to agent-1, allowing the other agent to reformulate the conjunctive goal in the same way so that when the plan is carried out, both agents will agree that it satisfies their respective goals. Figure 3 shows a similar reformulation of the conjunctive goal, but this time the plan is to spread the limited resource out in time.

The act of communication about goals is critical to cooperative behavior. This is true for two reasons. First, it is not possible to represent another agent's goals unless they are communicated in some way (we will take up a subtle variation of this point, inferring the goals of another, later). Second, once mutual goals are established, the planners must agree about what will satisfy their goals. One agent may generate a plan which will be rejected by the other, for example, if the agent in example 3 does not want to wait until tomorrow to eat pie. Cooperation requires that an agent know that another agent will be satisfied with a particular plan for achieving conjunctive goals.

Irrelevance of Goal Outcomes

To further make the point about the importance of communication and agreement on plans for the achievement of mutual goals, consider the following scenarios in which the outcomes differ:

6. Two children want a piece of pie. One suggests that they divide it in half. The other says that he or she wants a whole piece of pie.
7. Two children want a piece of pie. They talk about how to get one. Their mother brings each of them a piece.
8. Two children want a piece of pie. They pool their money and buy one. A dog snatches it and runs away.

Cases 6, 7, and 8 all involve mutual goals and depict actions taken in common toward those goals. All of these examples are considered instances of cooperation.

Case 6 is the case discussed earlier in which one agent rejects the plan of another agent to achieve a mutual goal. Figure 4 presents a goal analysis of this example. In this case, communication of the plan from agent-2 to agent-1 does not result in adoption of the plan by agent-1 as it did in example 2 (compare Figure 2). Note that this does not preclude further planning toward the mutual goal and thus we consider it still an instance of cooperation.

There is a definitional ambiguity here that results from the nature of the example and that is potentially confusing. Some might say that agent-1's action in example 6 is *noncooperative* because he or she did not go along with the plan. This assessment is based on a judgment that agent-1 could be satisfied with the pie division plan but that he or she is just being difficult. In the analyses presented here, it is assumed that plans are accepted and rejected on sound principles. Thus the rejection is part of a process of communication about plans for achieving common goals and is part of cooperative problem solving.

In case 7, the goal is achieved in the same way that it was in case 5, by the intervention of an agent external to the cooperating group. Interestingly, we say that the agents were cooperating in case 7 and not in case 5 because, in case 7, they engaged in problem-solving activity toward a mutual goal. A similar point is made in case 8, where the agents were engaged in problem solving toward a mutual goal but an external agent intervened to thwart the goal. Thus, just as acceptance or rejection of a plan by other agents is irrelevant to the cooperative nature of problem solving, the outcome of the planning process and the actions that are taken are also irrelevant.

Figure 4.

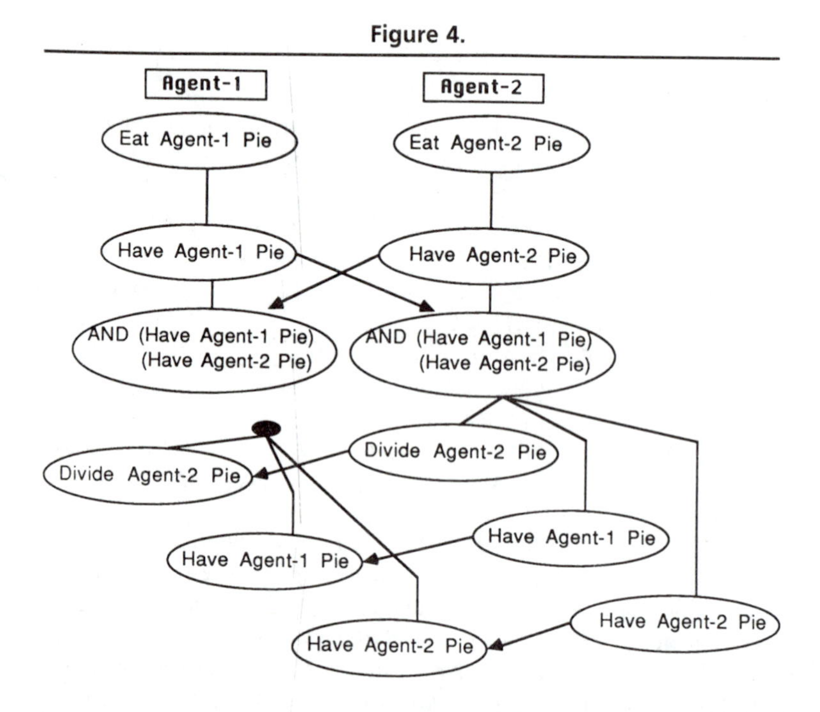

Situations 1–8 begin to suggest that shared goals and communication about plans for achievement of shared goals are central features of cooperative acts. They stress the fact that the outcome of goal-directed behavior is irrelevant to its cooperative nature since we have noncooperative examples where goals are achieved and cooperative examples where goals are not achieved. The examples above also suggest strongly that cooperation has something to do with the nature of the problem-solving process as much as it has to do with the existence of mutual goals.

Cooperating Without Sharing Goals

Consider next some examples where goals are not shared:

9. *Two children notice a piece of pie on the table but they are so stuffed that they don't give it a second look.
10. *One child wants a piece of pie. She takes a piece and offers some to her friend who refuses.
11. One child wants a piece of pie. She can't reach it. She asks her taller friend who gets it and gives it to her.
12. One child wants a piece of pie. She can't reach it. Her taller friend notices and gets it for her.

Figure 5.

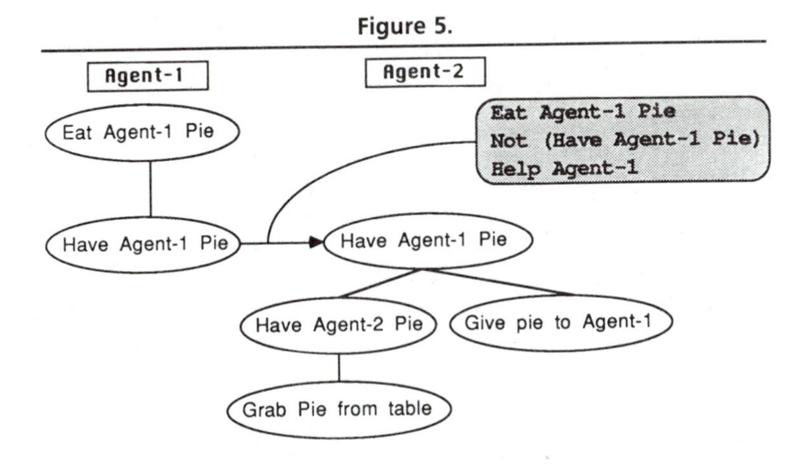

In case 9, the failure of either agent to possess a goal that involves the object they have noticed precludes them from cooperating. In case 10, one agent has a goal that is not shared by the other agent, and this also precludes cooperation. Before deciding that mutual goals are necessary for cooperative behavior, however, note that cases 11 and 12 show that agents which do not share goals can be judged as cooperative if one agent acts to achieve the goal of another.

Case 12 is problematic because it does not seem to share with the other cooperative examples the feature of communication between the agents concerning the achievement of a goal. Figure 5 shows a goal analysis of this case. Note that like the other cooperative cases, the agent who acts on behalf of another agent must incorporate the other agent's goal. In this example, however, there is not a conjunctive goal. Rather, agent-2 plans to satisfy the precondition that agent-1 must have the pie by getting the pie and transferring possession.

Communication About Goals and Plans

The issue of how agent-2, in example 12, takes on the goal of agent-1 is interesting and bears on differences among chapters in this book. The example states that agent-2 "notices" that agent-1 can't reach the pie. This is represented in Figure 5 as the state (Not [Have Agent-1 Pie]) which in conjunction with the known goal (Eat Agent-1 Pie) gives rise to a goal (Help Agent-1). The example has *prewired* the cooperation by representing agent-1's goal to eat pie as part of the initial knowledge structure of agent-2. (After this move we can assume cooperative planning processes or meta-planning knowledge for achieving the precondition for agent-1's goal.) But, how does agent-2 know what agent-1 wants?

There are at least three possibilities for the communication of agent-1's goal to agent-2. First (and least interesting), agent-1 could have told agent-2 what he or she wanted (example 11). This would make the example the same as previous cooperative examples, except that the resulting cooperative goal is not conjunctive. We can refer to this as *request-based* cooperation between agents. Second, agent-2 could infer agent-1's goal from the behavior of agent-1. This is also an act of communication, except that agent-1 now does not intend to communicate the goal. This type of inference would require that agent-2 know the kinds of behavior that count as evidence for another agent's goals. We can refer to this as *inference-based* cooperation. Another intriguing possibility is that agent-2 helps agent-1 because the situation is structured in a particular way—if agent-2 is a waitress and agent-1 is engaged in eating a meal that includes pie, for example. We can refer to this as *structurally-based* cooperation.

Requests, inferences, and system structure can all serve as ways to communicate about goals and coordinate cooperative behavior among agents. The contributors to this book describe mechanisms for all of these cases. In request-based systems, agents need not know anything about the behavior of other agents. They require only a communication network that allows requests to be passed efficiently. Agents must have the capability to initiate, process, and plan requests. Questions about such systems might include the structure of communication, the protocol of requests, autonomy of agents to accept or reject requests, and differing roles of agents as planners versus doers, or requestors versus recipients of requests. These are the issues discussed by chapters that emphasize distributed problem solving or that analyses distributed systems.

In an inference-based cooperative system, the agents must have access to information about the behavior of other agents in order to make inferences about their states and capabilities. Agents which initiate actions on behalf of other agents must know enough about the way other agents are structured to be able to undertake useful activity at appropriate times. Also, agents which have suspended problem solving to await the results of other agents' actions must be informed about the outcomes and effects of these actions. Questions concerning such systems might involve plan recognition problems, communication among agents about intermediate results, availability of information about the behavior of agents to other agents, or differing roles and capabilities of specialized agents. Chapters that discuss intelligent tutoring or that analyze learning situations address many of these issues.

Finally, agents in a structure-based cooperative system have predefined roles that can include initiating actions on behalf of the goals of other agents. Many of the human–machine interaction chapters stress the importance of structuring computational environments so that they help humans to achieve their goals. There is a recognition that *cooperation* can be passively engineered into the design of a human–machine system.

Independence of Planning, Communication, and Execution

In a final set of examples, consider the independence of planning, communication, and execution in cooperative situations:

13. *Two children want a piece of pie. They call each other on the phone and mention it. Then they hang up and go to their respective refrigerators for pie.
14. Two children want a piece of pie. They call each other up and figure out the recipe for apple pie. Then they hang up and each makes a pie.
15. One child wants some pie. An adult wants some coffee. They discuss where to go for the best of both. They drive to a coffee shop and get what they want.

Cases 13 and 14 both involve a mutual goal and in both cases the goal is achieved. Case 13 is not judged to be cooperative, however, because the problem solving that took place in service of achieving the goal (the *refrigerator plan*) was arrived at independently and was not influenced by the communicative act. Note that there is no communication resulting in the establishment of goals. The telephone call is independent of the planning mechanism utilized in achieving their goal. In case 14, on the other hand, actions taken by each agent to achieve the goal are independent but they are the result of a plan that was formulated during the communicative act. Case 14 is now judged to be a cooperative situation. In case 15, the initial goals are different and the agents both achieve their respective goals. Cooperative communication, however, was required to formulate the common plan.

Summary of the Examples

These examples reinforce the point that the communication process is the critical feature of cooperative interaction. It is not necessary that the agents share goals initially nor is it necessary that they come to share goals at a later time. It is not necessary that either agents' goals be achieved. It is only necessary that information about the goals of other agents be shared in such a way that common goals can be generated or mutual plans developed. This communication need not be direct, but can be indirect (via inference or situation structure). Researchers and developers of cooperative systems are really asking questions about the content and structure of communication among multiple agents engaged in goal-directed behavior.

ORGANIZATION OF THE BOOK

The book is organized into three parts. The first, *Observation and Analysis of Cooperative Systems,* contains analyses of empirical contexts in which cooperation exists and can be directly studied. This includes situations involving human–machine interaction, organizations and organizational behavior, teaching and learning, and *simple* conversation. Each chapter in the first part presents a theoretical analysis of one of these contexts. Together, they provide an overview of cooperation from the viewpoint of human cognition and human behavior, which is presently our best (and arguably our only) empirical example of the phenomenon of cooperative interaction.

The first chapter in this section, by Lucy Suchman, combines an empirical and theoretical examination of human–computer interaction. She begins by treating the concept as an implicit metaphor—"the notion that using a machine could be like [human] interaction"—and examining how human–human interaction and computer usage are alike and different. For Suchman, the essential unit of interaction is the face-to-face transaction, an event that is situated in the context of the moment as well as the culture and social system of the participants. She points out the context sensitivity of human interaction, a phenomenon which in linguistics is covered under the rubric of *pragmatics.* Drawing on the work of anthropologists (particularly ethnolinguists and ethnomethodologists like Garfinkle and Gumperz), Suchman argues that contextual factors are behind most of the empirical shape of human face-to-face interactions. Such contextual factors (and their implicit ties to cultural and social cues) are unavailable to the computer in a human–computer interaction, so the machine must rely on more syntactical and semantic factors in interpreting and participating in a dialog. Using an empirical case of transactions between people and a supposedly intelligent computer-based machine, she demonstrates that computer–human interaction is still fundamentally and qualitatively different than human–human interaction. Suchman's paper also presents a new and important methodology for analyzing human–computer interaction.

In the next chapter, Thomas Malone reverses the metaphor, and uses human organizational behavior as a metaphor for distributed computer systems. Rather than focusing on the smallest unit of human interaction (i.e., the face-to-face transaction), Malone instead examines the largest units, large-scale organizations and social systems with ill-defined boundaries, such as economic markets. Unlike traditional economic or sociological views, though, Malone approaches these large-scale organizations as information processing (i.e., cognitive) systems, in which cognitive constructs, like *goals,* are part of the basic definition of an organization. With this view, the main issues in analyzing human interaction concern the definition, coordination, and communication of these resources. He examines a range of macroscopic human organizations, and notes a general tradeoff between the flexibility of an information processing organization and its efficiency. This analysis of large-scale human interaction is then turned back onto the design of computer systems, which he approaches as organizations of information processing agents. In the end, he argues that a common framework can and should be applied to examining both human and computer-system cooperation.

The next chapter in the first section, by Denis Newman, looks more narrowly into a single model of one type of interaction, the *appropriation* model of learning. In this model, related to the work of the psycholinguist Vygotsky, "social interaction . . . [is] an essential component of cognitive change." Appropriation occurs when the actions of one agent in an interaction are used (appropriated) by the other in terms of a goal, plan, or meaning space that the original actor may not have understood or even known existed. When the appropriated action, which may have been deliberately appropriated for pedagogical purposes, is within some epsilon of the original agent's own state of understanding, then that agent will be able to infer or recognize the underlying model into which the action was appropriated. Newman takes this model in two directions. First, he analyzes two empirical instances of (human) teacher–student interaction to further refine and demonstrate the value of the model. Second, he applies the interaction-based appropriation model to question the assumptions behind computer-based tutoring and educational software. In the end, there are some surprising similarities between his conclusions from the educational domain and Suchman's from the personal interaction domain with regard to the criticality of situations and context in human interaction.

The final chapter in the first section is by Ray Gibbs and Rachel Mueller, which examines the most ubiquitous form of cooperative interaction, everyday human conversation. Their chapter provides an integrative overview to the large literatures on conversation from linguistics, psychology, and artificial intelligence. Conversational speech is an inherently interactive process, and one, they argue, that must be both implicitly and explicitly cooperative. They note that conversation is an interactional and therefore a social context, in which speakers pursue both communicational and social, interpersonal goals. These goals all require a negotiation of the common ground of shared beliefs and knowledge among the participants, which runs the continuum from culturally shared information (such as the referent of the word *dog* versus *perro*) to very locally shared information (such as what has transpired earlier in the conversation). They present both some computationally oriented mechanisms for managing this negotiation process, as well as relate this theory of conversation to the design of natural language computer systems and the analysis of human communication. The first section ends with a brief commentary by the editors on the common themes and issues raised by the chapters in that section.

The second section, *Design and Architecture of Cooperative Computer Systems*, contains contributions about the organization and design of distributed computing systems. As in the first section, each chapter examines the theoretical structure of such systems and the actions required of components of distributed intelligent computing systems. There is a deliberate complement between the approaches of the chapters in these first two sections. Even though computation has become a primary metaphor for the psychology of cognition, there is nonetheless a fundamental difference of perspective between the computational and behavioral sciences. The computer scientist is constantly concerned with problems of design—how to create a system with a certain set of behavioral properties. In this endeavor, purely explanatory studies of existing system designs can be, at best, a useful background for the design process. The behavioral scientist is, in contrast, constantly concerned with explaining the observable properties of one specific empirical system—the human being. The central problem for behavioral sciences is reconstructing the hidden design of the empirical system. In this endeavor, design studies provide, at best, useful examples of how underlying mechanisms may be related to observed properties.

A major part of the interest (to behavioral science) in the design orientation of computational sciences is the latter's use of abstract formalisms and methods to describe specific design problems and design approaches. These formalisms lead to languages in which design features and behavioral properties can be related. Such language can be appropriated to the empirical studies of human systems. This borrowing is, in fact, the basis for the wide use of the computational metaphor of cognition. Turing-style computation provides a convenient language in which underlying cognitive mechanisms and empirical behavior can be simultaneously discussed.

Just as unitary computer programs are analogous to single humans thinking and solving problems, distributed computer systems and programs are analogous to pairs or groups of humans interacting and cooperating. The computationally-oriented chapters in the second section therefore form a counterpoint to the human-based analyses in the first section.

The first chapter in section two is by David Woods, Emily Roth, and K. Bennett, which presents and applies a design framework which is termed *joint human–machine cognitive systems*. They are concerned with the design of interactive computer programs which optimize the cognitive performance of the higher order system that the person and machine comprise. Since the human component of the system is not subject to *engineering* (except by training), the primary focus is on how the machine component should be designed to meet this system-level performance criterion. They compare two models, a standard model of letting the machine do those canonical tasks where the human performs poorly and vice versa (which they call the cognitive prosthesis approach), and a new approach in which the machine is designed to support flexibly whatever tasks the person's needs supported *in the environment and problem at hand*. This approach is termed the cognitive instrument approach, and they argue that it leads to person–machine problem-solving interactions that are more typical of person–person problem-solving interactions. Much of the chapter consists of a detailed empirical examination of people interacting with an artificial intelligence expert system for trouble-shooting. Most expert systems, including the one they discuss, are built from the cognitive prosthesis approach, and their analysis demonstrates in detail why so many human end-users have such difficulty in working with expert system programs in real world situations. Another contribution of Woods et al. in this chapter is to develop a notation and methodology for charting and analyzing interactive problem solving.

The next chapter in the second section, by Edmund Durfee, Victor Lessor, and Dan Corkill, focuses on distributed artificial intelligence, a field concerned with design of cooperative systems that involve only intelligent machines. Durfee et al. describe techniques for coordinating decision making among nodes in a distributed problem-solving network. A distributed problem-solving network can be thought of as a multi-agent planning system where subtasks are formulated and dynamically allocated by the various nodes. Problems within this paradigm involve constraining decision making to the appropriate nodes and avoiding debilitating redundancies, node interactions, and resource tie-ups. After describing various approaches to distributed architectures, Durfee et al. focus on the role of prediction in coherent cooperative systems. Prediction is the ability of a node to anticipate the outcomes of processing at other nodes, and even the overall behavior of a network. Successful prediction comes from providing nodes with knowledge about the capabilities of other nodes and the communication patterns between nodes. Durfee et al. describe techniques for acquiring and updating such shared information during problem solving in a distributed network. Although this contribution is the only one which does not involve analysis of human agents, the requirements for distributed problem-solving nodes are strikingly relevant and familiar to the other chapters here.

The final chapter in section two is by John Goodson and Chuck Schmidt, and focuses on a similar issue to Woods et al. Goodson and Schmidt also consider the design of artificial intelligence systems that exhibit cooperative behavior and problem-solving human beings. They approach this problem from the perspective of distributed artificial intelligence (à la Durfee et al.), however, and arrive at some different conclusions than Woods et al. Goodson and Schmidt argue for decomposing the problem to be solved into functional units, identifying units which the machine can perform well and the human poorly, and building a machine problem solver for that unit. Up to this point, their approach is very much like the cognitive prosthesis approach which Woods et al. decry, but Goodson and Schmidt also specify the need to design a joint problem-solving method in which the human and computer share responsibility. Their implicit metaphor is perhaps one of industrial engineering, in which a one-person job becomes too complex, and is so divided into a two-person job which requires a complete redesign of the job itself. They discuss in detail an application of this design strategy in a real-time problem domain where the main task is vehicle tracking via remote sensing. In this process, they define a number of roles of the computer *vis-à-vis* the human in the system, and even go so far as to develop computational representations of each of these roles. The second section, like the first, concludes with a commentary and summary by the editors.

The third section, *Studies of Cognition and Cooperation,* contains studies of humans interacting with computer systems. This is the one key arena in which behavioral and design concerns come together in a single problem that is usually termed human factors engineering. As noted above, interactive computing is rapidly evolving to the point where people using computers expect and demand the computers to behave as people playing the same roles—teacher, consultant, advisor, helper—would behave. The concerns of the computer scientist and behavioral scientist overlap in this case. The computer scientist is trying to design a system that has at least some of the interactive properties of a person. The data and theories of human cognition and interaction are clearly of primary importance in both defining and solving this design problem. In fact, as some of the chapters in section three show, this design of cooperative computer interfaces can help define an experimental agenda for the behavioral sciences. The design of computer–human interaction also provides a rare arena in which the behavioral scientist may empirically test theories of human behavior in a design framework. A theory that leads to a system whose behavior approximates human behavior (or at least meets the system user's expectations) is in some degree validated. The *degree of validation* at the same time poses some difficulty for the behavioral sciences.

A main concern that arises from the first two sections of the book is the role that expectations play in cooperative interaction. Each agent must have some expectations about the other's goals and intentions, as well as about the other's behavior. The chapters in the first section demonstrate the importance of these expectations in learning, conversation, organizational behavior, and human–computer interaction. The chapters in the second section further concern the formal requirements for these expectations in computational systems:

- What kinds of expectations are necessary for what kind of cooperation?
- What expectations about role and behavior must be present?
- How should they be formalized and represented?

The chapters in the third section similarly center around the issue of expectation. In section three, both methodological questions (e.g., how can people's expectations about behavior be elicited?) and design questions (e.g., how can such expectations be represented and manipulated in a user interface?) are explored.

The first chapter in section three is by Arthur Grasser and Kelley Murray. Their chapter is primarily methodological, and focuses on ways of gaining data on the person's expectations, assumptions, and beliefs about the behavior of the computer during a human–computer interaction. They present a technique called a Question Answering methodology, one of the family of verbal protocol methods used extensively in cognitive science, and demonstrate its application to human interactions with a text editing system. Grasser and Murray are concerned with capturing the beliefs and expectations that are under both *normal* interaction as well as misdirected interactions or *errors*. They are also concerned with capturing and examining the differences in these among various learning states that a user might pass through—novice, intermediate, and expert. They point out that different interface design details (in terms of the optimal form) will depend on the type of user, goal of the interaction, and pragmatic context of the interaction. Such design decisions can only be made with knowledge of the expectations, plans, and beliefs that the humans will bring to the interaction. Thus, methodologies such as theirs are necessary components of the interaction design tool kit.

The second chapter in the last section is by Dana Kay and John Black. They are also concerned with mapping the knowledge that a person may bring to bear on an interaction with a computer system. Kay and Black, however, utilize the more traditional methodology of multidimensional scaling to build representations of knowledge for computer commands. By looking at computer naive, novice, and expert groups of subjects, the experimenters were able to trace changes in cognitive representations as users gained expertise with a computer system. They propose four distinct phases of learning, distinguished by the nature of conceptual representations utilized by human users to understand the system with which they are working. Any notion of cooperation through a learning period must recognize the changing nature of human representations of knowledge.

The third chapter in section three is by Robert Mack, and focuses again on the importance of identifying user expectations and accounting for them in the engineering of human–computer interaction. Mack looks at the expectations of new users of a computer system. Kay and Black noted in the previous chapter that in the first phase of learning, people utilize prior expectations about the behavior of a system that may be based on the natural language semantics of the system's command language or other features related to general knowledge. Mack emphasizes this point by looking exclusively at novice users. In a detailed analysis of new users' verbal explanations and predictions, Mack shows both the intrusion of related knowledge (i.e., typewriter knowledge used to understand a computer text editor) during learning and the simple causal structures (*one action, one outcome*) into which new users try to fit their expectations about system behavior. Mack goes a step further by describing a prototype training system that does not violate new user expectations. In a sense, such a system is *cooperative* in that it accommodates the cognitive representations of its users.

The final chapter in this section is by Marc Sebrechts and Richard Marsh, which combines the concerns of Kay and Black for changes in conceptual representations with the concerns of Mack for supportive training systems. Sebrechts and Marsh use a set of questions designed to elicit information about users' knowledge to examine the effects of different instructional materials. The instructional materials differ on the degree to which they support *conceptual elaboration* and *syntactic elaboration* of the material to be learned. They trace changes in conceptual models of the computer system in all of these conditions and provide a model, which they call *integrative modeling*, to explain them.

As the last chapter in the book, the editors offer some concluding remarks. Rather than summary, these remarks attempt a synthesis of the subject matter contained here. It was a working hypothesis during the formation of this book that important parts of a theory of cooperation will emerge from comparison of these different approaches. This hypothesis, actually more of a hope, has only partly proven true. Although many of the chapters discuss common issues, conflicting viewpoints on these issues are as often the rule as not, so no consensus on cognition, computation, and interaction has emerged. Instead, we try to identify in the concluding remarks the main components of theory of cooperation, and the main open research questions that must be answered before a consistent and concensus theory of cooperation can emerge.

References

Axelrod, R. (1981). The emergence of cooperation among egoists. *American Political Science Review, 75*(2), 306–318.

Chomsky, N. (1957). *Syntactic structures.* The Hague: Mouton.

Chubin, D. E., Rossinni, F. A., Porter, A. L., & Connolly, T. (Eds.). (1986). *Interdisciplinary analysis and research.* Mt. Airy, MD: Lomond.

Neeley, J. D. (1981). The management and social sciences literatures: An interdisciplinary cross-citation analysis. *Journal of the American Society for Information Science, 32,* 217–223.

Pylyshyn, Z. (1984). *Computation and cognition.* Cambridge: MIT Press.

Rigney, D., & Barnes, D. (1980). Patterns of interdisciplinary citation in the social sciences. *Social Science Quarterly, 61,* 114–127.

Small, H., & Crane, D. (1979). Specialties and disciplines in science and social science: An examination of their structure using citation indexes. *Scientometrics, 1,* 445–461.

I

Observation and Analysis of Cooperative Systems

Scott P. Robertson

Department of Psychology
Rutgers University
New Brunswick, NJ

INTRODUCTION

The first section of this book contains chapters that explore cooperation analytically and theoretically. The authors, all social scientists, are examining the question of what constitutes a *cooperative* system. Each is either trying to draw analogies between different types of systems or trying to show the cooperative nature of a particular system. Several different types of systems are analyzed. Lucy Suchman observes interactions with a copying machine, Thomas Malone discusses human organizations and computer systems, Denis Newman describes classroom learning situations and potential intelligent tutoring systems, and Raymond Gibbs and Rachael Mueller examine human conversation.

Two chapters, Suchman's and Newman's, utilize naturalistic observation of human behavior coupled with a theoretical analysis of what guides that behavior. Suchman, an anthropologist, observes the interaction of humans with a copying machine and analyzes the actions of each. The machine is an example of what Suchman calls an *interactive artifact,* a computationally based device that is designed to participate cooperatively in some activity. She notes that the development of this type of artifact has made communication theory relevant to human–machine interaction, and her analysis of actions in this situation makes use of "intentions" and "interpretations" of the human and machine participants. A primary finding of Suchman's is that failure of the interaction to go smoothly results from a lack of correspondence between the user's plan and the machine's plan for accomplishing goals. This basic idea appears in many forms throughout the book.

Newman, who is concerned with learning and tutoring, observes classroom interaction and comes to many of the same conclusions as Suchman. He suggests that educators think about the context in which knowledge is acquired. He notes that actions and concepts are always interpreted in context, and that students and tutors may be thinking of very different contexts when they produce and interpret behavior. Besides learning a procedure or a set of facts, students learn the goals that are appropriate in a particular context. This knowledge is acquired indirectly, by noticing when something is right and when it is wrong, or by trying to understand why certain behaviors are appropriate. In terms of a tutoring strategy, Newman points out that tutors and students should explicitly pursue common goals, instead of the tutor having a *teaching* goal and the student having a *learning* goal. In this paradigm, the tutor's job is to allocate responsibility for producing actions toward goals, not to tell the student what to do or to evaluate the student's actions. Thus we see that the problem of shared goals is critical in this context as well.

Malone draws upon the literature in organizational theory to make an explicit analogy between human organizations and computer systems. An organization is an observable system which relies on the coordinated behavior of many participants. Malone focuses on analyses of organizations as information processing systems and describes the various communication structures that have been identified within organizations. He notes that a central assumption in "market" models of organizational behavior is "mutual adjustment," or an agreement to share resources in order to achieve mutual goals. Thus, mutual goals again appears as an important component of cooperative systems. Malone also notes that market based systems are more decentralized and contain more autonomous components than other organizational systems, and he points out that these features are also characteristic of distributed computing systems. His suggestions for the organization of such systems are taken up by authors in Section II.

Suchman, Newman, and Malone all identify *communication* as a central component in the analysis of the behaviors that they discuss. Suchman explicitly notes that this focus on communication makes analysis of human linguistic behavior relevant to human–machine situations. In the last chapter in this section, Gibbs and Mueller bring an experimental approach to the analysis of human communication. They ask how indirect uses of language can be understood according to any theory of language that emphasizes formal analysis of speech acts or purely linguistic features. They argue that much of language comprehension is based on "conventional meaning," or intentional referents for utterances which are different from their formal referents but which are shared among speakers. They present experimental evidence from a psycholinguistic paradigm in support of their claim for the importance of convention, another form of sharing goals.

2
What is Human–Machine Interaction?

Lucy A. Suchman

System Sciences Laboratory
Xerox Palo Alto Research Center
Palo Alto, CA

INTRODUCTION

New technology has brought with it the idea that we no longer simply use machines, we interact with them. In particular, the notion of human–machine interaction pervades both technical and popular discussion of computers, their design, and their use. As an anthropologist and student of human interaction, I have become deeply interested in this notion of interactive artifacts; its use by those involved in computer research and design, its propagation out into the popular press and, most crucially, its basis in what actually goes on when people use computational machines.

Starting from the premise that *interaction*, or *communication*—I will use the two terms interchangeably—turns on the extent to which my words and actions and yours are mutually intelligible, we can look at the basis for beginning to speak of interaction, or mutual intelligibility, between humans and machines. In part, the notion of human–machine interaction derives simply from the increased reactivity of computer-based artifacts. Each action by the user effects an immediate machine response. And while the responses are entirely determinate, they may be combined and recombined into a large number of configurations. As a consequence, machine responses may be both occasioned by, and relatively particularized to certain user actions. These behavioral features alone suggest the comparison of using a computer to human interaction, and encourage attempts to increase the likeness.

A more profound basis for the comparison, however, is the fact that the means for controlling computing machines and the behavior that results are increasingly linguistic rather than mechanistic. That is to say, machine operation becomes less a matter of pushing buttons or pulling levers with some physical result, and more a matter of specifying operations and assessing their effects through the use of a common language. Increasingly, therefore, in describing what goes on between people and machines, designers and users employ terms borrowed from the description of human communication: *interaction, dialogue, conversation*, and so forth, terms that carry a largely unarticulated collection of intuitions about properties common to human communication and the use of computer-based machines. While the terms take on a narrow, technical meaning within the vocabulary of design, they also maintain the broader connotations of their ordinary usage.

INTERACTION AND INSTRUCTION

The reactive and linguistic properties of computer based artifacts alone suggest new directions in the design of user–machine interfaces. But more specifically, there is a strong sense in which the problem of designing and using artifacts is in fact a problem of communication. The designer of any system that is a tool must communicate the system's intended use to the user, and an appealing solution to this problem is an artifact that, as a feature of its use, tells about its use. This idea of a self-explicating artifact accords well, moreover, with the notion that using a machine could be like interaction.

The idea of interactive machines, in this sense, represents the latest solution to the longstanding problem of providing the user of a tool with instruction in its use. To a large extent, the designer of a machine just assumes that its use will be discovered through particular, situated inquires from the user. In physical design itself the designer anticipates certain questions from the user such that, in the event, an answer is there ready-at-hand. So, for example, the user's question "Where do I grab?" is answered by a handle fitted to the action of grabbing. In the case of the instruction manual, some further classes of inquiry are anticipated, and answers provided. The step-wise instruction set, which anticipates the question "What do I do next?", and the diagram, which anticipates "Where?", exemplify this approach. In either case, the questions anticipated and answered must be those that any user of the system might ask. And the occasion for both questions and answers is found by the user.

One obvious challenge in designing a system that would be interactive is to provide information that is fitted to user and occasion in some stronger sense than the instruction provided by a written manual. Every instructional system is guided by the general maxim that utterances should be designed for their recipients. As with any type of communication, the extent to which the maxim is observed in instruction is limited in the first instance by the resources that the medium of communication affords. Face-to-face human interaction is the paradigm case of a system for communication that, because it is designed for maximum context sensitivity, supports a response designed for just this recipient, on just this occasion. Face-to-face instruction brings that context sensitivity to bear on problems of skill acquisition. For example, the gifted coach draws on powers of language and observation, and uses the situation of instruction, in order to specialize his or her instruction for the individual student. Where written instruction relies upon generalizations about its recipient and the occasion of its use, the coach draws his or her pedagogical strength from exploitation of the unique details of particular situations.

A consequence of the coach's method is that his or her skills must be deployed anew each time. In contrast, an instruction manual has the advantage of being durable, reusable, and replicable. In part, the strength of written text is that, in direct contrast to the pointed commentary of the coach, text allows the disassociation of the occasion of an instruction's production from the occasion of its use. For the same reason, however, text affords relatively poor resources for recipient design.

The promise of interactive systems is a technology that can move instructional design away from the written manual in the direction of the human coach and of the resources afforded by face-to-face interaction. Efforts at building self-explicating machines in their more sophisticated forms now adopt the metaphor of the machine as an expert in its own use and the user as a novice, or student. One such case is Bluebonnet, an experimental system designed to assist novice users of a complex photocopier.[1] The system is something of a hybrid of old and new technologies, as the photocopier is controlled by a computer-based interface intended to act as an artificially intelligent "expert" in the machine's use. The goal of the system is that, rather than providing a compendium of information and leaving decisions of relevance to the user, information should be occasioned by and fitted to the user's inquiries. In order to provide, not just a set of instructions, but an occasioned response, the designer must now specify, not only the information, but how the system should recognize the situation for which that information is appropriate. Crucially, the relevant situations are constituted by the user's actions. Consequently, the system must in some sense be able to find their significance.

---

[1] Bluebonnet was developed, for research purposes, by Richard Fikes at Xerox Palo Alto Research Center in 1982-1983. For an interesting example of a computer based coach designed on quite different principles, see the West system as described in Burton and Brown (1982).

ACTION INTERPRETATION

The approach to action interpretation taken in the design of Bluebonnet is based on a view of purposeful action that is central to work on machine intelligence, and is deeply rooted in our common sense.[2] Briefly, the view is that the meaning of our actions is best understood as the reflection of our underlying plans. The strategy adopted in the design of interactive machines is to project the user's actions as the enactment of a plan, and then use the plan as a template for the actual action's interpretation.

In contrast to most occasions of human interaction, certain aspects of the event of using a machine provide grounds for imagining that one might safely predict, in some detail, just how the interaction will go:

- The interaction is instrumental;
- The possible goals of the interaction are constrained by the machine's functionality;
- The structure of the interaction is procedural, constituted by a sequence of actions whose order is partially enforced;
- The criteria of adequacy for each action can be specified.

Inasmuch as users' purposes are constrained by the machine's function, and their methods by its design, it seems reasonable to suppose that, in human–machine interaction, the machine's function and design should serve as adequate context for the interpretation of the user's actions. In the case of Bluebonnet, the user's responses to an initial series of questions are identified with one of a set of possible goals, the goal invokes an associated plan, and the enactment of the plan is prescribed as a step-wise procedure. The prescribed procedure then provides the system with a ready-made template against which certain of the user's actions can be located in the plan. And the location of the user's action in the plan determines what the system does in response.

The premise of the design is that the plan corresponds to the user's actions, and that the correspondence enables the interaction. However, observation of what actually happens when users encounter the system suggests that user and system each have a fundamentally different relationship to the plan. While the plan directly determines the system's behavior, the user is required to find the plan, as the prescriptive and descriptive significance of a series of procedural instructions. While the instructions, and the procedure that they describe, are the object of the user's work, they do not, strictly speaking, determine either the work's

[2] For a more extended discussion and analysis of this and other issues raised here, see Suchman, 1987.

course or its outcome. This difference in the relation of user and system to the design plan has consequences for their interaction.

Those consequences became apparent in looking closely at vidotapes of first-time users of the system.[3] The most general aim of the video analysis was to find the locus of "shared understanding" between the users and the system. More particularly, I wanted to compare the user's and the system's respective "views" of what happened over the course of events. In working to organize the transcripts of the videotapes, therefore, I arrived at the following simple framework:

Figure 1. The Analytic Framework

	The User's Actions		The Machine's Behavior	
	I Not available to the machine	II Available to the machine	III Available to the user	IV Rationale
T1				
T2				
T3				
.				
.				
.				

The framework revealed that the coherence of the users' actions was largely unavailable to the system, and something of why that was the case. Beginning with the observation that what the user was trying to do was, somehow, available to the researcher, one could ask how that was so. The richest source of information for the researcher, as a full-fledged "intelligent" observer, was the verbal protocol recorded in Column I. In reading the instructions aloud, the user located the problem that he or she was working on. The user's questions about the instructions identified the problem more particularly, and further talk provided the user's interpretations of the machine's behavior, and clarified his or her actions in response.

A second, but equally crucial, resource was visual access to the user's actions themselves. Of all of the user's actions, one could clearly see the very small subset, recorded in Column II, that were actually detected by

[3] The videotapes were made during March of 1983. Two people, neither of whom had ever used the system before, worked together in pairs. In the interest of their collaboration, each makes available to the other what she believes to be going on; what the task is, how it is to be accomplished, what has already been done and what remains, rationales for this way of proceeding over that, and so forth. Through the ways in which each collaborator makes available to the other her own sense of what is going on, she provides that sense to the researcher as well. An artifact of such a collaboration, therefore, is a kind of naturally generated protocol.

the system. From the system's "point of view," correspondingly, one could see how it was that those traces of the user's actions available to the system—the user's behavior seen, as it were, through a pin-hole—were mapped onto the system's plan, under the design assumption that, for example, button *x* pushed at this particular point in the procedure must mean that the user is doing *y*.

The framework proved invaluable for considering seriously the idea that user and machine were interacting. By treating the center two columns as the behavioral "interface," one could compare and contrast them with the outer columns, as the respective interpretations of the user and the design. This comparison located precisely the points of confusion, as well the points of intersection or "shared understanding."

The Basic Interaction

The system presents the user with a series of video displays composed of text and drawings (see appendix). Each display either describes the photocopier's behavior or provides the user with some next instructions. In the latter case, the final instruction of each display prescribes an action whose effect is detectable by the system, thereby triggering a change to the next display (see Figure 2).

Figure 2. The Basic Structure of an Interaction

	The User's Actions		The Machine's Behavior	
	I Not available to the machine	II Available to the machine	III Available to the user	IV Rationale
T1			DISPLAY 1	
T2	User reads instruction, interprets referents and action descriptions			
T3		ACTION A		
				Action A means that user has understood Display 1 and is ready to proceed to the next.
T4			DISPLAY 2	

Through the device of display changes keyed to actions by the user, the design accomplishes a simple form of occasioned response, despite the fact that only a partial trace of the user's actions is available to the system. To compensate for the machine's limited access to the user's actions, the design relies upon a partial enforcement of the order of actions within the procedural sequence. This strategy works fairly well, insofar as a particular effect produced by the user (such as closing a cover on the copier) can be taken to imply that a certain condition obtains (a document has been placed in the machine for copying), which, in turn, implies a machine response (the initiation of the printing process). This strategy of binding certain user effects to particular machine "responses" enables the appearance of instructions occasioned by the user's actions.

Among those user actions that are not available to the system is the actual work of locating referents and interpreting action descriptions; the system has access only to the product of that work, as the detectable Actions in Column II. Moreover, within the instruction provided by a given display are embedded instructions for actions whose effects are not detectable by the system. To anticipate the discussion of troubles that arise if one of these instructions is misconstrued, the error will go by unnoticed. Moreover, since the implication of a next display is that prior actions have been noted and that they have been found adequate, the appearance of Display 2 will confirm the correctness not only of Action A narrowly defined, but of all of the actions prescribed by Display 1.

In the following example from the video corpus, two users want to make two-sided copies of an article from a journal. The two-sided copying procedure requires an unbound document, however, so they must begin by making a master copy of their article, using the "Bound Document Aid" (see Appendix for machine diagram).

I [22:7-68, 189–196]

(A and B are proceeding from the display that establishes their goal as making two-sided copies of a bound document. "Quotes" indicate that the user is reading instructions from the display.)

	The User's Actions		The Machine's Behavior	
	I Not available to the machine	II Available to the machine	III Available to the user	IV Rationale
			DISPLAY 1	Tells user to wait while system selects procedure
A:	"To access the BDA, pull the latch labelled Bound Document Aid"::		DISPLAY 2	Instructions for using the bound document aid
	(Both users turn to machine)			
A:	(Points) Right there.			
B:	(Hands on latch)			
A:	"And lift up to the left." (looks to B, who struggles with the latch) "Lift up and to the left."			
B:	(Still struggling)			
A:	Okay::			
B:	Pu::ll, and lift up to the left. (Looks at picture) Oh, the whole thing [			
A:	Yea.			
B:	lift up and to the left.			
		RAISES DOCUMENT HANDLER		
			DISPLAY 3	Instructions for placing document and closing document cover
A:	"Place your original face down, (Passes journal to B) on the glass, centered			

	The User's Actions		The Machine's Behavior	
	I Not available to the machine	II Available to the machine	III Available to the user	IV Rationale
	over the registration guide." (looks to machine) Got that? (pause) Want me to read it again?			
B:	Um:: I'm just trying to figure out what a registration guide is, but I guess that's this, um:			
A:	(Looking over her shoulder) Yea:			
B:	centered over this line thingy here.			
A:	Okay, let me read it again. "Place your original face down on the glass, centered over the registration guide, to position it for the copier lens." Okay?			
B:	'Kay.			
A:	Okay. "Slide the docu- ment cover: left over your original, until it latches."			
[Portion omitted, in which they first mislocate, then locate, the document cover]				
		CLOSES COVER		
			DISPLAY 4	Instructions to start printing
B:	Okay, now, [			
A:	All right:: "Press the Start button"			
		SELECTS START		

One way of viewing the interaction of A and B in this example is as the adept completion of what the design attempts. Specifically, A further decomposes and re-presents the instructions provided by the system, such that they are fit precisely to B's actions in carrying them out. A is able to do this because of her sensitivity to what B is doing, including B's troubles.

The design premise here is that, by detecting certain of the user's actions (those in Column II), the system can follow her course in the procedure and provide occasioned instruction. The instructions are grouped such that the last action prescribed by each display produces an effect that is detectable by the system, and initiates the processing that produces the next display. Below is the same procedure, as specified by the designer to the program that controls the display of instructions to the user:

Step 1: Set Panel
[DISPLAY 1]
Step 2: Tell User "To access the BDA . . . Raise the RDH"
[DISPLAY 2]
Step 3: Tell User "Place original face down . . . Slide document cover left"
[DISPLAY 3]
Step 4: Make Ready.
Step 5: Tell User "Press Start." Requirements:
Panel Set (If not, try Step 1)
RDH raised (if not, try Step 2)
Document cover closed (If not, try Step 3)
Ready State (If not, try Step 4)
[DISPLAY 4]
Step 6: Complete printing Step (Sets CopiesMade) Requirements:
Printing State (If not, try Step 5)

Rather than proceeding through these instructions consecutively, the system starts with the last step of the procedure, Step 6 in this case, and checks to see whether it is done. The step is done if a check of the machine's state confirms that the conditions represented by that step's "requirements" have been met. When a requirement is found that is not met, a further set of specifications, tied to that requirement, send the system back to an earlier step in the procedural sequence. The system then displays the instructions tied to that earlier step to the user, until another change in state begins the same process again. Each time the user takes an action that the system can detect, in other words, the system compares the state of the machine with the end state, returns to the first unfinished step in the sequence, and presents the user with the instructions for that and any subsequent step.

Through this simple device of working backward through the procedure, the relevance of information is preserved across certain variations in circumstance. In Example II, having discovered that their original is larger than standard paper, A and B decide to redo the job. They return to the job description display to select the reduction feature, and then direct the machine to proceed:

II [22:223–255]
(Again A and B are proceeding from the display that establishes their goal as making two-sided copies of a bound document, this time with reduction. The document is still on the copier glass, the document cover is closed.)

	The User's Actions		The Machine's Behavior	
	I Not available to the machine	II Available to the machine	III Available to the user	IV Rationale
			DISPLAY 1	Selecting
B:	It's supposed to— it'll tell "Start," in a minute.			
A:	Oh. It will?			
B:	Well it did: in the past. (pause) A little start: box will:			
			DISPLAY 4	Ready to print
	There it goes.			
A:	"Press the Start button"			
		SELECTS START		
			STARTS	
	Okay.			

This time, on their instruction, to proceed, the system goes directly to the instruction to "Press the Start button," bypassing the instructions to raise the document handler, place the document on the glass, and close the document cover, all of which are irrelevant for the present situation in that the actions have already been taken. The result is that, while B predicts the system's behavior—specifically, that it will provide them with a "Start button"—on her recollection of an occasion (Example 1) on which it actually behaved somewhat differently, her prediction holds. That is, even though the system's behavior in fact changes on this occasion, it appears to behave in the "same" way. In human interaction, this accommodation to circumstance is fundamental, and largely taken-for-granted. The success of the system's achievement of that accommodation is evident in the accommodation's transparency to the users.

Things work in this instance just because a detectable state (the closed document cover) can be linked to a prior assumption about the user's intent with respect to a next action (ready to press start), regardless of the particulars of this time through. In the next instance, however, this very same strategy leads to trouble. In this case the users have completed the unbound master copy of their journal article, and have gone on to attempt to make their two-sided copies. The order of pages in the copies are found to be faulted (a fault not available to the machine, which cannot actually read the documents), so they try again. As in II, for the users this is a second attempt to accomplish the same job, while for the machine it is just another instance of the procedure. On this occasion that discrepancy turns out to matter.

III [22:582–608]
(Again proceeding from the display that establishes their goal as two-sided copying from a bound document.)

	The User's Actions		The Machine's Behavior	
	I Not available to the machine	II Available to the machine	III Available to the user	IV Rationale
B:	Okay, and then it'll tell us,			
			DISPLAY 1	Selecting procedure
	okay, and:: It's got to come up with the little Start thing soon. (pause)			
			DISPLAY 2	Instructions for using the bound document aid
	Okay, we've done all that. We've made our bound copies. (pause)			
A:	It'll go on though, I think. Won't it?			
B:	I think it's gonna continue on, after it realizes that we've done all that.			

In sequence II, the system's ignorance of the relation between one time through and the last did not matter, just because a check of the current state of the machine caused the appropriate behavior. Or, to put it another way, the current state of the interaction could be read from the state of the machine, independent of the embedding history of the larger course of events. Here, however, a check of the current state of the machine belies the users' intent. To appreciate what they are doing now requires that one recognize the relation between this time through and the last, and the system does not. So, while both users and system are, in some sense, starting the job again, there are two different "interpretations" of what, at this particular point, it means to do so. The users' actions are located in a history of trying to make two-sided copies, finding the pages to be out of order, and starting over. As far as they are concerned, they are still trying to make two-sided copies of a bound document, so they leave their job description as such. For the machine, however, the correct description of their goal this time, having made their master copy, is two-sided copying from an *unbound* document. The result is that what they tell the machine they are doing is not what they mean to do, and what they mean to do is not available from the current state of the world as the machine is able to see it.

COMMUNICATIVE TROUBLES

The design succeeds just in those cases where a detectable state can be linked to a prior assumption about the user's intent with respect to a next action. In those cases, the system can be engineered to provide the appropriate next instruction despite the fact that, for example, the machine cannot actually detect the presence or absence of a document on the glass, and lacks any sense of continuity across instances of the procedure. In other cases, however, this indifference leads to trouble. The problem is that a single user action does not carry sufficient information to allow the attribution of its intent. Goals, similarly, such as "the intention to produce two-sided copies of a journal article," while expressing a desired outcome and recommending a set of procedural instructions, do not carry sufficient information to enable the prediction of the user's situated actions. The latter turn on the uncertainties of the user's interpretation of the instructions, under particular and at least partially unforeseeable circumstances. A consequence of the uncertain relation of expressed intent to situated action is an unavoidable uncertainty to shared understanding.

One response to that uncertainty in human–machine communication is to compile a catalogue of frequently occurring misunderstandings and attempt to forestall them. An obvious source of potential trouble are points where the procedure requires an action whose purpose is obscure. The problem with the page order of the two-sided copies in the last example arose because the procedure requires at one point that a blank sheet of paper included with the copies not be discarded. Without fail, users' first inclination on finding a blank sheet in their output is to pull it from the stack and set it aside. This problem was part of a more general class of troubles, arising from procedural requirements that go against what is the "typical" case for machines of a certain kind. In these cases, the instructions are competing with the user's preconceptions. Take the following example:

IV [20:28–30]
(Another pair of users are proceeding from the display that establishes their goal as making four copies of an unbound document. They have read the instruction "Place all of your originals in the document handler.")

	The User's Actions		The Machine's Behavior	
	I Not available to the machine	II Available to the machine	III Available to the user	IV Rationale
		A PUTS SINGLE PAGE IN DOCUMENT HANDLER		
			DISPLAY 6	Ready to print
A:	"Press the Start button." Where's the Start button? (looks around machine, then to display)			
B:	(points to display) Start? Right there it is.			
A:	There, okay.			
		SELECTS START		
			STARTS DELIVERS COPIES	
A:	So it made four of the first?			Job complete
			DISPLAY 7	Removing documents
	(looks at display) Okay.			

From the system's point of view, there is no evidence here of any trouble or confusion. The system instructs the user to place her documents in the automatic document feeder, sees the user do so, instructs her to press start, the user does, and the machine produces four copies of a single sheet. To a human observer with any knowledge of this machine, however, A's question "So it made four of the first?" indicates a misunderstanding. Specifically, her question conveys the information that this in fact is not a single page document, but the first page of several. And in contrast to previous machines that require the placement of pages on the glass one at a time, copying a document of multiple pages with this machine requires loading the pages all at once.

As an assertion in the form of a question, A's statement not only formulates her view of what just happened, but requests confirmation of that formulation. Interactionally, her statement provides an occasion for the discovery of the misunderstanding. She even looks to the display for a response. The information provided there is sufficiently general, however—it simply says, "The copies have been made"—to support her assertion, rather than challenging it.

So at the point where the machine starts to print, A is making four copies of page 1 of her document, while the machine is just making four copies of the document in the document handler. This seems, on the face of it, a minor discrepancy. If the machine copies the document, why should it matter that it fails to appreciate more finely the document's status as one in a set of three?

The problem lies in what happens next:

V [20:32–35]

	The User's Actions		**The Machine's Behavior**	
	I **Not available to the machine**	**II** **Available to the machine**	**III** **Available to the user**	**IV** **Rationale**
			DISPLAY 7	Removing original from the document handler.
A:	(Takes first page out of document handler)			
		REMOVES ORIGINAL		
			DISPLAY 8	Removing the copies
	(holding second page over the document handler, looks to display) Does it say to put it in yet? (Puts second page into document handler)			
		REPLACES ORIGINAL		
			Return to DISPLAY 7	Removing originals

The strength of A's preconception of what is going on (repeating the procedure for each page) provides her with a logical next action (loading her second page into the document handler) in advance of any instruction. The form of her question, "Does it say to put it in yet?", anticipates the system's next utterance and treats the reference of her own utterance as obvious. That the instruction will appear, and what it will say, is not in question, only when. While she is going on to the next run of the procedure, however, the system is still engaged in the completion of the last. What remains are the removal of originals and copies from their respective trays:

VI [20:38–42]

	The User's Actions		The Machine's Behavior	
	I Not available to the machine	II Available to the machine	III Available to the user	IV Rationale
			DISPLAY 7	Removing originals
A:	"Remove the original—" Okay, I've re—I've moved the original. And put in the second copy.			

By simply transforming "remove" to "move," A adeptly makes the instruction to undo her last action into confirmation of her last action. The misunderstanding between A and the system at this point turns on just what the document in the document handler is, and how it got there. For A, a first page has been replaced by a second, a necessary step for the next run of the procedure. For the machine, there is just a document in the document handler, and its removal is required in order to complete the last run. The result is a deadlock wherein both user and system are waiting for each other, on the assumption that their own turn is complete, that their next action waits on an action by the other.

The problem here is not simply a failure of anticipation on the designer's part. On the contrary, in anticipation of this very problem the instruction for loading documents explicitly states that "all" the pages should be placed in the document handler. But the user's preconception prevails over the instruction, and the "all" is ignored. Given that trouble, the further difficulty arises from the fact that the misunderstanding goes by unnoticed at the point where it occurs. This happens because the user's actions are not faulted in terms of allowable procedure, but only in terms of her intent. The latter, while demonstrated in her talk and actions, is unavailable to the machine. Because those traces of the user's actions that are available to the machine are indistinguishable from the traces of some other procedure, the machine's response at the point where the user selects "Start," rather than flagging the trouble in her reading of the instruction and occasioning its repair, appears to provide confirmation.

The system can recognize the situation only as one of a class, not in its particulars. Specifically, the system does not have access to what the referrent of "all of the originals" is for any given case. It must take the number of documents in the document feeder, whatever that may be, as "all" that the user intends to copy. We might imagine getting around this problem by having the machine ask the user how many documents she means to copy, and then count to see that the number in the document handler and the number specified match. But imagine the comparable situation wherein, before we embarked on a collaborative task, I was required to sit down and specify to you what my goals were. You then scrutinized each of my actions to see if they fit the sequence of actions that you believed were the necessary means to my ends. Imagine further that you had no access to my words, only to a small subset of my behavior. At each point where something that I did failed to match your preconceptions about what my behavior should look like, given my stated intentions, you would challenge me and require that I explain myself. The collaboration itself, quite apart from the demands of my original task, would quickly become an onerous chore.

FOUNDATIONS OF INTERACTION

Successful communication relies crucially on situated interpretations of action. In Bluebonnet, as in other efforts to engineer human–machine interaction based on a planning model, the significance of actions is located in underlying plans. User's actions are interpreted by mapping the actions or, more accurately, the effects of certain of the actions detectable by the machine, onto a projected plan. It is the fit of the action to the plan that gives the action its significance.

The troubles in communication exemplified in the case above, however, raise questions about the adequacy of plan recognition for interaction. At the least, plan recognition seems not to address the question of how the significance of what we do, in some actual situation, is derived from the particular, concrete circumstances of our action. Plans, in other words, appear to have certain context-dependent, or *indexical* properties. Among philosophers and linguists, the term *indexicality* typically is used to distinguish those classes of expressions, like *this* and *that*, *here* and *now*, *I and you*, whose meaning is conditional on the situation of their use, from those such as, for example, noun phrases that refer to a class of objects, whose meaning is claimed to be specifiable in objective, or context-free terms.

But in an important sense, namely a *communicative* one, the significance of a linguistic expression is always contingent on the circumstances of its use. In this sense deictic expressions, place and time adverbs, and pronouns are just particularly clear illustrations of a general fact about situated language.

The problem of communicating instructions for action becomes clearer with this fact about language in mind. Like all efficient descriptions, instructions must be interpreted with respect to a collection of actions and circumstances that they never fully specify. The necessary incompleteness of instructions means that the significance of an instruction with respect to some particular, situated action must, in the end, be found by the instruction-follower. To an important extent, moreover, the methods for interpreting instructions, if they are to accommodate the unforeseeable contingencies of situated action, *must be ad hoc:*

> To treat instructions as though *ad hoc* features in their use was a nuisance, or to treat their presence as grounds for complaining about the incompleteness of instructions, is very much like complaining that if the walls of a building were gotten out of the way, one could see better what was keeping the roof up. (Garfinkel, 1967, p. 22)

The project of instruction writing is ill-conceived, in short, if its goal is seen as the production of action descriptions that guarantee a particular interpretation. What 'keeps the roof up' in the case of instructions for action is not only the instructions as such, but their situated interpretation. And the latter has all of the properties of *ad hocery* and uncertainty that characterizes every occasion of the situated use of language.

Communicative Resources

A central tenet in the social study of action is that the sources of action's significance, and the resources for action's interpretation, are interactional and situated. These aspects of human action are the subject matter of a recent branch of sociology that goes under the name of ethnomethodology.[4] While research in human–machine interaction has made extensive use of cognitive psychology, this other, essentially sociological, perspective has gone unnoticed. Acknowledging the role of individual predispositions in mutual intelligibility, the branch of ethnomethodology called conversation analysis focuses on the neglected other side of shared understanding; namely

[4] For a programmatic statement of this perspective, see Garfinkel and Sacks (1970) and Heritage (1984). For a more recent survey of its application to studying science, see Lynch, Livingston, and Garfinkel (1983).

the local, interactional work that produces intelligibility in situ, as an essentially collaborative achievement. Rather than the reliable recognition of intent, mutual intelligibility is understood to rely on the availability of communicative resources to detect, remedy, and at times even exploit the inevitable troubles that arise in constructing action's significance.[5] Everyday conversation is taken to incorporate the broadest range of such resources, with other forms of interaction (e.g., interviews, lectures) being characterizable in terms of particular resource limitations, or additional constraints. This view suggests that we might productively take human–machine interaction to be an extreme form of resource-limited communication, applying essentially the same methods to its analysis as those used in the study of human conversation.

As the basic system for situated communication, conversation is characterized by (a) an organization designed to support local, autonomous control over the development of topics or activities, and to maximize accommodation of unforeseeable circumstances that arise; and (b) resources for locating and remedying the communication's troubles as part of its fundamental organization. A primary mechanism for both local control and the identification and repair of trouble is the *conditional relevance* of conversational turns:

> By conditional relevance of one item on another we mean; given the first, the second is expectable; upon its occurrence it can be seen to be a second item to the first; upon its nonoccurrence it can be seen to be officially absent. (Schegloff, 1972, p. 364)

The first of two utterances that stand in a relationship of conditional relevance, in this local sense, sets up an expectation with respect to what should come next, and directs the way in which what does come next is heard. By the same token, the absence of an expected second part is a notable absence, and therefore takes on significance as well. In this way silences, for example, can be meaningful—most obviously, a silence following an utterance that implicates a response will be heard as belonging to the recipient of the utterance, and as a failure to respond. Similarly, a turn that holds the place of a response, but cannot be made relevant to the prior turn will be seen as a non sequitur, or as incoherent.

While conditional relevance is a constraint on inference, it is a weak constraint in the sense that it does not prescribe what counts as a response to a given action, only that whatever is done next will be viewed as a response. In fact, the range of actions in a second position that can be

---

[5] This observation has been documented extensively in the analysis of conversation; e.g., Jefferson (1972), Jordan and Fuller (1975), Sacks et al. (1978).

heard as a response is extended in virtue of the expectation that conditional relevance sets up, rather than constrained. That is to say, an action that is not in any explicit way tied to the action that follows will nevertheless be interpreted as a response, just in virtue of its position:

A: Are you coming?
B: I gotta work. (Goffman, 1975, p. 260)

The position of B's utterance as a response means that we look for its relevance to A's question. So in this case, B's statement can be heard as a negative reply, just as B's question can be heard as an affirmative reply in the following:

A: Have you got coffee to go?
B: Milk and sugar? (Merritt, 1977, p. 325)

The conditional relevance of questions and answers is illustrative of the kinds of structure to which conversation analysis is attuned. The general prescript is that, as participants in conversation use the local environment of their talk to provide for its mutual intelligibility, so must we as analysts. The coherence of a stretch of talk, whatever the prior agendas or retrospective accounts we might assign to it, must first and foremost be produced locally, then and there, through the contingent details of some actual occasion of interaction.

Managing Communicative Troubles

Because communication takes place in real environments, under real performance requirements on actual individuals, it is vulnerable to internal and external troubles that may arise at any time, from a misunderstanding to a clap of thunder (cf. Schegloff, 1982). Our communication succeeds in the face of such disturbances, not because we reliably predict what will happen, and thereby avoid any problems, or even because we encounter problems that we have anticipated in advance, but because we work, in situ, to identify and remedy the inevitable troubles that arise:

> It is a major feature of a rational organization for behavior that accommodates real-worldly interests, and is not susceptible of external enforcement, that it incorporates resources and procedures for repair of its troubles into its fundamental organization. (Sacks, Schegloff, & Jefferson, 1978, p. 39)

The resources for detecting and remedying problems in communication, in other words, are the same resources that support communication that is trouble free. The same option that provides for ordinary turn transitions, for example, affords the recipient of an utterance the occasion to assert that he or she has some trouble in understanding, or to request some clarification. Turn transition places provide recurring opportunities for the listener to initiate some repair, or to request clarification from the speaker. Alternatively, clarification may be offered by the speaker, not because the recipient of an utterance asks for it, but because the speaker finds evidence for some misapprehension in the recipient's response. The work of repair in either case includes calling the other's attention to the occurrence of some troublesome item, remedying it, and resuming the original line of action in which the troublesome item is embedded.

Given the lack of specific criteria for shared understanding in most cases, a crucial part of interactional competence is the ability to judge whether evidence of misunderstanding on a given occasion warrants the work required for repair (Jefferson, 1972). The decision whether to challenge a troublesome item or to let it pass involves, in part, a weighing of the relative work involved in the item's clarification, versus the foreseeable dangers of letting it go by. The risks of the latter are exemplified by the 'garden path' situation, where speakers fail to identify some communicative trouble at the point where it occurs, and only discover at some later point in the interaction that there has been some misunderstanding (cf. Jordan & Fuller, 1975; Gumperz & Tannen, 1979). At the point of discovery, the coherence of the interaction over some indefinite number of past turns may be called into question, and the source of the trouble may be difficult or impossible to reconstruct. In contrast to the routine problems and remedies that characterize local repair, such a situation may actually come close to communicative failure; i.e., it may require abandoning the current line of talk, or beginning anew.

A side sequence initiated by an assertion of misunderstanding or request for clarification sets up an exchange that the participants in most cases did not anticipate. In this and many other eventualities, the organization of everyday interaction is necessarily an ad hoc accomplishment. While any exchange can be reconstructed into an hierarchy of topics and subtopics, every instance of coherent interaction is in fact an essentially local production, accomplished collaboratively in real time, rather than "born naturally whole out of the speaker's forehead, the delivery of a cognitive plan" (Schegloff, 1982, p. 73):

> Good analysis retains a sense of the actual as a achievement from among possibilities; it retains a lively sense of the contingency of real things. It is worth an alert, therefore, that too easy a notion of 'discourse' can lose us that.
>
> If certain stable forms appear to emerge or recur in talk, they should be understood as an orderliness wrested by the participants from interactional

> contingency, rather than as automatic products of standardized plans. Form, one might say, is also the distillate of action and interaction, not only its blueprint. If that is so, then the description of forms of behavior, forms of discourse . . . included, has to include interaction among their constitutive domains, and not just as the stage on which scripts written in the mind are played out. (p. 89)

The organization of face-to-face interaction is the paradigm case of a system for communication that has evolved in the service of orderly, concerted action over an indefinite range of essentially unpredictable circumstances. What is notable about that system is the extent to which mastery of its constraints localizes, and thereby leaves open, questions of control and direction, while providing built-in mechanisms for recovery from trouble and error. The constraints on interaction in this sense are not determinants of, but are rather "production resources" (Erickson, 1982) for, shared understanding. The limits on available resources for accomplishing a shared agenda, and for detecting and remedying the troubles that that task poses, define the problem for human–machine interaction.

CONCLUSION

Whatever the common properties of using computer-based machines and human communication turn out to be, their discovery is better served by clarifying the distinction between human interaction and machine operation, than by their a priori identification. Amid ongoing debate over specific problems in the design and use of interactive machines, little question is raised regarding the bases for the idea of human–machine interaction itself. At the same time, recent developments in social studies regarding the foundations of human interaction have had remarkably little influence on the discussion of interactive machines. One aim of the research reported here is to suggest the mutual relevance of two fields of endeavor—research in human–machine communication, and social studies of human interaction—that today are largely unaware of each other.

Attempts to simulate interaction challenge those of us with an interest in human interaction to say just what we take that to be. By the same token, social studies have produced initial descriptions of human communication that are relevant to the problem of design. First, the mutual intelligibility that we achieve in our everyday interactions—sometimes with apparent effortlessness, sometimes with obvious travail—is always the product of local, collaborative work. Second, the apparatus of face-to-face communication, which supports that work, is designed to maximize sensitivity to situation particulars. Third, that apparatus incorporates resources for locating and remedying the communication's troubles as part of its fundamental organization. And finally, every occasion of human communication is embedded in, and makes use of, a mutually accessible linguistic and extralinguistic world.

Everything about our current communicative practices assumes such an embedding world. The designer of an interactive artifact must contend with the situatedness of interaction in two, related respects. First there is the fact that the significance of action, including but not limited to linguistic utterances, relies upon its circumstances. The user of a system deploys a full complement of observational, inferential, and language-using skills in order to construe—and misconstrue—the system's behavior according to the behavior's particular circumstances. At the same time, the system is mapping a restricted set of effects, left by the user's actions, onto a prescribed template of possible meanings. As a consequence of the asymmetrical access of user and machine to their situation, the ordinary collaborative resources of human interaction are unavailable. The real project of the designer of an interactive artifact, accordingly, is to engineer alternatives to interaction's situated properties.

The most plausible alternative—and the second way that designers must contend with the situatedness of interaction—is to substitute a description of what the situation of action will be, for access to the actual situation. The user is necessarily left to contend with the disparity between the idealized sequence of events, on which the design is based, and the action's actual course. The discrepancy of idealized behavior and situated action is not, in itself, the problem for human-machine communication. The problem is, rather, that designers tend to identify representations of action, like plans, with situated action. As a consequence, the question of the correspondence of idealized descriptions of human behavior to situated action is solved a priori, rather than defined as a problem for research. And only if that question is addressed directly can we hope to build our theories about human interaction on solid ground, and bring them usefully to bear on the design of a new technology.

REFERENCES

Burton, R., & Brown, J. S. (1982). An investigation of computer coaching for informal learning activities. In D. Sleeman & J. S. Brown (Eds.), *Intelligent tutoring systems*. London: Academic Press.

Erickson, F. (1982). Money tree, lasagna bush, salt and pepper: Social construction of topical cohesion in a conversation among Italian-Americans. In D. Tannen (Ed.), *Georgetown University Roundtable on Language and Linguistics*. Washington, DC: Georgetown University Press.

Garfinkel, H. (1967). *Studies in ethnomethodology*. Englewood Cliffs, NJ: Prentice-Hall.

Garfinkel, H., & Sacks, H. (1970). On formal structures of practical actions. In J. C. McKinney & E. A. Tiryakian (Eds.), *Theoretical sociology*. New York: Appleton-Century-Crofts.

Goffman, E. (1975). Replies and responses. *Language in Society, 5,* 257–313.

Gumperz, J., & Tannen, D. (1979). Individual and social differences in language use. In C. Fillmore et al. (Eds.), *Individual differences in language ability and language behavior*. New York: Academic Press.

Heritage, J. (1984). *Garfinkel and ethnomethodology*. Cambridge, England: Polity Press.

Jefferson, G. (1972). Side sequences. In D. Sudnow (Ed.), *Studies in social interaction*. New York: Free Press.

Jordan, B., & Fuller, N. (1975). On the non-fatal nature of trouble: Sense-making and trouble-managing in Lingua Franca talk. *Semiotica, 13,* 1.

Lynch, M., Livingston, E., & Garfinkel, H. (1983). Temporal order in laboratory work. In K. Knorr-Cetina & M. Mulkay (Eds.), *Science observed*. London: Sage.

Merritt, M. (1976). *Resources for saying in service encounters*. Unpublished Ph.D. dissertation, Department of Linguistics, University of Pennsylvania.

Sacks, H., Schegloff, E., & Jefferson, G. (1978). A simplest systematics for the organization of turn-taking for conversation. In J. Schenkein (Ed.), *Studies in the organization of conversational interaction*. New York: Academic Press.

Schegloff, E. (1972). Sequencing in conversational openings. In J. Gumperz & D. Hymes (Eds.), *Directions in sociolinguistics: The ethnography of communication*. New York: Academic Press.

Schegloff, E. (1982). Discourse as an interactional achievement: Some uses of "uh huh" and other things that come between sentences. In D. Tannen (Ed.), *Georgetown University roundtable on language and linguistics*. Washington, DC: Georgetown University Press.

Suchman, L. (1987). *Plans and situated actions: The problem of human–machine communication*. Cambridge, England: Cambridge University Press.

APPENDIX

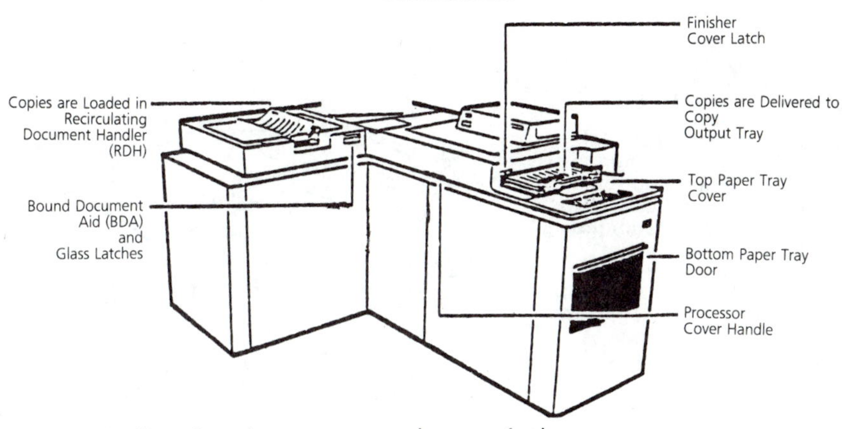

Describe the document to be copied:

Is it a bound document? Yes **No**

Copy both sides of each sheet? Yes **No**

Is it on standard size (8.5" × 11") paper? **Yes** No

Is it on standard thickness paper? **Yes** No

Quality of original: darker than normal **normal** lighter than normal

About how many images are to be copied?

1	2	3
4	5	6
7	8	9
	0	Clear

Describe the desired copies:

Number of copies:

1	2	3
4	5	6
7	8	9
	0	Clear

The currently described job will take about 1 minute.

Use standard paper? Yes No

Staple each copy? Yes No

Put images on both sides? Yes No

Reduce size of images? No 35% smaller 25% smaller 2% smaller

PROCEED HELP

DISPLAY 0

OVERVIEW:

You need to use the Bound Document Aid (BDA) to make an unbound copy of your original. That copy can then be put into the Recirculating Document Handler (RDH) to make your collated two-sided copies.

INSTRUCTIONS:

Please wait.

THE MACHINE

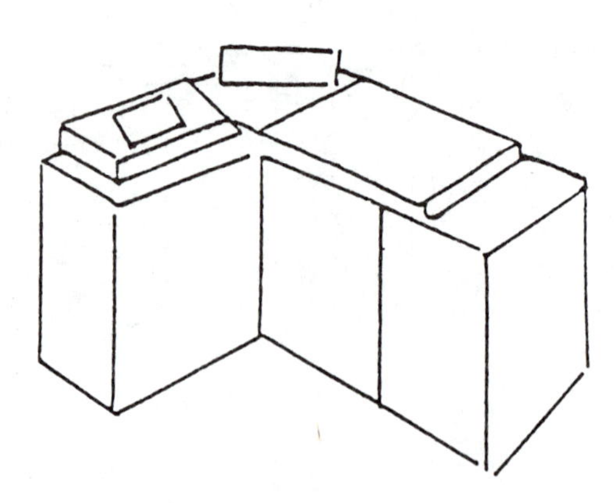

Change Task Description

Help

DISPLAY 1

OVERVIEW:

You need to use the Bound Document Aid (BDA) to make an unbound copy of your original. That copy can then be put into the Recirculating Document Handler (RDH) to make your collated two-sided copies.

How to access the BDA:

To access the BDA, pull the latch labelled bound document aid,

And lift up and to the left.

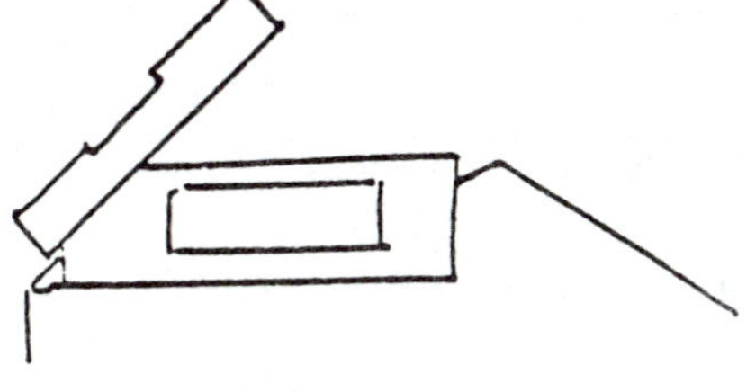

INSTRUCTION:

Pull the latch labelled bound document aid. (To release the RDH.)

Raise the RDH. (To enable placement of the bound document on the glass.)

Change Task Description

Help

DISPLAY 2

OVERVIEW:

How to close
the document cover:

To close the
document cover,
grasp the cover,
and slide it firmly
to the left.

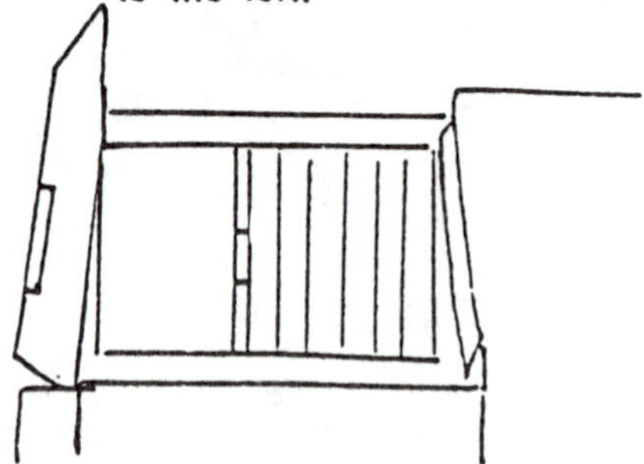

INSTRUCTION:

Place your original
face down on the glass,
centered over the
registration guide.
(To position it for the copier lens)

Slide the document cover
left over your original
until it latches.
(To provide an eye shield
from the copier lights)

DISPLAY 3

OVERVIEW:

THE COPIER

ASSUMPTIONS:

The first page to be copied is on the glass.

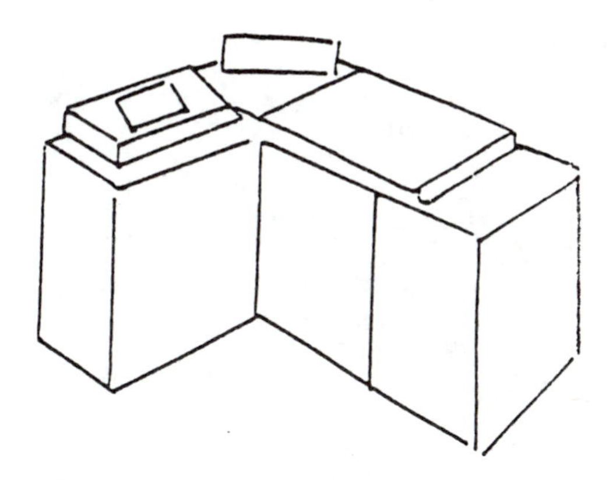

INSTRUCTION:

Press the Start button (to produce a copy in the output tray)

START

DISPLAY 4

OVERVIEW:
You can use the
Recirculating Document
Handler (RDH) to make
your copies.

THE MACHINE
This is the RDH

INSTRUCTIONS:

Place all of your originals
in the RDH,
first page on top.
(so that the RDH can
automatically feed each
sheet into the copier.)

Change
Task Description

Help

DISPLAY 5

OVERVIEW:
You can use the
Recirculating Document
Handler (RDH) to make
your copies.

THE MACHINE

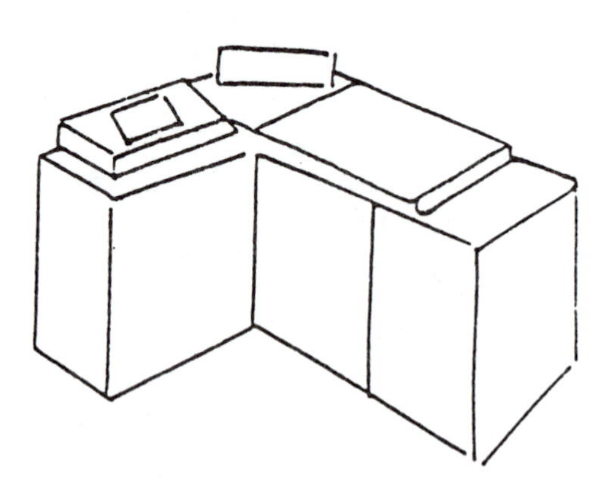

ASSUMPTIONS:
The document to be
copied is in the RDH.

INSTRUCTIONS:

Press the Start button.
(to produce 4 copies
in the output tray.)

Change Task
Description

Start

Help

DISPLAY 6

THE MACHINE

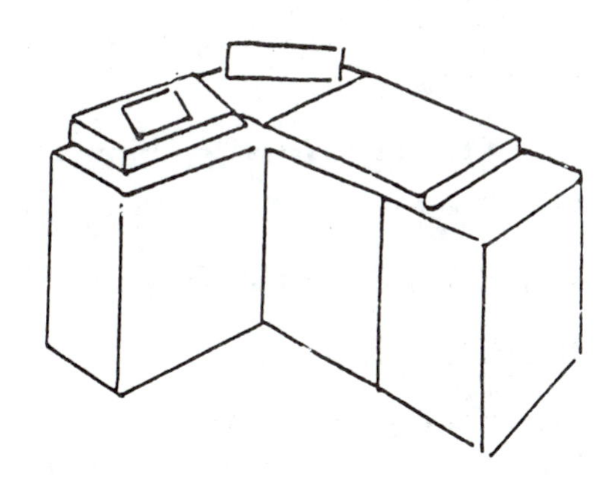

ASSUMPTIONS:
The copies have been made.

INSTRUCTIONS:

Remove the originals from the RDH.

Change Task Description

Help

DISPLAY 7

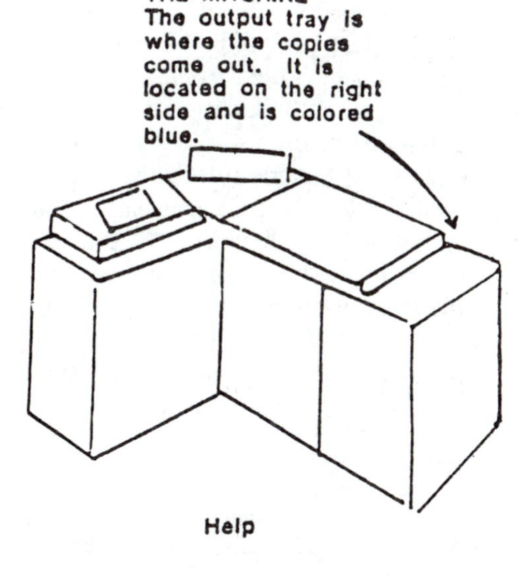

THE MACHINE
The output tray is where the copies come out. It is located on the right side and is colored blue.

ASSUMPTIONS:
The copies have been made.

INSTRUCTIONS:

Remove the copies from the output tray.

Change Task Description

Help

DISPLAY 8

3

Organizing Information Processing Systems: Parallels Between Human Organizations and Computer Systems*

Thomas W. Malone

Sloan School of Management
Massachusetts Institute of Technology
Cambridge, MA

In this chapter, I will describe some striking parallels between the organizational forms that occur in computer programs and those that occur in human organizations. In addition to suggesting analogies between these two kinds of information-processing entities, I will show how concepts and results from the fields of computer science and organizational design can each be translated into the other, and I will give examples of new implications for each field from such translations.

One goal of this chapter is to develop a framework for using ideas about human organizations as a source of inspiration in the design of computer systems. As Mead and Conway (1980) remark, "Analogies with human structures may help to suggest the kinds of behavior we might achieve in computational structures. . . . The design of computers and of algorithms has yet to show the ingenuity reflected in human organizations" (p. 264, 265). One good reason for studying large scale organizations for this purpose, instead of individual human brains (e.g., Hillis, 1985; McClelland & Rumelhart, 1985; Touretzky & Hinton, 1985) is that organizations are much more open for inspection than brains. Because our introspective

*Most of this chapter was written while the author was at the Xerox Palo Alto Research Center (PARC), Palo Alto, CA. An earlier version of the chapter was distributed as a Xerox PARC working paper (April 1981, revised August 1982). The chapter benefited from the comments of a number of people, including John Seely Brown, Richard Fikes, Austin Henderson, Paul Ricci, Steve Smith, Lucy Suchman, and Shoshanah Zuboff. Long discussions with Michael Cohen, in particular, were the source of a number of insights reflected here.

access to our own minds is largely serial, introspection is distinctly unhelpful in designing highly parallel computational structures. Human organizations, on the other hand, are highly parallel in a very visible way. Observations of, and in some cases detailed theories about, human organizations can thus be used to design highly parallel computer systems. A number of recent papers (Chandrasekaran, 1981; Wesson, Hayes-Roth, Burge, Stasz, & Sunshine, 1981; Kornfeld & Hewitt, 1981; Smith & Davis, 1981; Fox, 1981; Lesser & Corkill, 1981) have discussed social systems as a metaphor for designing computer programs. Fox (1981), in particular, has begun to develop a framework similar to the one developed here. The analysis here, however, separates the different levels of coordination structures more clearly, highlights the trade-off between flexibility and efficiency, and gives several new examples of cross-disciplinary transfers.

Another goal of this chapter is to move toward analytically derivable principles that must hold for any information processing system. This analysis, by illuminating the fundamental information processing constraints inherent in any organization, suggests how to design new forms of human organizations. Most interestingly, this approach can provide a principled basis for predicting or designing new kinds of human organizations that take advantage of the unprecedented information-processing capacities of computers (e.g., see Malone & Smith, 1988; Malone, Yates, & Benjamin, 1987).

Finally, this chapter will hint at how some of the concepts traditionally used to describe human organizations can be made more rigorous using computational metaphors (e.g., Malone, 1987; Malone & Smith, 1988). The success of computational metaphors in helping to understand individual thought processes argues for the fruitfulness of this approach in understanding organizations as well. In this endeavor, the current paper is similar in spirit to earlier work on organizations by March and Simon (1958), Cyert and March (1963), Galbraith (1973, 1977), Cohen (1981, 1984), and others. In describing human organizations as information processing systems, I do not intend to imply that they are "nothing but" information processing systems. I will argue, however, that a very important part of what happens in human organizations can be explained in terms of information processing, and that an understanding of this aspect of organizations provides an essential foundation for understanding other facets of organizations such as power relationships and personal satisfaction.

Almost all the issues I will discuss here are also involved in individual human information processing. Though I will not dwell on the point in this chapter, one interesting possibility is that organizational concepts like those discussed here may also be useful in modeling the behavior of individual minds (see, e.g., Minsky, 1986; Dennett, 1978).

This chapter is intended primarily as an introduction to a novel way of looking at both human and computer organizations and seeing common structures in both. Accordingly, it presents a broad overview of promising directions for further exploration and brief descriptions of several forays along these paths. If it is successful, it will raise as many questions as it answers and stimulate as many ideas as it provides. Since these ideas are intended to be of interest to researchers in both computer science and organizational behavior, I have attempted to provide enough background to make the paper accessible to both groups.

THE FRAMEWORK

One dictionary definition of *organize* is to "arrange systematically for harmonious or united action." Thus a group of agents is an *organization* if (1) they are connected in some way ("arranged systematically"), and (2) their combined activities result in something better (more "harmonious"[1]) than if they were not connected. In other words, an organization consists of:

1. a group of *agents*,
2. a set of *activities* performed by the agents,
3. a set of *connections* among the agents, and
4. a set of *goals* or evaluation criteria by which the combined activities of the agents are evaluated.

To *organize*, then, is to

1. establish (either explicitly or implicitly) the goals of the organization,[2]
2. segment the goals into separate activities to be performed by different agents, and

---

[1] Mesarovic, Macko, and Takahara (1970, p. 124) define *harmony* among a group of agents as the existence of an element which simultaneously optimizes their separate evaluation functions. The definition I am using here is somewhat weaker, and—I believe—more consistent with the intuitive notion of harmony. This new definition is that a group of agents is in harmony (with respect to a given evaluation function) if the value of the combined activities of the agents is greater than the value of their separate activities would be if they were not connected.

[2] There may well be a great deal of ambiguity about what the goals of an organization really are (e.g., see Cyert & March, 1963, chap. 3). Different observers and different members of a given group may have different ideas about the goals of the group (e.g., to maximize profit or to provide employment), but one is not justified in calling a group an "organization" without at least some implicit notion of how to evaluate its success.

3. connect the different agents and activities so the overall goals are achieved.

Alternative Ways of Segmenting a Problem

In general, problems should be divided so that tasks or resources that interact strongly with each other are placed together. The more self-contained the separate segments are, the less need there is for coordination. For example, if all the people working on design, manufacture, and marketing for a new product are in the same division of a company, this division is relatively self-contained and will have relatively few needs for coordination with the rest of the company. Similarly, if a computer system is designed to keep track of inventory in a warehouse and of customer billing records, these two tasks should probably be performed by separate programs, since they will have relatively few interactions with each other.

There is a tension, however, between the need to minimize coordination costs by creating self-contained tasks and the need to take advantage of economies of scale by sharing resources among several tasks. For example, in the case of organizations, there may be large savings from using the same manufacturing facilities for several different products. Or in the case of computer programs, it may be very helpful to use the same procedures for reading and writing disk files in several different programs. But sharing resources among several tasks requires coordination. If two products are made using the same factory facilities, their needs may sometimes conflict. And if two different programs use the same disk accessing procedures, a change that makes the procedure more convenient to use in one program may make it less convenient in the other program.

These issues have been analyzed in some detail for organizations (see, for example, Galbraith, 1973; Mintzberg, 1979; Malone & Smith, 1988; Malone, 1987). The questions are usually phrased in terms of the desirability of grouping by *product* (or geographical region, type of customer, etc.) or grouping by *function*. The best solution for a given organization depends on the relative importance of coordination costs and economies of scale in the technology used by the organization. Organizations in rapidly changing and complex environments (e.g., high technology companies) often use product groupings to minimize their needs for coordination. Organizations in relatively stable environments often use functional groupings to take advantage of economies of scale in their technologies. This trade-off between flexibility and efficiency is a central issue in organization design and will appear again below. It is possible, as Malone and Smith (1988) show, to explain a number of major historical changes in American business structures in terms of changes in the costs and relative importance of these dimensions.

Most computer programs use a kind of multilayered organization that includes, in a sense, both product and functional groupings. As I will discuss in more detail below, this structure more closely resembles that of a marketplace than that of a single hierarchical organization.

Functions of Coordination

Two of the most important kinds of interaction among parts of a problem involve sharing scarce resources and sharing intermediate results. Thus two of the most fundamental components of coordination are (a) *allocation of scarce resources*, and (b) *communication of intermediate results*. Even though information about intermediate results may sometimes be a scarce resource (e.g., because of limited communication capacity or because of intentional withholding of information to enhance power), it is still useful to distinguish these two components of coordination.

A number of aspects of coordination can be analyzed in terms of these two components. For example, *synchronizing interdependent activities* involves both of these components. If one activity requires as input the results of other activities, then synchronizing the communication of intermediate results is required. If nondivisible scarce resources (such as assembly line time or input/output channel time) must be shared then allocating these resources requires spreading demands out over time.

As another example, these two functions of coordination include the three kinds of interdependencies in organizations identified by Thompson (1967): (a) *pooled interdependence,* when several parts of an organization make discrete contributions to the whole and are each supported by the whole (e.g., separate doctors in the same medical clinic), (b) *sequential interdependence,* when one activity depends on another but not vice versa (e.g., one stage of a manufacturing process produces the component parts for the next stage), and (c) *reciprocal interdependence,* when each activity depends for its inputs on the other's outputs (e.g., an airline maintenance department receives planes to be repaired from the operations department, and the operations department, in turn, receives repaired planes from the maintenance department). Pooled interdependence, as Mintzberg (1979) notes, is essentially a requirement for the first coordination function described here—allocation of shared resources. Both sequential and reciprocal interdependence are subsumed under the second function of coordination—communication of intermediate results.

Smith and Davis (1981) discuss two forms of cooperation in distributed problem solving: (a) task sharing, and (b) result sharing. Result sharing is communication of intermediate results. Task sharing includes both segmenting the problem and allocating scarce processing resources (in this case, the processors themselves).

Kinds of Connections for Coordination

In order to fulfill the allocation and communication functions of coordination, agents must be connected to each other. We will be primarily concerned here with information connections rather than physical connections. In order to communicate intermediate results, information must be transferred over these *information links*, which may run in either or both directions. Even though I will use both unidirectional and bidirectional links in this chapter, a more detailed analysis such as that used in Information Control Nets (Ellis, 1979; Ellis, Gibbons, & Morris, 1979; Cook, 1979) would decompose all bidirectional links into unidirectional links at different times in a process. One way to think about information links between agents is as *messages* passed between them. This is the approach taken by object-oriented programming languages such as Smalltalk (Goldberg & Robson, 1983; see also Hewitt, 1977).

In order to allocate shared resources, activities must be able to transfer control over the shared resources. This ability often leads to the ability to prescribe behavior, as well. Therefore, it is sometimes useful to distinguish *control links* as a special kind of information link that carries information the recipients are *motivated* to follow as instructions. Of course, there are many other possible distinctions that can be made among kinds of links, depending on the kinds of information conveyed and on how the recipients are motivated to respond to the information.

Processes for Coordination

The links I have described can be used to analyze (or invent) arbitrarily complex coordination processes, but there are a few very common processes. In this section, I will describe three of these common structures. The three processes are taken from Mintzberg (1979), who, in turn, drew heavily from Simon (1957), March and Simon (1958), and Galbraith (1977). The three fundamental coordination processes (illustrated in Figure 1) are *mutual adjustment, direct supervision*, and *standardization*. They are arranged here in order of decreasing flexibility and increasing efficiency.

Mutual adjustment. Mutual adjustment is the simplest form of coordination. It occurs whenever two or more entities agree to share resources in order to achieve some common goal. In the process of agreeing to collaborate and in carrying out the joint project, the entities must usually exchange a lot of information and make many adjustments in their own behavior, depending on the behavior of the others. In coordination by mutual adjustment, no party has any prior control over the others, and decision making is a joint process. Coordination in peer groups and in markets is usually a form of mutual adjustment. Other examples are the

Figure 1. Three fundamental coordination processes.

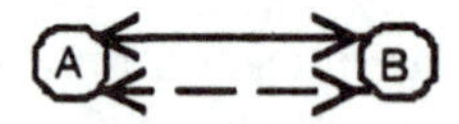

(a) Mutual adjustment

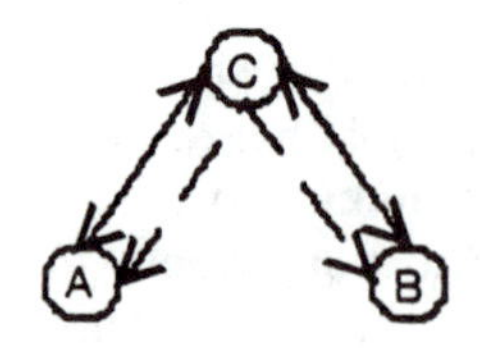

(b) Direct supervision

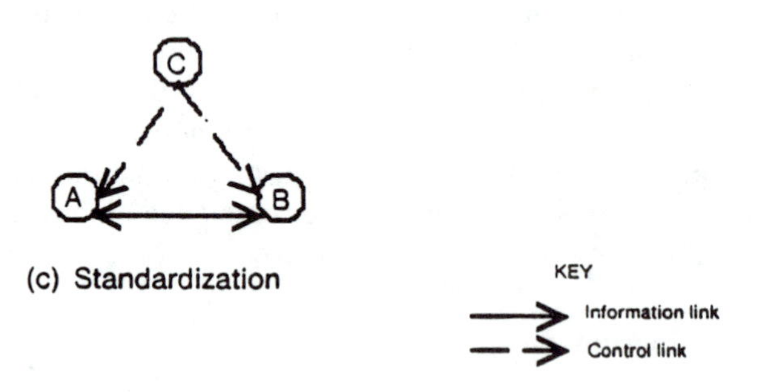

(c) Standardization

"Worm" program described by Shoch and Hupp (1982) and the "contract net protocol" described by Smith and Davis (1978, 1981; Smith, 1980), where independent computer processors negotiate about collaborating on common tasks.

Direct supervision. Another very common form of coordination is direct supervision. This form occurs when two or more entities have already established a relationship in which one entity has some control over the others. This prior relationship may have been established by mutual adjustment (as when an employee or subcontractor agrees to follow certain kinds of directions from a supervisor in exchange for control over a valued resource such as money) or by standardization (as when a procedure in a computer program has been standardized to operate only when called by another procedure and then only with the parameter values supplied by the calling procedure). In this form of coordination, the supervisor controls the use of shared resources (e.g., human labor, computer processing time, money) by the subordinates and may also prescribe certain aspects of their behavior. In many cases the supervisor communicates intermediate results between the subordinates, but there may also be information links directly between subordinates who share a common supervisor.

By eliminating the needs for all subordinates to know all aspects of a situation, and for there to be a consensus about all courses of action, direct supervision usually substantially reduces the amount of communication and information processing necessary to coordinate an activity. This increase in efficiency, however, may decrease certain kinds of flexibility and satisfaction.

Standardization. When a supervisor directs subordinates to perform certain actions in the current situation, the supervisor is coordinating by direct supervision. When, instead, the supervisor establishes standard procedures for subordinates to follow in a number of situations, the supervisor is coordinating by standardization. Standardized organizational procedures of this kind (e.g. expense reporting procedures) are very common in human organizations. And computer programs consist entirely of standardized procedures, so standardization is, in an important sense, the only way different parts of a computer system are coordinated.

As will be discussed in more detail below, standardization generally allows a significant reduction in the amount of information transfer necessary compared to direct supervision. For instance, Figure 1 (c) shows only one information link directly between A and B in place of two links connecting each of A and B with C in Figure 1 (b).

Systems for Coordination

Using the fundamental coordination processes—mutual adjustment, direct supervision, and standardization—as building blocks, elaborate systems for coordination can be constructed. Two of the most prevalent systems for achieving coordination are *hierarchies* and *markets*. As I will describe below, these two systems are not strict alternatives. Instead, hierarchies are a very important special case of the kinds of structures that can be formed by multiple contracting in a market.

Hierarchies. In very small groups, most tasks can be coordinated by mutual adjustment among the group members. However, as the size of the groups (or the number of subtasks) grows, the number of information links necessary and the amount of information to be exchanged becomes prohibitive very quickly. By using two levels of direct supervision, a large group can be divided into subgroups. If the subgroups are well chosen, as discussed above in the section on segmenting problems, most of the information transfer necessary can occur within subgroups and the few interactions between subgroups can be handled by the supervisors of the subgroups. This structure can, of course, be repeated to build up many levels of hierarchy.

Many human organizations have this parts-within-parts hierarchical structure, at least in the formal reporting relationships, and many computer programs are also organized in this way with layers of procedures composed of other procedures.

Hierarchies have a number of useful organizational properties besides reducing the overall information flow necessary. For example, Simon (1981) argues that complex systems are much easier to describe and understand and much more likely to evolve from simple systems when they are hierarchically structured than when they are not. Simon marshalls examples from sociology, biology, and chemistry in support of his arguments.

Using examples from computer science, Mead and Conway (1980) argue that, in the design of VLSI circuits, "the organization of real estate on silicon dictates a hierarchical communication system for any devices that must support global communication" (p. 292). What Mead and Conway actually show (pp. 314–317) is that a hierarchically organized random-access memory has a much faster response time (for large memory size) than a memory with a single linear bus. In other words, they do not show that hierarchies are optimal in the space of all possible organizations, they only show that a hierarchy is superior to a linear bus. Arguments such as those given by Simon (1981) suggest that a more rigorous proof of optimality over all possible structures might be given, but the discussion of lateral relations below suggests some other promising possibilities.

Augmented hierarchies. Even though hierarchies have a number of information processing and other advantages over nonhierarchical systems, hierarchies, too, have their limitations. It is a truism in the study of organizational behavior that there are many "informal" communication paths in an organization that are not shown on the formal organization chart. In fact, it seems likely that most organizations would be unable to function at all without large amounts of this nonhierarchical communication. The problems solved by most organizations are, to use Simon's (1981) term, "nearly decomposable," but they are only *nearly* so. If all the interactions between subgroups in different parts of a hierarchy had to be channeled up to a level where the two groups shared a common supervisor, there would almost certainly be severe information processing bottlenecks in most human organizations (see Mintzberg, 1979, pp. 46–53). As Galbraith (1973) points out, the overloads on ordinary hierarchical channels are most severe in organizations facing complex and rapidly changing environments.

In addition to using informal communication paths, there are several formal ways hierarchies can be augmented to deal with these problems (see Galbraith, 1973, for a more detailed discussion):

1. First, as discussed above, standardized procedures can be established for communications between different subgroups. When there are

standard procedures for salespeople from the marketing department to receive expense reimbursements from the accounting department, then this intergroup communication does not have to go up through the hierarchy to a level that supervises both departments.

2. Second, the hierarchical information processing capacities of the organization can be augmented by such devices as computerized management information systems. As traditionally used, these systems increase the efficiency of vertical information flow in the hierarchy.
3. Finally, hierarchies can be augmented by the explicit development of lateral relationships between subgroups in different parts of a hierarchy. Galbraith (1973, 1977) provides a detailed analysis of the kinds of structures that can be used for this purpose. In increasing order of cost, these structures include: direct contact, liaison roles, task forces, teams, explicit integrating roles, linking managerial roles, and matrix designs. In other words, these lateral relationships begin with information contact between groups, progress through various degrees of commitment to joint problem solving, and end with a matrix organization where one person may report simultaneously to two different supervisors in different parts of the hierarchy. As Galbraith describes, these lateral relationships are not without their own problems, but they seem essential to the successful functioning of large and complex organizations.

It is interesting to note, in this connection, that Minsky (1986), in a theory of mind modeled on human societies, also postulates as a crucial part of his theory the existence of a kind of lateral relationship (called K-lines) in hierarchies. In fact, in his scheme, these lateral structures form another hierarchy that is orthogonal to the original hierarchy.

Resource allocation and information dissemination in hierarchies. In strict hierarchies, information flows occur only among the subordinates of a common superior and the superior. Resources are allocated from the top down, dividing and subdividing the available resources at each level (e.g., see Crecine, 1982, for a discussion of this process in the U.S. federal government). In augmented hierarchies, there are more lateral information flows, and there may be some lateral transfers of resources as well.

Markets. Even though markets are often not thought of as a form of organization, they are one of the most pervasive of all mechanisms for coordinating the activities of many people (see, e.g., Simon, 1981). The fundamental coordination mechanism in a marketplace is mutual adjustment. In any kind of mutual adjustment, two agents, each of whom controls some scarce resources (e.g., labor, money, etc.), agree to share some of their respective resources to achieve some mutual goals. For example, when I buy a loaf of bread from a grocer, the grocer and I have agreed to reallocate the use of his bread and my money in order to achieve our respective goals of reducing my hunger and increasing his profits. All forms of mutual adjustment can be considered to be markets where valued resources are exchanged, either with or without explicit prices. Formal markets are a particularly explicit form of mutual adjustment where the exchanges take place after competitive bidding to establish prices.

Once a contract has been made, there is a sense in which the buyer becomes the supervisor of the supplier. Within the limits of my implicit contract with the grocer, for example, I can direct him to supply me with a loaf of bread. In this sense, then, buyers can exercise coordination through a form of direct supervision. If I buy bread from one store, ham from another, and then hire someone to make sandwiches, I have, in a sense, coordinated the activities of the bread suppliers (all the way back to the farmer who raised the wheat), the ham suppliers (again all the way back to the farmer), and the sandwich preparer. It should be clear that, in this sense, all economic activity is coordinated at various levels by markets for goods and services.

Many-to-many structures. One central feature of markets is that, in general, each producer buys from a number of suppliers and each supplier sells to a number of producers. This many-to-many structure is reflected in most computer programs, as well. A procedure in a computer program usually calls a number of subprocedures, each of which may also be used by several other calling procedures.

In fact, many computer systems have the kind of multilayered structure shown in Figure 2 (e.g., see Parnas, 1979). A number of procedures at one level (say, the procedures in an inventory program) will be written in terms of a set of primitive procedures at another level (say, the facilities of a high-level language like Pascal). Each procedure in the program will, in general, use many facilities of the language (like facilities for reading and writing files), and each facility will be used by many procedures. But each facility of the language is, in turn, defined in terms of more primitive procedures provided by the hardware instruction set of the computer being used, and the same many-to-many relationship holds there, as well.

This many-to-many structure of multilayered computer systems is a generalization of the matrix design used in organizations. Instead of reporting to only two supervisors, however, groups (or procedures) in a many-to-many structure can "report to" any number of supervisors.

A one-layer version of this structure is used for certain kinds of support services in human organizations. For example, members of legal service groups or typing pools provide support for many different groups in an organization. Thus, even though they may officially report to only one supervisor, they have information and control links to many temporary supervisors throughout the organization. For reasons that are not entirely

Figure 2. "Many-to-many" structure in a computer system.

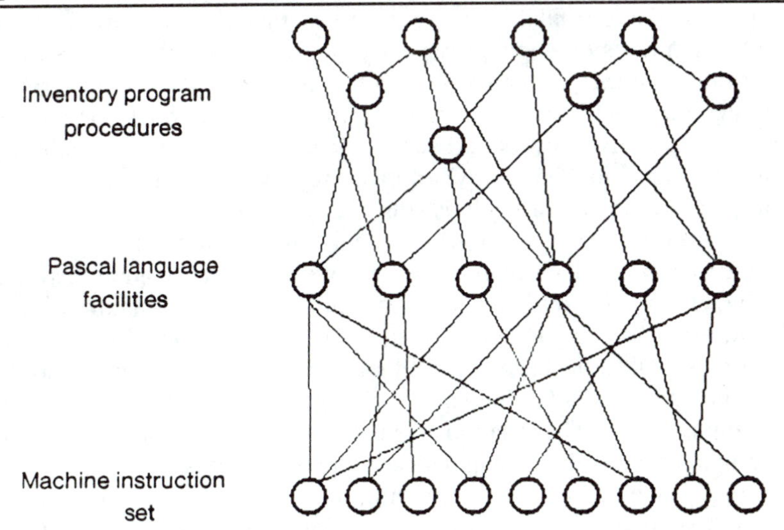

clear, this many-to-many structure seems to be used in human organizations mostly for staff services that are subsidiary to the primary goals of the organization, and seldom for the central "line" of the organizations.

Resource allocation and information dissemination in markets. The many-to-many structures established by markets are a way of transferring both resources and information in very nonhierarchical patterns. A central feature of resource allocation in markets is the use of (sometimes implicit) competitive bidding to establish prices. Under some fairly weak assumptions (such as that prices drop when there is an excess of a commodity and that quantities produced drop when prices drop), this mechanism is guaranteed to clear markets, that is, to find buyers for all products that are offered for sale (see Simon, 1981, p. 39; Samuelson, 1948, chap. 9). Under some much stronger assumptions (such as perfect competition and utility maximization by the agents), this mechanism leads to a Pareto optimal distribution of resources, that is, a situation where no agent could be made better off without making another one worse (see Simon, 1981, p. 39; Debreu, 1959). As discussed briefly below, these ideas from economics can apply to computer systems with market-like organizations, as well as to human markets. At the very least, some of the decentralized resource allocation mechanisms analyzed by economists (see Hurwicz, 1973, Reiter, 1986; for reviews) can inspire similar kinds of mechanisms in highly parallel distributed computer systems. If the mechanisms are sufficiently similar, the economic theorems about optimality and convergence properties of the mechanisms may be adapted to the computer systems as well.

It is often remarked (e.g., see Hayek, 1945) that one of the advantages of markets as coordination mechanisms is that they require very little information transfer—only prices and (as is often not mentioned) product descriptions, if they are not already known by the agents. For markets to operate efficiently, however, this information must be widely disseminated among prospective buyers and sellers. In actual markets, an elaborately structured "invisible hierarchy" of brokers, advertising media, market researchers, and so forth accomplish this dissemination (see, e.g., Hirshleifer, 1973). These mechanisms appear to be a fruitful source of metaphors for solving similar problems of information dissemination in computer systems.

Markets vs. hierarchies. As we saw in the sandwich-making example above, a market can be used to coordinate different activities by a form of direct supervision. In fact, most hierarchies in human organizations are constructed by a set of employment contracts negotiated through a market for human labor. The relationship between employee and supervisor is thus, in a sense, a relationship between a producer of labor and a consumer of labor. What, then, are the essential differences between markets and hierarchies as coordination systems? One eventual goal of the present approach is to clarify the fundamental dimensions on which markets and hierarchies differ. For the present, however, it seems that there are at least the following differences:

1. *Agents in a marketplace are more autonomous than those in a hierarchy.* That is, agents in a marketplace have, in principle, a greater degree of choice about whether they perform an activity or not (even though in practice, these choices may be severely limited by the alternatives available). This decentralization of decision-making power is also usually associated with a decentralization of information as well (e.g., see Hurwicz, 1973; Arrow & Hurwicz, 1960). For example, firms typically reveal less information to their customers, than subsidiaries reveal to their owners.
2. *Markets can establish many other forms of relationships besides tree-structured hierarchies.* For example, Figure 3 shows three kinds of coordination structures that might be established across a market. The first structure is a linear sequence of subcontracting relationships without any reduction of complexity through grouping. The second is a vertically integrated hierarchy to accomplish the same task as the linear sequence. The final structure is a hierarchical grouping of subcontracting relationships.

Thus a market is a decentralized mechanism for establishing relationships. Some of the structures created by multiple contracts in a market are hierarchies, and others are not.

Figure 3. Example of coordination structures in a market.

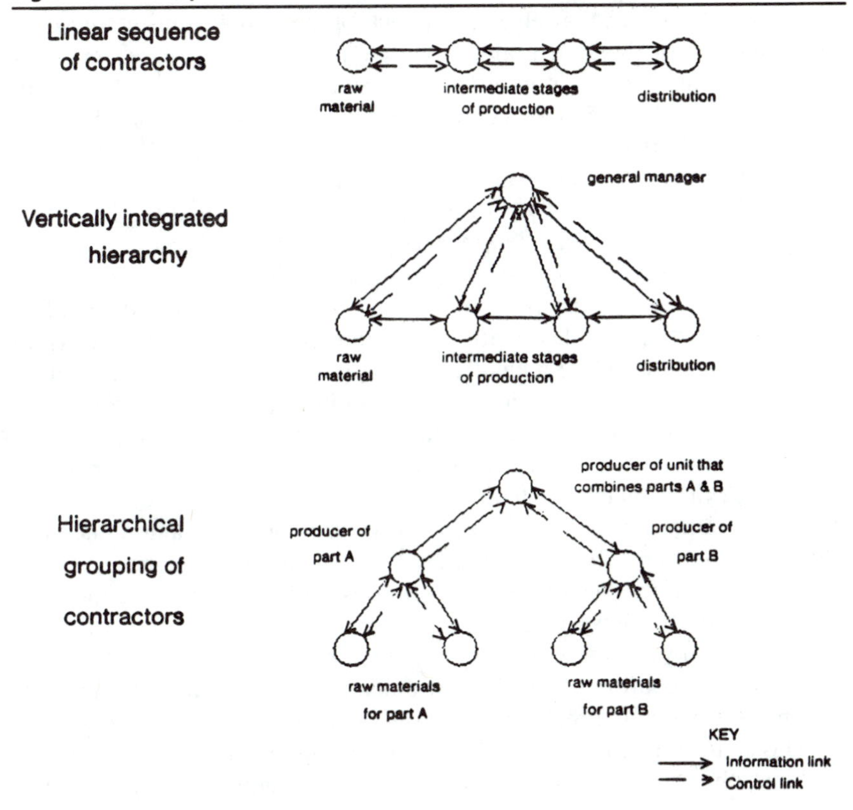

Now, what are the relative advantages of markets vs. preestablished hierarchies as coordination systems? In the perfectly efficient markets of contemporary neoclassical economic theory, the optimal mix of short- and long-term contracts would be established, with some of the long-term contracts forming hierarchical firms. Perfect markets would thus be both optimally flexible and optimally efficient. One of the primary barriers to the optimality of markets is the cost of establishing contracts. Williamson (1975, 1979, 1980, 1981a, b) has begun to elaborate a detailed approach to analyzing these "transaction costs." A central theme of this approach is that activities will (and should) be coordinated within a firm, rather than across a market, when the transaction costs of market coordination are too high. Williamson describes how two human characteristics (bounded rationality and opportunism) often interact with two environmental characteristics (uncertainty/complexity and small numbers of potential traders) to produce high transaction cost. The essence of the argument is that, when there are only a few potential traders in an uncertain and complex market, traders must either protect themselves from each other's opportunism with complex contracts or they must take expensive risks. In either case, the transaction costs are high. Within a firm, however, opportunism can be more easily controlled by internal auditing and long-term evaluation of organizational members (see also Ross, 1973; Grossman & Hart, 1983).

For instance, suppose an automobile manufacturer is considering contracting with an outside firm to make its automobile bodies. If the two companies try to establish a long-term contract, it will have to be extremely complex to protect both parties against uncertain future events such as dramatic increases in the price of steel or large drops in the demand for cars. Even with a complex contract, any unexpected condition that is not provided for may still be a chance for one party to take advantage of the other. To protect itself against these possibilities, the automobile manufacturer can try to buy the car bodies from the outside supplier under a series of short-term contracts. One of the dangers here, however, is that the outside firm will be able to drive competitive suppliers out of the market (e.g., through acquiring specialized knowledge about the unique needs of this manufacturing firm) and then supply the automobile bodies under monopolistic terms. All these considerations lead to high transaction costs, since both parties have to protect themselves against the other's possible opportunism by complex contracts or price premiums or both. If, on the other hand, the automobile bodies are made by an internal division of the automobile manufacturer, the risks from opportunism are greatly reduced.

This argument depends crucially on the assumption of *opportunism* (which Williamson defines as "self-interest seeking with guile"). Without opportunism, the argument gives no basis for believing that complexity and uncertainty favor a hierarchical organization over a market organization. Even in a complex and uncertain market, nonopportunistic traders can modify contracts in a "fair" way as situations change without necessarily incurring higher transaction costs than would be the case in an equivalent hierarchy.[3] While nonopportunistic traders may be rare in human markets, there is no reason at all why computer programs cannot be constructed with "honest" participants in a market-like organization (see examples below). Fox (1981) fails to take this fact into account in his interpretation of Williamson's argument and therefore claims (mistakenly, I believe) that uncertainty favors hierarchies over market-like organizations in computer programs as well as in human organizations. Even if, as Fox suggests, modules in a program become "self-motivated" in the sense of deciding

[3] By "equivalent hierarchy," I mean a hierarchy with a similar task division. A more detailed version of this argument is being developed in collaboration with Michael Cohen.

for themselves how to spend their time, they need not do so in an opportunistic way.

Thus Williamson's argument about an important source of high transaction costs in human markets appears not to hold for markets composed of nonopportunistic agents such as might be created in a computer system. There may still be other sources of high transaction costs, however. For instance, the information costs of repeatedly finding agents with specific products available to sell (e.g., consultants) may sometimes be greater than simply establishing a long-term contract with (e.g., hiring) a producer of a product (see Baligh & Richartz, 1967; Malone & Smith, 1988; Malone, 1987, for an analysis of some of these nonopportunistic coordinating costs).

Alchian and Demsetz (1972) explore another barrier to the optimality of markets—the difficulty of measuring individual contributions to the output of a team. They argue that, in cases where individual contributions are difficult to measure, hierarchies are often better able to reduce "shirking" than markets. This argument, like Williamson's, is only partly applicable to systems of nonopportunistic agents such as in computer systems. Computer programs can certainly be designed to not "shirk" responsibility, but for parts of programs where performance measurement (or credit assignment) is difficult, market metaphors in computer programs are less likely to be useful.

Criteria for Evaluating Coordination

Ultimately the success of a coordination scheme is judged by how well the organization meets whatever specific goals the coordination is designed to achieve. But there are two general criteria for evaluating how well the coordination scheme itself performs. They are: *flexibility* and *efficiency*.

Flexibility vs. efficiency. In both organizational design and computer system design, one of the crucial trade-offs is between flexibility and efficiency.[4] One example of this trade-off was mentioned above in the discussion of coordination costs vs. economies of scale in segmenting organizations by product or by function. Another example of this trade-off is illustrated in computer systems by, on the one hand, compiled procedures that are highly optimized for efficiency in a specialized task, and, on the other hand, flexible intelligent systems that use generalized methods to plan actions and solve problems in many different specific situations. In organizations, this trade-off involves moving between, on the one hand,

[4] Of course, flexibility and efficiency trade off against each other only at a given "technical frontier." An organization with substantially superior components can be both more efficient and more flexible than one with inferior components.

highly bureaucratic organizations with formalized procedures for dealing with almost all eventualities, and on the other hand, very loosely structured organizations that depend on massive amounts of informal communication and mutual adjustment to adapt to rapidly changing complex environments.

It is sometimes useful, as Malone and Smith (1988) point out, to view flexibility and efficiency as two different versions of the same measure with different time spans: what we usually call efficiency is efficiency in the short term; and what we call flexibility is really efficiency in the long term—over a wide range of different possible situations.

To illustrate these points, let us consider some examples. The first example is of a flexible artificial intelligence program called Hearsay-II (Erman, Hayes-Roth, Lesser, & Reddy, 1980) that was designed to understand human speech. In part because speech understanding is a complex and poorly understood task, the designers of the Hearsay-II system chose to create a number of independent program modules called *knowledge sources* (KSs) that communicate with each other through a global data base called a *blackboard*. The blackboard is divided into regions corresponding to partial hypotheses at different levels of speech understanding: syllables, words, phrases, etc. Each knowledge source scans the blackboard for conditions under which it may contribute to the problem-solving activity. When such a condition arises, the KS makes its contribution, usually by inserting new information at another place on the blackboard. Processing time is allocated among KSs by a heuristic scheduler that calculates a priority for each action and, at each time, executes the waiting action with the highest priority.

This system achieves its flexibility, in part, by having all communication of intermediate results between KSs be implicitly accomplished through the global blackboard. The designer of an individual KS (and the KS itself during its execution) need not be explicitly concerned with which KS created the information it uses or with which other KSs will use the results it creates. Thus KSs can be added, deleted, or radically modified, relatively independently of each other, and the flow of control between KSs is heavily dependent on the specific problem being solved. The allocation of processing resources is relatively independent of the KSs themselves, being accomplished by the scheduler and by a few control-oriented KSs (e.g., WORD-CTL, WORD-SEQ-CTL; see Erman et al., 1980).

While this flexibility is highly desirable in complex and poorly understood tasks, it can be unnecessarily inefficient when the task is simpler or better understood. In fact, for the limited cases used to test the Hearsay-II program, a less flexible "precompiled" program called Harpy (Lowere, 1976; Lowerre & Reddy, 1980) was superior in both speed and accuracy (Erman et al., 1980). Most of the knowledge used in Harpy is precompiled into a network structure that represents all possible utterances in a limited vocabulary. This structure is searched in each instance using a heuristic form of dynamic programming to find the path that best matches the spoken utterance.

Summary

Table 1 summarizes the framework I have presented for analyzing coordination. The trade-off between flexibility and efficiency occurs at several levels in the framework. When segmenting a problem into parts, there is a trade-off between functional organizations, which take advantage of economies of scale (thus improving efficiency), and product organizations, which reduce the number of interactions to be coordinated (thus improving flexibility). When choosing coordination processes, there is a trade-off between flexible but often inefficient mutual adjustment and inflexible but efficient standardization, with direct supervision being an intermediate case. When choosing a coordination system, the trade-offs between flexibility and efficiency are more complex, since they depend on the details of how transaction costs differ for different kinds of contracts and on other characteristics of the agents such as opportunism.

This framework integrates previous work by a number of different organizational theorists. As noted above, Thompson (1967) presented a slightly less parsimonious description of the functions of coordination. A number of researchers, including Ellis (1979) in a slightly different sense than that used here, have distinguished between information and control links. Mintzberg (1979) integrated previous work by Simon (1957), March and Simon (1958), and Galbraith (1973) to distinguish the three processes for

Table 1. Framework for Organizational Analysis.

Level of analysis	Alternatives
Segmenting the problem	product function
Functions of coordination	allocation of scarce resources communication of intermediate results
Connections for coordination	information links control links
Processes for coordination	mutual adjustment direct supervision standardization
Systems for coordination	hierarchies augmented hierarchies markets
Criteria for evaluating coordination	flexibility efficiency

coordination. Finally, Galbraith (1973, 1977) clarified the relationship between hierarchies and augmented hierarchies, and Williamson (1975, 1980) clarified the relationship between markets and hierarchies. Malone and Smith (1988; Malone, 1987) used a formal model based on assumptions at the micro level of message processing to clarify tradeoffs between macro-level structures such as various kinds of hierarchies and various kinds of markets.

My contribution here has been to integrate these different levels of analysis into a coherent framework and to show how the framework can also be applied to other information processing systems such as computers.

IMPLICATIONS FOR COMPUTER SYSTEM DESIGN

The above analysis of hierarchical and nonhierarchical coordination systems has a number of implications for the design of computer systems. First, it suggests that, as the problems to be solved by an information processing system become more complex and less well understood, they become less and less susceptible to purely hierarchical solutions. Second, the organizational concepts mentioned above may suggest ways of organizing nonhierarchical interactions in computer systems. Two such examples are given below. One example, based on an earlier version of this chapter, has already been implemented in prototype form.

Market Metaphors in Computer Systems

Several existing computer programs embody aspects of market metaphors. For example, the Hearsay program described above has a form of hierarchical information transfer that resembles a market. But the resource allocation decisions in the Hearsay program are all still centralized in the scheduler and the two control-oriented KSs. A few recent systems have begun to use genuinely decentralized market-like mechanisms for allocating processors to tasks. For example, Smith and Davis (1978, 1981; Smith, 1980) have developed an explicit market-like protocol with which processor nodes in distributed problem-solving systems establish "contract nets" among themselves for subdividing tasks. A similar system has subsequently been developed, implemented, and analyzed by Malone, Fikes, Grant and Howard (1986, 1988). In these systems processors with tasks to be done ("clients") send out "announcements" of the tasks; other processors ("contractors") respond with bids; and the task is "awarded" to one of the contractors. The allocation of tasks to processors is decentralized in the sense that "clients" can rank contractors on the basis of the characteristics of the bids and "contractors" can rank contracts on the desirability of the tasks required. Malone and Smith (1988) show, based on very rough parameter estimates, that this decentralized scheduling method appears superior to a centralized alternative in the environment in which their system is implemented.

"Entrepreneurial" resource allocation in artificial intelligence programs. Another analogy between computer programs and markets is suggested by the computer language Ether (Kornfeld, 1979). Before presenting the analogy between this system and a market, a brief description of the system is in order. Computation in Ether is done by computational elements called *sprites*. Each sprite recognizes certain kinds of messages (those in its *interest set*). As part of the computation triggered by receiving a message in which it is interested, a sprite can create new sprites and broadcast more messages. There are two types of messages: *assertions* and *goals*.

Kornfeld and Hewitt (1980) analyze this language using the metaphor of a scientific community. A system of sprites can be compared to a community of scientists working in parallel, broadcasting their results and goals. Resource allocation in the system is performed by *sponsors* who provide processing time to sprites working on goals created by the sponsors. When a sprite creates subgoals for other sprites to achieve, it must divide its processing time among the subgoals it creates based on some notion of how important each subgoal is.

In some ways a market is an even more suggestive metaphor for this system than a scientific community. Assertions are equivalent to products being offered for sale. Goals are equivalent to bids to buy a certain product. In the resource allocation scheme described for the current system, no products are ever produced except under prior contract from a sponsor. Sponsors subdivide the processing power they control among their subgoals until either a goal is successfully achieved or some energy limit is reached.

One implication of a market metaphor for this system is that, if more than one sponsor is interested in the same goal, they should be able to pool their resources in trying to achieve it. A more interesting implication is that there could also be a kind of adaptive resource allocation using *entrepreneurs*. These entrepreneurs could be sponsors who use heuristics and a knowledge of current goals to fund products that are not yet called for by the goals of any other sprites. If a product is later desired by another sprite (or sprites) who have resources to sponsor its production, these sprites can instead purchase it from the entrepreneur for the amount of resources they would have been willing to devote to it themselves. (Their determination of the value of the product could be done by some heuristic like the priority setting scheme in Lenat's (1976) agenda mechanism.) If the total revenue obtained in this way is greater than the cost of production, then the entrepreneur makes a "profit." If not, the entrepreneur sustains a "loss." This scheme embodies a simple kind of learning because entrepreneurs who use good heuristics will acquire more resources to invest and thus control more of the system's processing.

Note that in this scheme, price has no necessary effect on demand or supply in the short-run, and thus the classical economic theorems (see Samuelson, 1948) do not even guarantee market clearing, much less optimality. Instead, prices have a long-run (evolutionary) effect on the allocation of resources by different kinds of heuristics (see Simon, 1981; Nelson & Winter, 1982).

IMPLICATIONS FOR ORGANIZATIONAL DESIGN

One result of this analysis for organizations is a clearer focus on the price that must be paid for flexibility. It is common in some organizational theories to extol the virtues of flexible organizations. But one should not lose sight of the fact that there is a cost in terms of time and communication for this flexibility. For well-understood and relatively stable environments, just as for well-understood computer tasks (like Harpy), less flexible but more efficient organizations are probably preferable.

On the other hand, the dramatically decreasing cost of computer hardware means that the use of computers will make more and more kinds of organizational flexibility cost-effective. Galbraith (1973, 1977) talks about how traditional management information systems can be used to make the *vertical* information flows in an organization more efficient, thus increasing the organization's ability to respond quickly and flexibly to changing situations. Curiously, Galbraith hardly mentions the possibility of using computers to enhance the *lateral* information flows in an organization. The use of electronic mail, computer conferencing, and similar systems seems likely to have at least as profound an effect on increasing organizational flexibility through improved lateral communication as through improved vertical communication. One such possibility is suggested by the blackboard concept in the Hearsay-II system.

An Organizational Blackboard for Adhocracies

Mintzberg (1979) describes an emerging new form of organization which he calls *adhocracy* (after Toffler, 1970). This is the form of organization that is necessary for sophisticated and continued innovation such as might be found in a high technology manufacturing company, a consulting firm, or an advertising agency. It requires the joining of experts from different disciplines into smoothly functioning and often changing ad hoc project teams. To continue to be innovative and flexible, this kind of organization cannot depend only on hierarchical bureaucratic channels of information flow. It must also rely heavily on unplanned lateral communications.

This kind of unplanned, nonhierarchical information flow is also one of the key features of the blackboard in the Hearsay-II program. Here, knowledge sources can communicate with each other without either the sender or the receiver having to know ahead of time which other knowledge source the information is coming from or going to.[5] But there are clearly limits to the amount of this nonhierarchical information that can be useful, in either computer or human organizations. In the case of human organization, it would be impossible for each individual to scan all the information that passes through the organization (books, memos, telephone calls, etc.) to find the few pieces of information that are relevant to his or her job.

In conventional organizations, much of the filtering of this information is accomplished by the hierarchy and by standardized procedures. If a computer-based information system is to facilitate unplanned lateral communication, it must provide for some other way of accomplishing this filtering for all the nonprivate information handled by the system, including electronic mail and other documents. There are a number of possibilities for this filtering (see also Malone, Grant, Turbak, Brobst, & Cohen, 1987; Hiltz & Turoff, 1978; Stevens, 1981):

1. *Filtering by classification.* The Hearsay-II blackboard has a classification scheme for all "messages" between knowledge sources based on a simple two-dimensional subdivision of the problem being solved. In electronic mail and computer conferencing systems, named distribution lists and discussion topics provide a similar way for senders to address their messages, not just to specific individuals, but to a group of people who are identified only by the kind of information in which they are interested. The more accurately this classification system reflects the structure of the work being performed, the more useful it will be in aiding communication. There are a number of interesting questions about how this classification system should be designed and maintained. Should specific people be responsible for specific parts of the system? From where in the organizational hierarchy should these people be chosen? Is there a way to automatically aid the emergence and change of a group consensus about the structure of the classification system?

---

[5] This content addressability is also a property of ordinary bulletin boards, and of many other kinds of mass media such as newspapers, magazines, and books. In a sense, these media constitute a kind of invisible hierarchy with very many information links but no direct control links. Many people report to the editors of these organs, and then the editors, in turn, filter and disseminate the information to many other subordinates.

2. *Filtering by quality.* One of the primary functions of editors in traditional media like books and journals is to screen material, not only for its appropriate classification, but also for its quality. Information sources that have been prescreened for quality are more worthy of attention by others. Therefore a useful computerized information system should provide a way for this editing function to be performed for certain categories of the data base. The same questions about who and where in the organization this should be done arise here as they did for classification.

In addition to providing for filtering by classification and by quality, computer systems have the promise of providing arbitrarily complex automated procedures to scan a data base of messages or other documents and select items that would be of interest to a specific individual. An *active interest filter* like this needs to contain a model of the kinds of information in which its user is likely to be interested.

Malone, Grant, Turbak, Brobst, & Cohen (1987) describe precisely such a system that uses techniques from artificial intelligence such as frame inheritance networks and production rules to help people specify rules for automatically finding, filtering, prioritizing, and classifying messages that are of interest to them. The system is based on a rich network of semistructured message types. For example, the *meeting announcement* message type contains fields for "time," "place," "topic," and "sponsor," as well as for any unstructured information (such as a speaker's abstract) that the sender of the message wants to include. By making it easy for senders to compose messages using these semistructured message types, the system avoids the unsolved problems of natural language understanding while still allowing receivers to specify much more sophisticated rules for describing their interests than would be possible with simple keyword searches.

If a global "organizational bulletin board" like this is to be truly useful in the offices of the future, there are a number of other problems that need to be solved. Perhaps the most important of these problems involve how to allocate resources (including people's time) in such a scheme. People who think they might find useful information in such an electronic bulletin board would clearly be motivated to consult it. But why should people bother to insert information into a bulletin board like this? In some cases, there will be clear motivations to publicize something like a meeting or a public report. But in other cases, internal memoranda in one group may be of vital, but unsuspected, interest to people in another group. Automatically inserting all such information into a public data base raises many serious privacy problems.

And in still other cases, a nontrivial amount of effort may be required to answer direct queries or to contribute unsolicited information that might be of interest to people in other parts of the organization. Part of the solution may be to create a social environment in which formal or informal rewards, like money or status, accrue to people who contribute useful information to public computer data bases, much as they do to people who contribute to noncomputer databases like books and journals.

CONCLUSIONS

In this chapter I have outlined a framework of systematic analogies between two different kinds of information processing entities: computer programs and human organizations. Any information processing system must segment its tasks into parts and then coordinate the solutions of the separate parts. I suggested a framework for analyzing different ways of segmenting and coordinating tasks, and I described one of the most important design trade-offs in information processing systems—that between flexibility and efficiency. To demonstrate the usefulness of this framework for facilitating the transfer of ideas between the two fields, I gave examples of new implications in each field based on concepts from the other field.

One of the most promising research directions suggested by this paper appears to be developing computer programs that embody organizational metaphors such as the market-like systems described above. Such systems can be useful both as programming tools in their own right and as formal models for analyzing human organizations.

REFERENCES

Alchian, A. A., & Demsetz, H. (1972). Production, information costs, and economic organization. *American Economic Review, 62*, 777–795.

Arrow, K. J., & Hurwicz, L. (1960). Decentralization and computation in resource allocation. In R. W. Pfouts (Ed.), *Essays in Economics and Econometrics* (pp. 34–104). Chapel Hill, NC: University of North Carolina Press. (Reprinted in K. J. Arrow & L. Hurwicz (Eds.), *Studies in resource allocation processes* (pp. 41–95). Cambridge, England: Cambridge University Press, 1977).

Baligh, H. H., & Richartz, L. (1967). *Vertical market structures*. Boston: Allyn and Bacon.

Bobrow, R. J., & Brown, J. S. (1975). Systematic understanding: Synthesis, analysis, and contingent knowledge in specialized understanding systems. In D. G. Bobrow & A. Collins (Eds.), *Representation and understanding: Studies in cognitive science* (pp. 103–130). New York: Academic Press.

Chandrasekaran, B. (1981). Natural and social system metaphors for distributed problem solving: Introduction to the issue. *IEEE Transactions on Systems, Man, and Cybernetics, SMC-11*, 1–4.

Cohen, M. D. (1984, June). Conflict and complexity: Goal diversity and organizational search effectiveness. *American Political Science Review, 78*, 435–451.

Cohen, M. D. (1981, August). The power of parallel thinking. *Journal of Economic Behavior and Organization, 2*, 287–306.

Cook, C. L. (1979). *Streamlining office procedures* (Tech. rep. no. SSL-79-10). Palo Alto, CA: Xerox Palo Alto Research Center.

Crecine, P. (1982). *A positive theory of public spending* (working paper). Carnegie-Mellon University.

Cyert, R. M., & March, J. G. (1963). *A behavioral theory of the firm.* Englewood Cliffs, NJ: Prentice-Hall.

Debreu, G. (1959). *Theory of value: An axiomatic analysis of economic equilibrium.* New York: Wiley.

Dennet, D. C. (1978). Artificial intelligence as philosophy and as psychology. In D. C. Dennet (Ed.), *Brainstorms* (pp. 109–126). Cambridge, MA: MIT Press.

Ellis, C. A. (1979, August). *Information control nets: A mathematical model of office information flow.* Paper presented at proceedings of the ACM Conference on Simulation, Measurement, and Modeling, Boulder, Colorado.

Ellis, C. A., Gibbons, R., & Morris, P. (1979). Office streamlining. *Proceedings of the International Workshop on Integrated Offices*. Versailles, France: Institut de Recherche d'Informatique et d'Automatique.

Erman, L. D., Hayes-Roth, F., Lesser, V. R., & Reddy, D. R. (1980). The Hearsay-II speech-understanding system: Integrating knowledge to resolve uncertainty. *Computing Surveys, 12*, 213–253.

Fox, M. S. (1981). An organizational view of distributed systems. *IEEE Transactions on Systems, Man, and Cybernetics, SMC-11,* 70–79.

Galbraith, J. (1973). *Designing complex organizations.* Reading, MA: Addison-Wesley.

Galbraith, J. (1977). *Organization design.* Reading, MA: Addison-Wesley.

Goldberg, A., & Robson, D. *Smalltalk-80: The language and its implementation.* Boston: Addison-Wesley.

Grossman, S., & Hart, O. (1983). An analysis of the principal-agent problem. *Econometrica, 51*, 7–45.

Hayek, F. (1945, September). The uses of knowledge in society. *American Economic Review, 35*, 519–530.

Hewitt, C. (1977). Viewing control structures as patterns of passing messages. *Artificial Intelligence, 8*, 323–364.

Hillis, D. (1985). *The connection machine.* Cambridge, MA: The MIT Press.

Hiltz, S. R., & Turoff, M. (1978). *The network nation: Human communication via computer.* Reading, MA: Addison-Wesley.

Hirshleifer, J. (1973, May). Where are we in the theory of information? *American Economic Review, 63*, 31–39.

Hurwicz, L. (1973, May). The design of resource allocation mechanisms. *American Economic Review Papers and Proceedings, 58*, 1–30. (Reprinted in K. J. Arrow and L. Hurwicz (Eds.), *Studies in Resource Allocation Processes* (pp. 3–40). Cambridge, England: Cambridge University Press, 1977.)

Hurwicz, L. (1986). On mechanisms. In S. Reiter (Ed.), *Studies in mathematics: Vol. 25. Studies in mathematical economics*. Washington, DC: Mathematical Association of America.

Kornfeld, W. A. (1979). *Using parallel processing for problem solving* (A. I. Memo No. 561). Cambridge, MA: Massachusetts Institute of Technology, Artificial Intelligence Laboratory.

Kornfeld, W. A., & Hewitt, C. (1981). The scientific community metaphor. *IEEE Transactions on Systems, Man, and Cybernetics, SMC-11*, 24–33.

Lenat, D. (1982). AM: Discovery in mathematics as heuristic search. In R. Davis & D. B. Lenat (Eds.), *Knowledge-based systems in artificial intelligence*. New York: McGraw-Hill.

Lesser, V. R., & Corkill, D. D. (1981). Functionally-accurate, cooperative distributed systems. *IEEE Transactions on Systems, Man, and Cybernetics, SMC-11*, 81–96.

Lowerre, B. T. (1976). *The HARPY speech recognition system.* Unpublished doctoral dissertation, Computer Science Dept., Carnegie-Mellon University.

Lowerre, B. T., & Reddy, R. (1980). The HARPY speech understanding system. In W. A. Lea (Ed.), *Trends in speech recognition*. Englewood Cliffs, NJ: Prentice-Hall.

Malone, T. W. (1987). Modeling coordination in organizations and markets. *Management Science, 33*, 1317–1332.

Malone, T. W., Fikes, R. E., Grant, K. R., & Howard, M. T. (1987, April). *Market-like load sharing in distributed computing environments*. Massachusetts Institute of Technology, Center for Information Systems Research Working Paper #139.

Malone, T. W., Fikes, R. E., Grant, K. R., & Howard, M. T. (1988). Enterprise: A market-like task scheduler for distributed computing environments. In B. A. Huberman (Ed.), *The ecology of computation*. Amsterdam: North Holland.

Malone, T. W., Grant, K. R., Turbak, F. A., Brobst, S. A., & Cohen, M. D. (1987). Intelligent information sharing systems. *Communications of the ACM, 30*, 390–402.

Malone, T. W., & Smith, S. A. (1988, May–June). Modeling the performance of organizational structures. *Operations Research, 36*, 421–436.

Malone, T. W., Yates, J., & Benjamin, R. I. (1987). Electronic markets and electronic hierarchies. *Communications of the ACM, 30*, 484–497.

March, J. G., & Simon, H. A. (1958). *Organizations.* New York: Wiley.

McClelland, J. L., Rumelhart, D. E., & the PDP Research Group. (1986). *Parallel distributed processing: Explorations in the microstructure of cognition* (Vols. 1–2). Cambridge, MA: MIT Press.

Mead, C., & Conway, L. (1980). *Introduction to VLSI systems.* Reading MA: Addison-Wesley.

Mesarovic, M. D., Macko, D., & Takahara, Y. (1970). *Theory of hierarchical multi-level systems.* New York: Academic Press.

Minsky, M. L.(1980). K-Lines: A theory of memory. *Cognitive Science, 4*, 117–133.

Minsky, M. L. (1986). *Society of mind.* New York: Simon and Schuster.

Mintzberg, H. (1979). *The structuring of organizations.* Englewood Cliffs, NJ: Prentice-Hall.

Nelson, R., & Winter, S. (1982). *An evolutionary theory of economic change.* Cambridge, MA: Harvard University Press.

Parnas, D. L. (1979, March). Designing software for ease of extension and contraction. *IEEE Transactions on Software Engineering, SE-5*, 128–137.

Reiter, S. (1986). Information and incentives in the (new)^2 welfare economics. In S. Reiter (Ed.), *Studies in mathematics: Vol. 25, Studies in mathematical economics.* Washington, DC: Mathematical Association of America.

Ross, S. The economic theory of agency: The principal's problem. *American Economic Review, 63*, 134–139.

Samuelson, P. A. (1948). *Foundations of economic analysis.* Cambridge, MA: Harvard University Press.

Shoch, J. F., & Hupp, J. A. (1982). The "Worm" programs—early experience with a distributed computation. *Communications of the ACM, 25*, 172–180.

Simon, H. A. (1957). *Administrative behavior* (2nd ed.). New York: Macmillan.

Simon, H. A. (1981). *The sciences of the artificial* (2nd ed.). Cambridge, MA: MIT Press.

Smith, R. G. (1980, December). The contract net protocol: High-level communication and control in a distributed problem solver. *IEEE Transactions on Computers, C-29*, 1104–1113.

Smith, R. G., & Davis, R. (1978, July). *Distributed problem solving: The contract net approach.* Paper presented at the proceedings of the Second National Conference of the Canadian Society for Computational Studies of Intelligence, University of Toronto, Toronto, Ontario.

Smith, R. G., & Davis, R. (1981). Frameworks for cooperation in distributed problem solving. *IEEE Transactions on Systems, Man, and Cybernetics, SMC-11*, 61–69.

Stevens, C. H. (1981). *Many-to-many communication* (Tech. Rep. No. 72). Cambridge, MA: Center for Information Systems Research, Massachusetts Institute of Technology.

Strassman, P. A. (1980, December-January). The office of the future: Information management for the new age. *Technology Review*, pp. 54–65.

Thompson, J. D. (1967). *Organizations in action.* New York: McGraw-Hill.

Toffler, A. (1970). *Future shock.* New York: Bantam Books.

Touretzky, D., & Hinton, G. E. (1985). Symbols among the neurons: Details of a connectionist inference architecture. *Proceedings of the Ninth International Joint Conference on Artificial Intelligence.*

Wesson, R., Hayes-Roth, F., Burge, J. W., Stasz, C., & Sunshine, C. A. (1981). Network structures for distributed situation assessment. *IEEE Transactions on Systems, Man, and Cybernetics, SMC-11*, 5–23.

Williamson, O. E. (1975). *Markets and hierarchies.* New York: Free Press.

Williamson, O. E. (1979). Transaction cost economics: The governance of contractual relations. *Journal of Law and Economics, 22*, 233–261.

Williamson, O. E. (1980). The organization of work: A comparative institutional assessment. *Journal of Economic Behavior and Organization, 1*, 6–38.

Williamson, O. E. (1981a). The economics of organization: The transaction cost approach. *American-Journal of Sociology, 87*, 548–575.

Williamson, O. E. (1981b). The modern corporation: Origins, evolution, attributes. *Journal of Economic Literature, XIX*, 1537–1568.

Williamson, O. E. (1981c, February). *The economics of organization: The transaction cost approach.* Discussion paper no. 96, Center for the Study of Organizational Innovation, University of Pennsylvania.

4
Cognitive Change by Appropriation*

Denis Newman

Bolt, Beranek & Newman
Cambridge, MA

When novices have to learn to do something entirely new, they have to learn, not just the specific procedures, but the overall structure and purpose of the new activity. This is true whether it is an office worker learning to do word processing, a second grader learning a subtraction algorithm, or a nursery-schooler learning to draw a picture.

The need to learn both the goal and the procedure at the same time presents interesting theoretical puzzles about how learning can happen. It also presents difficult problems for the design of educational software which goes beyond simple drill and practice. This paper outlines a theoretical approach to how cognition changes in social interaction, presents examples of how cognitive change works in classrooms, and concludes with implications for the design of educational environments. The ways that teachers provide interactive support for a student's new interpretation can be observed in an essentially interactive process we call *appropriation*.

LEARNING SOMETHING ENTIRELY NEW

A very simple hypothetical example of human–computer interaction illustrates some features of how people can learn to do new things. A novice is working at a terminal to some large, old-fashioned computer system. It is his first session since a colleague showed him how to log on and use the editor. He has been playing around with the editor but now wants to go have lunch. He remembers that you get off the system by typing "logout," so he does that. The computer responds "LOGOUT is not an editor command."

The first thing that we can imagine is that the novice is confused because he wasn't trying to give an editor command. He was trying to log out and

*The author thanks Peg Griffin and Seth Chaiklin for comments on earlier drafts.

go have lunch. But there are some very important facts about the overall structure of the system that the novice might conclude from such an error message. The editor has its own set of associated goals, such as saving and exiting. The computer system has a compartmentalized structure, and the user has to be somewhere else to execute the logout command.

Even unfriendly systems can help novices learn as long as the novice works very hard at interpreting the system's response to his actions. In this case, the novice would have to arrive at the insights about the system's structure by a process of inference, because the error message did not say them explicitly. Because the user was talking to the editor, that program *presupposed* that he was trying to give an editor command and took "logout" as a bad case of one. So, once the novice is able to interpret the system's presupposition, he can come to realize important facts about the system's goal hierarchy.

The art and science of interface design has come a long way in the last decade (Norman & Draper, 1986). The point of the example, however, is not to argue for better user interfaces but rather to point out that, even under adverse conditions, *nonartificial* human intelligence is often able to learn about properties of systems through seeing how the system interpreted its actions. Adults and older children are able to reason about the way that one actor's perspective on the interaction can be embedded as part of the other actor's perspective (Bruce & Newman, 1978; Newman & Bruce, 1986). Such reasoning is very common in human interaction and can be the basis for learning about how the other person (or system) interpreted one's actions.

A particular kind of embedding found in interpersonal as well as human–machine interactions is a process we call *appropriation* (Newman, Griffin, & Cole, 1989). The process is illustrated in the initial example. The machine *appropriates* the novice's actions into the interpretive system that it is working from. The interpretive system may include goals as well as procedures that the novice is unaware of. The novice ends up playing a role in a system he did not entirely understand when he first took the action. This interactive process of appropriation is worth exploring further, because it is a pervasive feature of instructional interactions. Under proper conditions, even very young children can undergo cognitive change as a result of having their actions appropriated. It is a process that is particularly relevant when learning things that are entirely new.

TWO VIEWS OF COGNITIVE CHANGE

The concept of appropriation is a central concept in the theory of cognitive change that Peg Griffin, Michael Cole, and I have been working with (Newman, Griffin, & Cole, 1989). The theory has social interaction as an essential component of cognitive change. The process of appropriation is critical to the explanation of how novices can learn from experts without undergoing explicit instruction. A brief outline of our approach to cognitive change sets the stage for several examples of appropriation in instructional interactions which illustrate the conditions that allow the process to work for young children as well as adults.

The general framework for the theory of cognitive change in interaction is illustrated in Figure 1, which contrasts two approaches—the traditional cognitive science approach, and what I will call the ZPD approach. When most cognitive scientists have conceptualized cognitive change, the student is in some state of understanding at time 1, some experience or teaching occurs, and, consequently, the student is turned into an expert at time 2. Even in cases of explicit instruction, it is usually assumed that some constructive cognitive process occurs *in the novice* resulting in the change. In contrast, the ZPD theory has the expert and novice student accomplishing the task interactively at time 1, and the student (now an expert) accomplishing it without expert help at time 2. Vygotsky (1978) called this interactive locus of change a *zone of proximal development* (ZPD), meaning an interactive system in which the novice is able to work interactively at the level he would, in the near future, be able to work on his own. In this construction zone, the interactive process involves gradually handing over

Figure 1. Two approaches to cognitive change.

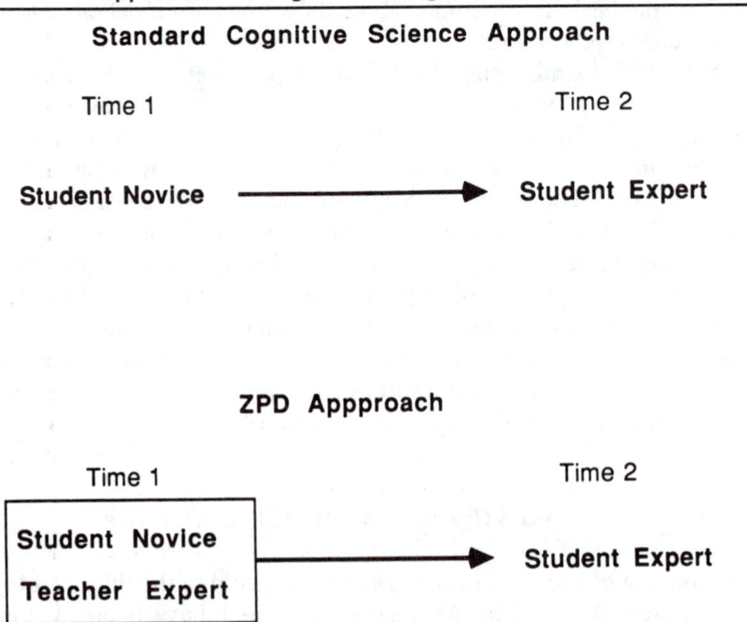

more responsibility for the task to the student. It is assumed that a constructive process occurs in such an interaction; that is, some of the process which constructs new knowledge can be occurring in the *interaction* rather than exclusively in the head of the student. In Vygotsky's terms, the activity moves from the interpersonal plane to the intrapersonal plane.

I have presented these two characterizations of cognitive change in order to highlight a difference in orientation that can lead researchers along quite different paths with respect to their objects of study, their theoretical puzzles, and the practical work of designing instruction. What I have characterized as the standard view focuses attention on cognitive processes that are internal. There is a great concern with the forms of representation and the specific mental models of the learners. The alternative view focuses attention on the social interactions themselves. Without denying the importance of internal processes, there is equal concern with how understandings can be established in an interaction, the consequences of misinterpretations, the process of collaboration. Interactive processes are not reduced to individual processes but are examined for their special features.

A theoretical puzzle that arises in the standard approach was expressed most clearly by Fodor (1980) in his critique of Piaget's stage theory. A higher order logic can derive lower order ones but cannot be derived from them. Fodor's solution was to assume that the higher stages are somehow innately in place to be triggered by the environment. The Fodor paradox is a paradox only for theories which assume that stage 2 has to arise directly from stage 1. The diagram above makes clear that, for the ZPD theory, the state of the student at stage 2 is continuous with the state of the interaction at time 1 but not necessarily with the state of the novice at that time. Nor does the ZPD theory place the entire burden of the construction of the new structures on the process of internalization, as Bereiter (1985) has argued. In a ZPD, the interactive processes operate externally to construct complex structures without depending on continuity with whatever simpler structures may have been useful for the novice's initial entry into the zone. Because the construction happens in the interaction with the expert, children do not have to do the building of more complex structures inside their heads. Nor do they have to leave the zone and internalize in isolation. If the construction occurs in the interaction, it need not happen all over again "in the head." This is not to say that internalization is automatic and nonproblematic. But internalization does not have to bear the whole burden of a constructivist theory.

The two ways of looking at how change happens have very different implications for how one might go about designing and carrying out instruction. In the standard approach, it is very important to understand the novice's mental model of the domain with its potential conceptual bugs and other misconceptions (Gentner & Stevens, 1983). Then one creates lessons of other learning materials that will move the novice from this initial state to the goal state. In the ZPD approach, one doesn't really have to know too much about what the novice thinks he or she is doing at time 1. The expert has to find some way for him or her to play at least a minimal role in the accomplishment of the task and give feedback in terms of the expert understanding of the task: what the goal is, what's relevant, why his or her move was not optimal and so on.

In instructional interactions, both the novice and expert are necessarily somewhat ignorant of each other's mental state. This situation is not necessarily a problem. The novice does not have to have the expert understanding of what it is all about. All the novice needs to be able to do is produce some move that in some way contributes (or can be understood as an attempt to contribute) to the task. The expert does not have to know exactly what the novice thinks he or she is doing as long as he or she can appropriate what the novice does into the joint accomplishment of the task.

The notion that a detailed mental model of the novice may not be essential is at odds with much of the work in cognitive development and cognitive science, where an important thrust of research has been to understand the learner's point of view (Gentner & Stevens, 1983). For example, a major issue in the design of intelligent tutoring systems is the specification of a "student model" on the basis of which the tutoring system can construct appropriate remediation (Wenger, 1987).

It would be foolish to argue that knowledge of the novice's mental model leads to poor instruction. This kind of information is certainly used in instructional interactions (Clark & Peterson, 1986). The argument here is that, first, in many cases such knowledge is unnecessary and second, it is very often impossible to obtain because of the complexity of the domain and the wide variety of incoming misunderstandings or because the teacher has to work with several children simultaneously. Considering these practical problems, we ought to explore ways that children are able to learn from experts who are simply providing them with an alternative interpretation for their actions.

EXAMPLES OF TEACHER–CHILD APPROPRIATION

The computer example at the beginning of the paper illustrates several features of appropriation. These can be specified in terms of a number of steps.

1. There are two systems engaged in joint activity (an expert and a novice), each with its own understanding of the task they are engaged in (in

this case the computer is the expert and if the anthropomorphization is allowed, it understands its system).
2. The novice makes some move that arises out of his own understanding.
3. The expert appropriates that move or attempts to appropriate that move into the system as she understands it and provides feedback about the move in terms of the expert system.
4. The novice comes to understand, in retrospect, what his action meant in the expert system and learns something about that system.

From the standard cognitive science perspective, the key to how this works is contained in the fourth step in which the novice comes to some conclusion about the nature of the task he is doing with the expert on the basis of the feedback he received from the expert. But a narrow focus on the internal psychological processes can lead us to ignore important aspects of the social processes that may contribute to the production of those understandings in the course of the expert–novice interaction, i.e., in steps 2 and 3 above. The social interactions between a human teacher and child can be much more intricate than many computer–user interactions and deserves careful study before we make conclusions about how to design educational computer systems.

In the initial example, the computer was relatively dumb. All it could do was give its interpretation of the user's command. It does not have any educational goals around which to carry on a dialog with the user in any interesting way. Human teachers, on the other hand, have a myriad of interesting strategies for engaging children in joint activity. Two documented cases of appropriation presented below show how a teacher can provide support for the child's process of understanding. The teacher, unlike the inanimate computer, actively creates opportunities for reinterpretations to arise. These are instructional interactions which help the child reflect on their previous actions in terms of the expert understanding of the task.

Gearhart and Newman (1980) studied videotapes of a nursery school classroom. It was a small class of four boys and one girl, all around 3 years old, and their teacher. During one 18-minute period, children were at a table drawing on paper with crayons and marking pens. These children were at the very beginning of learning to draw pictures and many things that we might take for granted are not part of their system of activity. For our current discussion a critical difference was that these children were not creating representations of things as an intentional action. That is, with one exception, they did not start out to draw anything in particular. When they started out they would choose a crayon and start to make marks or lines. As the drawing progressed, children at the table would talk to each other, and, just as they imitated what each other said, they would also imitate what each other drew.

Since the children did not have a plan for their drawing from the beginning there was always an interesting negotiation between the teacher and the children as to whether the drawing was finished. Children would announce they were finished when other more interesting activities caught their attention. The teacher would sometimes ask them to go back and do some more. But when she was satisfied that they had put in a good effort, she would have a little discussion about what they had done before hanging it on the wall. She would generally initiate the discussion with an open-ended question like, "Can you tell me about your picture?" and follow up with more specific questions like, "What is this?" The child would answer with a simple description like, "That's a moon," or, "It has two mountains on two orange circles." Then the teacher would compliment the child for a nice drawing and write his or her name on it.

In themselves, the question–answer sequences that constitute these finish discussions are unremarkable. What makes them interesting is their relation to what the child had just done. Since the children did not set out to draw anything in particular, the discussion is a chance to formulate, perhaps for the first time, what the picture is a picture of. The children were able to provide a description in response to the teacher's question, even though the response may have been fairly arbitrary. As long as the child could provide responses to the local demands of the questions, the teacher was then able, through gestures, follow-up questions and selective acknowledgements, to direct the discussion toward her interpretation of the picture as a representation with *some* particular content. In the terms of the current discussion, the teacher appropriated the drawings into her own scheme of relevancies and carried on discussions that presupposed that scheme.

The children can be learning something very important from these discussions, which were a routine part of drawing and painting activities. After participation in a number of drawing sessions, a child may come to anticipate the finish discussion and begin to produce a drawing that he will be able to talk about. One child, who was more advanced than the others in his social and language development, did talk about "making a house" as he was drawing it. The idea that drawing has the goal of producing a representation is not obvious to children from the outset. It appears to be something learned in interaction with experts.

A second case of appropriation is one in which we purposefully set up an appropriation process in order to teach children a strategy for solving the Piagetian combinations task (Piaget & Inhelder, 1975). As part of a larger study of the relation between classrooms and laboratory settings (Newman, Griffin, & Cole, 1989) we had children go one at a time to the library corner of the classroom, where our research assistant presented them with a task involving stacks of little cards, each stack with a picture of a different movie star on it. Our "laboratory" task was for the child to make all the pairs of movie stars that he could with no pairs the same. This is a relatively difficult task for fourth-graders to solve systematically. Typically, children of this age go about the problem *empirically*, by which Piaget meant that they make whatever pairs they can think of in no particular order until they cannot think of any more. But they lack any sense of certainty that they have all that are possible or even the sense that there is some definite set that constitutes "all." A strategy that is more typical of adolescents is what Piaget calls *intersection*. In this method, the children systemically pair 1&2 1&3 1&4 2&3 2&4 3&4. When they get to the end of the series, they have a clear sense that they have all the pairs. Our sample was consistent with Piaget's age norms in that only 3 out of the 27 children in the class used the intersection strategy on the first trial.

Our interest, however, was in teaching the intersection strategy, not in testing the children's operational level. There were three trials in which the tutor asked the child to make all the pairs. Each trial added another item so the initial trial used four stacks of cards; the second, five stacks; and the last trial, six stacks. Between trials, the researcher conducted a little tutorial in the form of a checking session. When the child made all the pairs he could, the tutor asked, "How do you know you have all the pairs?" The child usually answered, with a hint of frustration, "I can't think of any more." The tutor then asked, "Could you check to see if you have all the pairs?" The child usually said little and the tutor said, "Well, I have a way to check. Do you have all the pairs with Mork (or the first star on the left)?" From there she proceeded through a checking procedure, asking about each star in relation to every other star, going systematically from left to right across the row of stacks of movie star cards. As soon as the child got the idea of the checking procedure, the tutor let him complete it on his own.

The tutorial was quite successful, considering that 17 children (out of 27) used the intersection strategy in the second or third trials. This is in spite of the fact that the strategy for producing pairs was never directly taught. What the tutor did was appropriate the column of pairs the child had produced and talk about it in terms of the strategy she had in mind. The tutor's question, "How do you know you have all the pairs?" presupposes that the child was trying to get *all* the pairs. This may have been a false presupposition, but it was strategically useful. The question treated the child's column of pairs *as if* it had been produced in an attempt to get all the pairs. The tutor then invoked the intersection procedure as a means to fix up the child's "failed attempt to produce all the pairs." That is, she appropriated the child's pair-making, making it a part of an example of how to achieve the stated goal. It appears that, when their own empirical production of pairs was retrospectively interpreted in terms of the intersection schema, children began to learn the researcher's meaning of *all the pairs* and thereby discovered the goal considered as a definite set that can be specified even before pair-making is initiated.

A critical difference between the teachers in both the nursery school and fourth grade examples, and the computer in the initial example, is that the teachers provide a second task that engages the children in thinking about the first task. In the computer example, the computer simply provided feedback from its point of view. In the drawing lesson, the teacher engaged the children in a question–answer activity about their drawings. The questions presupposed that the children had been engaged in creating a representation. In the combinations tutorial, the tutor engaged the children in checking the results of their initial prediction of pairs. This checking activity presupposed that there was a definite answer to whether they had *all* the pairs. The feedback for their actions is not just a statement but an activity that they can engage in somewhat independently of the first activity.

Once the children engage in the second task, the presupposition becomes a premise for their actions. Since the second activity reflects on the products of the first activity, the process of reinterpretation is not something that has to go on entirely in the child's head. The reinterpretation is implicit in the second activity which is accomplished interactively. Because the expert is initially providing structure and support for the second activity, the child does not have to consider explicitly the difference between his own understanding of the product of the first activity that he had when he originally produced it and the new interpretation that the expert is offering. He simply engages in the second activity.

This is not true for the computer novice who is trying to log out of his computer so he can go get lunch. Here the novice has to engage in more-or-less explicit reasoning to infer that there are contexts in which certain commands are not appropriate. From a developmental point of view, there is considerable evidence that children increase in their ability to reason about the point of view of another person (Newman, 1986). Our observations in classrooms point to the ways the teachers can circumvent the need for that kind of reasoning in constructing learning activities.

IMPLICATIONS FOR EDUCATIONAL PRACTICE

Theories that reduce learning to an individual cognitive activity may be missing interesting social processes by which expertise is transmitted from an expert to a novice. Our classroom observations lend credibility to the idea that expertise follows out of the interaction with an expert who appropriates the novice's actions into the expert system. To use Vygotsky's terminology, the expertise exists first on the interpersonal plane before it is internalized to the intrapersonal plane. The implication of this for pedagogical theories is that we have to look outside the learner, as well as inside, if we are going to capture the full process.

By focusing on how teachers support children's cognitive change, we can begin to understand some of the social processes involved in instruction. The particular focuses of attention in this chapter are cases where there is a substantial gulf between the expert and novice understandings regarding the overall structure and goal of the activity. In cases where the novice has a good understanding of the goal and, for example, is just learning a new procedure for what he or she has done many times before, direct instruction may be quite feasible. But in many cases where a substantial gulf exists, an appropriation process may work successfully as long as the novice is in a position to take some action that can be appropriated. Furthermore, appropriation may be *necessary* in cases where a gulf exists, even when the expert has a reasonable picture of the novice's mental model, since direct instruction about the new understanding will be uninterpretable by the novice. The expert will still have to engage the novice in activities that help to bridge the cognitive gulf.

An implication of this analysis is that we can question the assumption that "intelligent" educational software must attempt to replace the teacher by simulating a diagnostic process in which the teacher is supposed to build up a student model as a basis for remediation. Our observations of how teachers support students' reinterpretations suggest a new way of looking at the form of the dialogue between the student and the machine. By analogy, we can ask how we can create computer-mediated tasks that get the child to reflect on and reinterpret his initial actions. Some interesting examples of this approach are already running. For example, ALGEBRALAND, developed at Xerox PARC (Brown, 1984), provides a trace of the students' reasoning process which itself can become an object of consideration and discussion between the teacher and student (Collins & Brown, 1987; Pea, 1987). The teacher may still have an essential role in the classroom use of such programs. Because the trace provides a new way for the student to look at the reasoning process, a human teacher may be needed initially to interpret the students' prior actions in terms of the trace. Attempting to replace the human teacher, which is often the orientation of intelligent tutoring systems (Anderson, Boyle, & Reiser, 1985), may not be the best strategy for educational software development in the near future. The study of teacher–student interaction, especially as technological tools are incorporated into the interaction, should be a priority.

REFERENCES

Anderson, J. R., Boyle, B. J., & Reiser, B. J. (1985). Intelligent tutoring systems. *Science, 228*, 456–461.

Bereiter, C. (1985). Toward a solution of the learning paradox. *Review of Educational Research, 55*, 201–226.

Brown, J. S. (1984). Process versus product: A perspective on tools for communal and informal electronic learning. In S. Newman & E. Poor (Eds.), *Report from the Learning Lab: Education in the electronic age*. New York: Educational Broadcasting Corporation, WNET.

Bruce, B. C., & Newman, D. (1978). Interacting plans. *Cognitive Science, 2*, 196–233.

Clark, C. M., & Peterson, P. L. (1986). Teachers' thought processes. In M. C. Whitroch (Ed.), *Handbook of research on teaching* (3rd ed.). New York: McMillan.

Collins, A., & Brown, J. S. (1987). The computer as a tool for learning through reflection. In H. Mandl & A. M. Lesgold (Eds.), *Learning issues for intelligent tutoring systems*. New York: Springer-Verlag.

Fodor, J. (1980). On the impossibility of acquiring "more powerful" structures. In M. Piattelli-Palmarini (Ed.), *Language and learning: The debate between Jean Piaget and Noam Chomsky*. Cambridge, MA: Harvard University Press.

Gearhart, M., & Newman, D. (1980). Learning to draw a picture: The social context of an individual activity. *Discourse Processes, 3*, 169–184.

Gentner, D., & Stevens, A. L. (Eds.). (1983). *Mental models*. Hillsdale, NJ: Erlbaum.

Newman, D. (1986). The role of mutual knowledge in perspective taking development. *Developmental Review, 6*, 122–145.

Newman, D., & Bruce, B. C. (1986). Interpretation and manipulation in human plans. *Discourse Processes, 9*, 149–196.

Newman, D., Griffin, P., & Cole, M. (1989). *The construction zone: Working for cognitive change in school*. Cambridge, England: Cambridge University Press.

Norman, D. A., & Draper, S. W. (1986). *User centered system design: New perspectives on human–computer interaction*. Hillsdale, NJ: Erlbaum.

Pea, R. D. (1987). Cognitive technologies for mathematics education. In A. H. Schoenfeld (Ed.), *Cognitive science and mathematics education*. Hillsdale, NJ: Erlbaum.

Piaget, J., & Inhelder, B. (1975). *The origin of the idea of chance in children*. New York: Norton.

Vygotsky, L. S. (1978). *Mind in Society: The development of the higher psychological processes*. Cambridge, MA: Harvard University Press.

Wenger, E. (1987). *Artificial intelligence and tutoring systems: Computational and cognitive approaches to the communication of knowledge*. Los Altos, CA: Morgan Kaufmann Publishers, Inc.

5

Conversation as Coordinated, Cooperative Interaction

Raymond W. Gibbs, Jr. and Rachel A. G. Mueller

Program in Experimental Psychology
University of California, Santa Cruz
Santa Cruz, CA

INTRODUCTION

When two or more people interact with one another, they must somehow coordinate their actions to achieve various goals. Whether two people have to coordinate while dancing, each partner assumes that the other will reliably follow the appropriate movements, or if two nations are discussing ways of eliminating nuclear weapons, there must be some agreement as to what each party will do and when. There can be disastrous consequences if there is a breakdown of the coordination between any two agents. Imagine the difficulties that would arise if car drivers did not coordinate which side of the road they drive on. It matters little to anyone whether or not to drive on the right or left side of the road, provided the others do likewise. But if some people drive in the left lane and others in the right, everyone is in danger of collision. Without coordination of some sort, people would be unable to attain the simplest of goals in everyday life.

Coordination is also extremely important when people use language. In some ways the coordination required during conversation is fairly obvious, as in turn-taking, where people have to figure out who speaks when and for how long (cf. Sacks, Schegloff, & Jefferson, 1974). But in other aspects of conversation the coordination necessary is less apparent and more difficult to determine. When a speaker says something to an addressee, there must be some coordination of the speaker's and addressee's mutual beliefs. When Al says to Beth *I'm real hungry, why don't we pizza tonight*? he assumes that Beth is capable of figuring out how this utterance is intended (i.e., "let's go and eat some pizza," see Gerrig, 1986). This belief on Al's part constitutes part of the set of mutual knowledge between Al and Beth and enables them to leave some things unsaid when they talk with one another. But how do Al and Beth know what information and beliefs they mutually share, and how does this shared understanding influence the way they speak and listen?

The issue we examine in this chapter concerns how people coordinate with one another to efficiently and successfully communicate in conversation. Our premise is that people must assume a certain degree of cooperation with one another, as well as coordinate their shared knowledge and beliefs to achieve social, communicative goals. The coordination that is necessary for conversational exchanges imposes certain constraints on the cognitive processes underlying the social activities of speaking and listening. A description of these constraints should be specified as part of any psychological theory of meaning and communicative interaction. Our plan is to discuss some of what is known about coordination and cooperation in language use, and then to demonstrate the importance of intentionality and mutual knowledge in psychological theories of conversational meaning.

COORDINATION AND COOPERATION

Consider a problem of communication between Bob and Betsy (adapted from Lewis, 1969, p. 5).

> Bob and Betsy want to meet each other. They will meet if and only if each goes to the same place. It matters little to either of them where (within limits) Bob goes if he meets Betsy there; and it matters little to either of them where he goes if he fails to meet Betsy there. They must each choose where to go. The best place for Bob to go is the place where Betsy will go, so he tries to figure out where Betsy will go and to go there himself. Betsy does the same. Each chooses according to his or her expectation of the other's choice. If either succeeds, so does the other; the outcome is one both Bob and Betsy desired.

Lewis (1969) has provided a nice analysis of coordination problems like this and has suggested four types of coordination devices used to solve such difficulties. The first is *explicit agreement*, where Bob and Betsy explicitly agree where to meet. The explicit agreement would give them each grounds for believing that both will mutually expect to go to meet, for example, at Betsy's home. The second coordination device is *salience*. Because Bob is known not to know anything about the town except where Betsy lives, the most salient solution is for each to meet there. Bob and Betsy could each be confident of this solution and could mutually expect to meet at Betsy's home. The third device is *precedence*, which says that Bob and Betsy realize that they last met at Betsy's home and so this is the most appropriate, and most salient, solution. Finally, there is the device of *convention*. Bob and Betsy have met so regularly at her house that it has become for them a convention to meet there. Whenever they want to meet, then, Bob and Betsy can mutually expect each other to go to her house, because that is the place where they conventionally meet.

Of these four devices, Lewis argued that *convention* serves as the best solution to recurrent coordination problems. A *convention* is a regularity in behavior evolved by two or more people to solve recurrent coordination problems. The recurrent problem of deciding which side of the road to drive on is solved by the convention that, in the United States, we drive on the right side of the road. We all could just as easily drive on the left side, as is done in England and some other countries, but the convention which everyone in the community of United States drivers both know about, adhere to, and expect everyone else to adhere to is to drive on the right side of the road.

For our purposes, the significant part of Lewis's argument is that language is largely a system of conventions that have evolved as solutions to certain coordination problems in communicative settings. For instance, when a speaker wants to refer to a four-legged animal that barks, he or she will use *dog*, because it is the convention in their linguistic community to call these animals this name. Speakers will comfortably use conventional words or phrases in many social situations because they can assume that hearers share this knowledge. The recurrence of many communicative situations has resulted in a variety of conventional linguistic routines to facilitate understanding. When making requests, for example, speakers use sentence forms that are generally known and well accepted, such as *Can you tell me the time*? (See Gibbs, 1981, 1984, 1986). Listeners easily understand these conventional expressions as requests, even though these are questions syntactically. The use of conventional language enables conversants to determine each other's plans and goals and to respond as the situation dictates.

Coordination in language use requires cooperation and a brief look at most any conversation illustrates some of the ingredients of what it means to speak cooperatively. In his seminal work on logic and conversation, Grice (1975) provided the first framework for a theory of cooperation in conversation. Grice proposed that each participant in a conversation obeyed a *cooperative principle*—"make your conversational contribution such as required, at the stage at which it occurs, by the accepted purpose or direction of the talk exchange in which you are engaged" (Grice, 1975, p. 45). Grice more specifically argued that participants can assume that each other obey certain *maxims*, which follow directly from the *cooperative principle*. The maxims are

QUANTITY: Don't provide more or less information than is required for the current purposes of the exchange.
QUALITY: Speak the truth.
RELATION: Be relevant.
MANNER: Be clear.

When we engage in conversation, we can assume that each person is acting cooperatively to sustain our joint activity and that we can do this by following the maxims. Although speakers mostly observe the *cooperative principle* and the conversational maxims, they do not always do so. Consider the following example (Grice, 1975, p. 51):

A: Smith doesn't seem to have a girlfriend these days.
B: He has been paying a lot of visits to New York lately.

On the surface, B's response violates the maxim of Quality and would appear not to uphold the *cooperative principle*. What is interesting about this example, though, is that the failure to satisfy a maxim is *deliberate*, but does not necessarily hinder communication, because a speaker may intend the violation to be part of the message. In this case, we have to assume that B's response is somehow relevant to what A uttered. According to Grice, in order to preserve this assumption, we infer (or rather A as the addressee infers) the proposition that Smith has, or may have, a girlfriend in New York. Grice calls this kind of inference a *conversational implicature*, and we say that B implicates the proposition just mentioned in virtue of what he said along with various background knowledge and beliefs shared with A. An *implicature*, therefore, is an inference generated in the course of a conversation to preserve the assumption that participants are obeying the maxims. Speakers produce and listeners interpret implicatures in light of the overriding assumption that both are trying to speak the truth, be relevant, avoid obscurity, and so on.

The problem of understanding conversational utterances where speakers deliberately flout one or more maxims has been widely studied, particularly in regard to how people comprehend figurative language. Consider how a listener might comprehend a sarcastic remark, such as *You're a fine friend* (meaning "You are not a good friend."). The speaker's sarcastic utterance does not uphold the maxims of Quality and Relevance, given the contradictory evidence that the addressee has not been a good friend. Implicit in Grice's characterization of conversational implicature is a set of computations which are likely to be used to work out an implicature, in this case "You're not a good friend." Presumably, a listener must first analyze the semantic or literal meaning of the sentence, compare this to the context to see if it is appropriate for the situation, and then, finding it inappropriate, derive its intended, nonliteral meaning by determining an interpretation which would fulfill the *cooperative principle*. According to Grice (1975, 1978), this is most easily done in the case of sarcasm by assuming the opposite of the sentence's literal meaning.

Psychological experiments have been conducted to test aspects of Grice's view (see Gibbs, 1982, 1984, 1986, for reviews) and have found that listeners or readers can actually comprehend the intended meanings of many nonliteral utterances as quickly as, if not more quickly, than corresponding literal expressions. Such findings strongly suggest that people can process the conversational implicatures underlying figurative discourse without having first to determine these utterance's literal meanings. Social and linguistic contexts provide an interpretive framework, so that people can understand what speakers mean without having to go through a stage of computation where the literal meaning of an utterance is computed and then rejected in favor of some other nonliteral interpretation. This experimental evidence against the computational process-model inherent in Grice's theory does *not* in any way damage Grice's main point that conversation is a cooperative venture influenced, in part, by the adherence to or flouting of a set of conversational maxims. The psychological findings do, nonetheless, show how contextual information constrains the comprehensive process involved in understanding the meanings of utterances where their intended interpretations differ in some way from their literal meanings. In fact, it is precisely *because* speakers and listeners are coordinated and share knowledge about the rules of language and each other's personal and cultural beliefs that they usually have such easy access to a speaker's meaning.

Context, then, plays an obviously important part in the comprehension process. But what constitutes the context used in interpreting conversational meaning? Certainly, context includes a listener's knowledge of the physical-social surroundings and knowledge of the speaker's probable beliefs. Psychological studies of language processing over the past 15 years have repeatedly shown that context influence both the process and product of interpreting discourse (cf. Bransford, 1979; Clark & Carlson, 1981). Our argument, though, is that the listener must not only utilize a context which includes information about what the speaker believes, but must also recognize that the speaker knows that he or she knows this as well. Consider again the earlier conversation:

A: Smith doesn't seem to have a girlfriend these days.
B: He has been paying a lot of visits to New York lately.

Speaker B means to inform the listener of the belief that Smith probably had a girlfriend in New York. Successful interpretation of B's remark demands that the listener make this inference about what the speaker meant. But recognition of the speaker's intention is of a special kind, what Grice (1968) called *m-intention*. An m-intention is a speaker's intention to produce an effect in the listener by means of the hearer's recognition of that intention. Speaker B wants the listener to recognize the intention that Smith might have a girlfriend in New York, in part by means of the listener's recognition of that intention. If the listener successfully recognizes this intention, the authorized inference will have been drawn (Clark, 1977). Any other inferences drawn, e.g., that the speaker believes Smith does not have a girlfriend and is doing something else in New York, are unauthorized or not m-intended. As Grice (1975) argued, the *cooperative principle* is a device for enabling listeners to draw only inferences authorized by the speaker. Any other inference produced does not adhere to the *cooperative principle*.

Most theories of comprehension in psychology, linguistics, and AI tend to ignore this crucial distinction between the intended meaning of a message that a speaker meant for a listener to draw as part of the message, and the implications of a message, which often are not strictly authorized or intended by the speaker. Nonetheless, listeners normally distinguish authorized and unauthorized inferences and interpret the meanings of conversational utterances accordingly. A few examples will help to illustrate the difference between authorized and unauthorized inferences (see Clark, 1977).

John is sitting in the living room, reading a book, when Sally walks in, leaving the door open. John says to Sally, *It's cold in here*. He means for Sally to understand this as a polite request to close the door, and she understands him to mean just that. Sally has therefore drawn an authorized inference, one which she believes John meant for her to infer.

Barbara and Sandra are talking on the telephone about their plans for the coming weekend. Sandra asks Barbara, *Are you going to the big party tomorrow night*? And Barbara replies, *Peter is going to be there*. Although Barbara's answer is not a direct response to the question, she intends Sandra to understand it to mean "Definitely, no!" because of the recent breakup of her relationship with Peter. Sandra knows this information, something which is mutually shared between Barbara and Sandra, and Sandra correctly infers that Barbara intends her to use this shared knowledge in understanding the response to her original question. The inference that Sandra draws then is authorized. Note that, if Sandra incorrectly believed that Barbara *wanted* to reestablish her relationship with Peter, then she probably would draw the unauthorized inference that Barbara meant "Definitely, yes!" by her response.

People do, of course, draw unauthorized inferences. Roger is having dinner one evening with his wife Emily and at one point asks her, *Did you prepare the lasagna differently tonight?* intending this as a serious question, because he thinks the lasagna is particularly delicious that night. Emily, on the other hand, believes that Roger is being critical of the dinner and responds by saying, *You don't have to eat it if you don't want.* This shows that Emily has clearly made an unauthorized inference about Roger's question; Roger did not intend for Emily to draw a negative impression about his comment.

These examples highlight that it is not difficult to see what distinguishes authorized from unauthorized inferences. Each authorized inference relies on a speaker–listener agreement about how the language can be used and about what each of them respectively knows and believes. Speakers and listeners keep track of these mostly implicit agreements and take them into account when speaking and understanding the meanings of their conversational utterances.

EVIDENCE FOR THE RECOGNITION OF INTENTIONS

There is much research in both psychology and AI to support the notion that people normally attempt recovery of the authorized intentions behind speakers' utterances. Psychological experiments usually examine shifts in memory as evidence for inferences subjects made during the encoding or comprehension of sentences. Many studies have shown that people are very likely to remember a pragmatic implication of an utterance rather than the utterance itself or what it directly asserts or logically implies. Johnson, Bransford, and Solomon (1973), for instance, used a recognition memory task to show that subjects hearing sentences like (1a) and (1b) embedded in a brief story were more likely to falsely recognize similar sentence containing pragmatic implications of the presented sentences, as in (2a) and (2b).

(1a) He dropped the delicate glass pitcher.
(1b) He broke the delicate glass pitcher.
(2a) He was pounding the nail.
(2b) He was using the hammer.

Schweller, Brewer, and Dahl (1976) found that memory shifts in a cued-recall task were consistent with the illocutionary force (or the speaker's intended meaning) pragmatically implied by an input sentence. Thus, subjects hearing sentences like (3a) often recalled them as (3b).

(3a) The housewife spoke to the manager about the increase in meat prices.
(3b) The housewife complained to the manager about the increase in meat prices.

Similarly, Schweller et al. found that subjects falsely recognize a pragmatically implied perlocution (or intended effect upon the listener) of an utterance. For example, sentences like (4a) were often falsely recognized as (4b), but very rarely as (4c).

(4a) The English professor told his students a dull story about Jane Austen.
(4b) The English professor bored his students with a dull story about Jane Austen.
(4c) The English professor amused his students with a dull story about Jane Austen.

In each of these different experiments, subjects encoded inferences that went beyond the explicit information contained in the sentences, but which were in accord with the speakers' probable intentions in making their utterances. It is the intention behind the speaker's utterance that is encoded and represented in memory, not the surface form of literal words used to convey the intention. Jarvella and Collas (1974) provided additional evidence for this idea. They had pairs of subjects act out several dialogues. After each acted version, the subjects either judged whether particular sentences from a comparison dialogue were the same or different from the acted dialogue, or, in another experiment, chose between two intentions suggested for some of the acted sentences. The results showed that subjects best recognized the test sentences which conveyed the intentions of utterances previously stated.

Although many of these experimental findings have been interpreted as supporting a schema-theoretic view of language comprehension (cf. Alba & Hasher, 1983), the results of these studies just as easily confirm the primacy of intention recovery in discourse processing. This perspective suggests that speakers formulate their words to activate specific knowledge structures to enable listeners to recognize particular intentions. Speakers and writers will purposely design their output in such a way as to induce listeners and readers to infer connections between ideas (cf. Haviland & Clark, 1974; Kintsch & van Dijk, 1978) and understanders will incorporate these inferences into their memory representations to increase the plausibility or coherence of a set of ideas (Owens, Bower, & Black, 1978; Thorndyke, 1976). Of course, listeners must also know that their activated knowledge is mutually shared with their speakers; otherwise, people can never be sure that the interpretation given to an utterance is the correct or authorized one.

A number of researchers in artificial intelligence have similarly stressed the importance of intentions in building dialogue systems (Allen & Perrault, 1980; Appelt, 1985; Cohen & Perrault, 1979; Cohen, Perrault, &

Allen, 1982; Carbonell, 1978; Robertson, Black, & Johnson, 1981). Most of these systems make use of ideas from speech act theory and emphasize access to a speaker's underlying psychological plans and goals as fundamental elements for both planning and understanding utterances. Analysis of dialogue between people and both real and simulated programs show that people implicitly assume that the person (or machine) with which they are interacting can compute their wants and goals on the slenderest of evidence, and use these wants and goals to determine the intended meaning of utterances in dialogue (Cohen et al., 1982). Allen and Perrault (1980), for example, describe a dialogue system where agents attempt to recognize the plans of other agents and then use this plan in deciding what response to make. Consider the following brief exchange between a patron and a clerk at a train station.

Patron: The 3:15 train to Windsor?
Clerk: Gate 10.

Allen and Perrault's system models what occurs when one agent A (patron) asks a question of another agent B (clerk) which B then answers. A has a *goal* to acquire some information, and constructs a plan (*plan construction*) that involves asking B a question whose answer will provide the information. A then executes the plan, asking B the question. B then attempts to infer A's plan from the question (*plan inference*). In this plan, there may be goals A cannot achieve without assistance, which are the obstacles in A's plan. B can accept some of these obstacles as his or her own goals and constructs a plan to achieve these goals, thereby removing the obstacles present for A. B subsequently formulates his response to best achieve his goal.

In general, Allen and Perrault's (1980) model explicitly models how a listener infers the plan of the speaker and cooperates with it. Their system assumes, though, that speakers only have a single goal (i.e., getting specific information) and that listeners automatically cooperate to help attain that goal. Speakers can also craft their utterances to satisfy multiple goals at different communicative levels. In their microanalysis of a naturally occurring conversation, Hobbs and Evans (1980) describe how a single utterance can actually accomplish multiple goals. Thus, with a single statement a speaker can, for instance, try to protect his or her reputation, retort to an earlier remark, and have the floor, all at the same time.

A complete account of understanding conversational meaning, therefore, requires processing statements on multiple levels, reflecting the idea that many speakers actually use their utterances to achieve various goals and intend for listeners to recognize these multiple intentions. Robertson, Black, and Johnson's (1981) computer program, called MAGPIE, was designed to use a goal-based model of conversation to simulate the multiple intentions underlying a dialogue between a husband and wife, as in the following exchange.

Wife: Why were you out so late last night?
Husband: I went out bowling with the boys.
Wife: I thought you hated bowling.

MAGPIE traces the conversational goals in a dialogue by analyzing each utterance on multiple levels. The wife's first utterance, for instance, can be motivated by four interacting goals—seek information about the location of the husband last night, to seek information about the husband's motivation for staying out late, to express anger over the husband's absence, and to regain equilibrium in the relationship once again. Each of these goals work together to produce the wife's initial utterance, and the husband's response is, in part, an attempt to satisfy some of these goals. Thus, when the husband responds, he satisfies the wife's goal of seeking information about his location last night, but the goal of seeking information about the husband's motivation for doing so remains active. This, in turn, results in the wife's second statement. One of the advantages of MAGPIE is its ability to capture the fact the conversational utterances are often made to satisfy multiple goals, reflecting, we believe, the complexity of people's intentions in speaking with one another.

CONTEXT AND THE PROBLEM OF MUTUAL KNOWLEDGE

Our discussion has emphasized the importance of coordination and cooperation in recovering speaker's intentions in conversation. Psychological theories of language use must explicitly model these pragmatic factors to accurately describe the cognitive processes by which discourse is comprehended. The question remains, however, whether listeners are really able to coordinate what they mutually know and believe with speakers to recover specific communicative intentions. Lewis' (1969) coordination devices are, in essence, part of what constitutes the context for comprehension of speakers' utterances in conversation. Although most theories of language comprehension depend upon context for successful description of human communication, few have attempted to fully define and specify the features which characterize context. Context has usually been viewed as any and every type of information which is present at a given time in a given situation for a given individual. Nonetheless, this all-encompassing characterization of context leaves the definition empty.

Clark and Carlson (1981) have tried to refine the notion of context by identifying two types of context. Suppose, for instance, that some person is in an experiment which examines response times for deciding whether a string of letters is a meaningful word. The process of identification, response time, and error production can be influenced by different types of things. One type includes what that person knows about the phonemic structure of words, knowledge about the category of the word, and knowledge of which sentences, if any, preceded the word. Another influence on the person's behavior in the experiment includes his or her thoughts about the job, boredom with the experiment, and tired eyes. The first group of influences are considered *intrinsic* for comprehension, because these aspects of the context are potentially necessary for the comprehension process and therefore need to be included in a theory which accounts for the comprehension process. The second type of context is generally considered to be *incidental* to the comprehension process, since this type of information only indirectly influences processing and therefore does not need to be included in a theory of comprehension. Clark and Carlson (1981) more specifically propose that the intrinsic context which is necessary in order for a listener to understand what a speaker means at a particular moment is the *common ground* that the listener and speaker share during their conversation. Common ground specifically includes mutual knowledge as well as mutual beliefs and suppositions.

But, how can participants reliably establish that they mutually know something? This issue has been widely discussed within linguistics and philosophy (see Bennett, 1976; Kempson, 1975; Stalnaker, 1978), with some researchers going so far as to abandon the idea that mutual knowledge plays any significant role in conversational analysis (Wilson & Sperber, 1982). One of the difficult problems with this question concerns the infinite regress that soon results when trying to determine that speaker and addressee mutually know some proposition p. By definition, a speaker S and an addressee A mutually know a proposition P if and only if

S knows that P
A knows that P
S knows that A knows P
A knows that S knows that A knows P
. . . and so on ad infinitem.

It is highly unlikely that listeners can compute such an infinite regression in a finite period of time. However, Clark and Marshall (1981) have convincingly argued that mutual knowledge can be represented as an unanalyzable concept of the form:

A and B mutually know that p, if and only if some state of affairs G holds such that

a. A and B have reason to believe that G holds.

b. G indicates to A and B that each other has reason to believe that G holds.
c. G indicates to A and B that p.

G is called the basis for the mutual knowledge that p. Essentially, if A and B make certain assumptions about each other's rationality, they can use certain states of affairs as a basis for inferring the infinity of conditions all at once. There is no need to confirm each and every one of the infinity of conditions, even though in practice this can be attempted. Clark and Marshall's *mutual induction scheme* prevents the infinite number of iterations usually seen as a necessary consequence of establishing mutual knowledge.

The ground that are used for inferring the mutual knowledge that p include what Clark and Marshall call *co-presence heuristics*. These include linguistic co-presence, where the listener takes as common ground all of their conversation up to and including the utterances currently being processed. A second source for common ground is physical co-presence. Here the listener takes as common ground what he or she and the speaker are currently experiencing and have already experienced. The final source of evidence is community membership, which includes information that is universally known in a community, such as the fact that everyone is assumed to know a certain language and certain information about that community. Normally, mutual knowledge is established by some combination of physical or linguistic co-presence and the mutual knowledge based on community membership. Whatever the case, these *co-presence heuristics* show that mutual induction can occur without having to rely on an infinite number of pieces of evidence. We see then that mutual knowledge can be readily established and is a necessary element in what it means to speak in a coordinated, cooperative manner.

Some debate still remains over the role of mutual knowledge in language comprehension. Despite Clark and Marshall's demonstration to the contrary, Sperber and Wilson (1982) continue to doubt whether mutual knowledge can be easily identified during moment-by-moment processing, because of the infinite recursion involved in describing it. Given the practical problem identifying what is mutually known between speaker and addressee, Sperber and Wilson question its necessity for comprehension. Mutual knowledge, in their proposal, is not a *prerequisite* for comprehension, but is a product of comprehension.

Harris, Begg, and Upfold (1980) have presented some initial results suggesting the importance of shared knowledge for successful verbal communication. Their experiment utilized a task which is similar to the television game "Password." Subjects were identified either as senders or receivers. The senders were given a target word, such as *table* or *chair*, and instructed to provide clues to the receivers until recognition occurred. The specific context within which the senders produced clues differed, in that some senders were told that the receivers had been given the category from which the target was an exemplar, for instance *articles of furniture*, while other senders were told that the receivers had not been given such information. The senders were also told whether or not the receiver was given a list of possible target responses from which to choose. These four possible contexts for the senders were varied with the same possible contexts for the receivers, so that, sometimes, the senders' and receivers' conditions were matched, and sometimes they were crossed. Harris et al. expected that senders who are misinformed about the amount of information the receivers have about a target, either too much or too little, will have problems communicating with the receiver. The results of their experiment showed this to be true. If senders correctly assumed the context from which the receiver was to recognize the target, then successful communication occurred. These findings again emphasize that shared knowledge between speakers and addressees is critical for effective communication.

The need for the coordination of speaker's and addressee's knowledge is particularly apparent for the use and understanding of reference. For instance, if a speaker says, *Mary got a car with a sun-roof*, the speaker, when designing the utterance, must have assumed that that car was not singularly identifiable from the common ground. However, if the speaker instead said, *Mary got the car with a sun-roof*, that car must have been assumed to be identifiable given the common ground. In general, speakers will design their utterances to use definite reference for objects which are a part of the common ground, and indefinite reference for objects which are not (Clark & Marshall, 1981).

Olson (1970) proposed a model of definite reference in which speakers design their utterances to differentiate among an *array* of possible referents. Based primarily on perceptual information, the addressee infers the intended referent from the array. If the referent array is perceptually present and explicitly in the speaker and addressee's common ground, then the problem of communication becomes one of specifying which referent is intended. Krauss and Weinheimer (1964, 1966) examined speaker's and addressee's strategies for identifying referents from an array of novel figures. In their study, a speaker was instructed to repeatedly refer to one of the novel figures while she could not see her addressee. The addressee's task was to correctly identify the referent from the array of figures. On the initial trials, the speaker used elaborate descriptions in order to successfully communicate the target referent. However, on successive trials, the speaker progressively shortened the description while still successfully conveying the identity of the referent. The speaker was able to use the strategy of employing a shorthand description based on the knowledge that he or she shared earlier descriptions of the same referent with the addressee. This suggests that the earlier descriptions of the referent became part of the common ground which was then used for subsequent references. In Lewis' (1969) terms, these earlier descriptions set the precedence for the subsequent referring expressions. Common ground, which is accumulated during some communication, is the basis for the continued success of that conversation. Furthermore, when the speaker was instructed to describe the target figure on a tape-recorder for some new listener, the descriptions were not as shortened. This suggests that speakers take into account whether common ground is or is not present when describing a referent.

However, the array, which includes the intended referent, does not always present itself clearly to the speaker and addressee. How does the addressee determine the correct referent if the array is not explicit in that conversational context? This is not specified in Olson's model; however, Clark and Marshall (1981) propose that the determination of the intended array, in addition to the target referent, may be based on common ground. One strategy which may be employed by the speaker to specify the target array is the use of a gesture. A demonstrative reference, in particular, includes a gesture by the speaker to indicate the domain of the referent as well as to identify the referent.

An experiment which highlights the influence of common ground to understanding demonstrative reference was performed by Clark, Schreuder, and Buttrick (1983). In their experiment, subjects were shown a photograph of several groups of like objects. The experimenter made a gesture towards that photograph and then asked the listeners a question which required that they choose one of the objects as the target referent. So, for instance, if a subject was shown a photograph of *several* types of flowers and asked "How would you describe the color of this flower?" subjects would have to choose the one flower which they thought the speaker was referring to. Responses to the questions were examined and categorized as either implicit acknowledgements of understanding of the referent, requests for confirmation, or requests for clarification. The results of their study showed that, when target items in the photograph were clearly more prominent than the distractor items, subjects more often made an immediate choice of the referent than when the target items in the photograph were only slightly more prominent.

Clark et al. (1983) interpret these findings in support of the notion that listeners select referents based on what they judge to be most salient with respect to the their common ground. This suggests that Lewis' (1969) use of salience as one aspect of the context for comprehension has some psychological reality. Further results, from a task which asked subjects to explicitly choose the referent which the speaker was pointing to, were similar to people's choices based on perceptual salience and to people's choices based on common ground. They conclude that mutual knowledge about perceptual salience, speaker's goals, assertions, and presuppositions is required for understanding of demonstrative reference.

We are presently conducting our own series of experiments testing whether mutual knowledge plays an operative role in language use. Our aim is to see if people's assessment of what is mutually known between themselves and an addressee influences both the production and comprehension of questions in conversation. In our first study, subjects read stories which contained conversations between two people. Each story was written to reflect one of three possible degrees of mutual knowledge states between the characters. The different degrees of mutual knowledge are as follows:

1. Both speaker and addressee know some proposition (belief), and both knows that the other knows.
2. Speaker knows some proposition, but doesn't know if addressee knows the proposition.
3. Speaker knows some proposition and knows that addressee doesn't know the proposition.

Each individual story were written three times to reflected the three mutual knowledge states. Subjects were presented one of the three versions of each story, along with all three possible completions, and were instructed to decide which of the three completions was most appropriate for that story context. The completion was a question posed by the speaker which reflected one of the three levels of mutual knowledge. For instance, consider the following incompleted story and three possible completions.

> Fred and Robin, who work at the same office, are on their coffee break. Fred mentions that he needs some new clothes and that he hopes that Macy's will be having a sale soon. After Fred leaves, Robin notices a newspaper ad which states that Macy's will be having a men's wear sale that evening. Robin returns to the office, sees Fred, and says . . .
>
> (a) Do you think you want to do some shopping at Macy's tonight?
> (b) Do you want to go to the sale that Macy's is having tonight?
> (c) Did you find out that there will be a sale at Macy's tonight?

It would be inappropriate to assume that Fred knows that there is a sale that evening, so Robin would not say, *Do you think you want to do some shopping at Macy's tonight*? Instead, the question which best reflects the mutual knowledge of the story is, *Do you want to go to the sale that Macy's is having tonight*?

If people take into account what is mutually known between speaker and addressee, then subjects in our study should choose the completion that best reflects the degree of mutual knowledge between speaker and addressee in each story. The results of our experiment showed that subjects most frequently chose story completions which corresponded to the degree of mutual knowledge presented in the story. That is, subjects reading the stories from the first group most often chose (a) sentences, subjects reading the second group of stories most frequently chose (b) sentences, and subjects reading the third group of stories chose (c) sentences a majority of the time. These findings demonstrate that people do take into account the degree of mutual knowledge between the speaker and addressee when selecting specific questions to ask. We are currently examining whether people utilize information about the mutual knowledge state during moment-by-moment comprehension of utterances in conversation.

Although the issue of audience design is clearly relevant to common ground, this feature has rarely been explicitly included in process models of language use. More recently, Gibbs (1986) has investigated speakers' design and listeners' comprehension of indirect requests in conversation. Specifically, Gibbs tried to show that speakers formulate their requests to anticipate potential obstacles for addressees in complying with requests, and that the form of an indirect request influences how it is interpreted. In a first experiment, subjects were given scenarios and asked how they would make a request in that context. Each context specified a particular obstacle facing the addressee in complying with the request, as in the following example.

> Tracy and Sara were tired of eating at their college's dining hall. So they went downtown to find something exciting to eat. They decided to go to Tampico's. Sara wanted an enchilada, but was unsure whether the restaurant had them. The waitress came up to take their order and Sara said to her . . .

Note that, in this situation, the main obstacle for the addressee (i.e., the waitress) in complying with Sara's request was whether the restaurant actually served enchiladas. Presumably, then, Sara would take this into account in formulating her utterance and would say something like, *Do you have any enchiladas*? The results showed that the requests generated by subjects did indeed take into account the obstacles present in the scenarios. This suggests that people design their utterances to best capture the knowledge and beliefs of their audience. The results of a reading-time study showed that how well speakers' requests specify the potential obstacles for addressees influences the speed with which these utterances are understood. People process indirect requests that adequately specified the reasons for an addressee not complying with a request faster than they understand indirect requests that did not specify such obstacles. In general, it seems clear that speakers' and listeners' assessments of each other's plans and goals in conversation have a direct influence on the processes by which requests are produced and comprehended.

CONCLUSION

The main theme of this chapter has been that linguistic communication is an interactive, negotiated process in which there are implicit agreements between speakers and listeners. These agreements constitute the common ground or set of mutual knowledge and beliefs which enable speakers to anticipate the contextual requirements of the listener by designing his or her utterance to address these requirements. Formulating utterances in this manner facilitates listeners' recovery of the meanings intended by speakers. This view departs radically from many psychological models of communication which see words as vehicles that speakers select to express certain ideas, and which listeners, in turn, unpack to extract the verbal message. Such models usually assume that everyone has the same basic knowledge and belief systems and that words simply label pieces of knowledge, which themselves refer to objects outside in the world.

Our approach suggests that the determination of conversational meaning is a process requiring cooperation and coordination on the part of speaker and listener to achieve various social, interpersonal goals. We believe that viewing conversational meaning as coordinated, cooperative interaction imposes several constraints on psychological models of meaning and language use. The first is a constraint on the computational processes involved in understanding the meanings of speakers' utterances. Because speakers and listeners share certain knowledge and beliefs, which constitute the context for comprehension, listeners can recover speakers' intentions without having to first perform complex compositional analyses on the literal meanings of sentences. The common ground shared by conversants enables listeners to directly infer the implicatures speakers intend for hearers to recover.

A second constraint is that the shared knowledge between speakers and listeners also allows listeners to easily distinguish inferences which are authorized and intended from other possible implicatures that are unauthorized. Again, mutual knowledge provides much of the contextual framework so listeners immediately determine what inferences are authorized and constrains the generation of other, unauthorized inferences. Of course, people do, on occasion, derive inferences which are not specifically intended by speakers. But this reflects some of the difficulties in correctly assessing what set of knowledge and beliefs make-up the common ground between conversants. As we have shown, nonetheless, establishing mutual knowledge is not computationally improbable and is a realistic, necessary part of the process of understanding conversational messages.

Finally, mutual knowledge allows speakers to tailor the wording of their utterances to meet the various plans and goals of their addressees. Speakers take into account precisely what they know about their addressees and what their addressees know about their (the speakers') own beliefs and knowledge in deciding how to say what they intend. We have reviewed evidence suggesting that listeners are cognizant of how well speakers do this, which affects the process of understanding speakers' utterances.

Each of these constraints on the processes of speaking and listening should be included as part of any psychological theory of meaning and language use. Many of the pragmatic properties of communication we have described (e.g., cooperative principles and maxims, intention recognition, mutual knowledge, audience design, etc.), cannot be dismissed as belonging to the sociological domain because, as we've described, these factors impose constraints on the psychological computations necessary to produce and understand conversational utterances. Models of interactive systems, whether they concern people talking with other people or people talking in some natural dialogue with computers, must explicitly model the beliefs and intentions of the agents, as well as the beliefs and knowledge that the agents have of each other.

REFERENCES

Alba, J., & Hasher, L. (1983). Is memory schematic? *Psychological Bulletin, 93*, 203–231.

Allen, J., & Perrault, C. (1980). Analyzing intention in utterances. *Artificial Intelligence, 15*, 143–178.

Appelt, D. (1985). *Planning English sentences*. Cambridge, England: Cambridge University Press.

Bennett, J. (1976). *Linguistic behavior*. Cambridge, England: Cambridge University Press.

Bransford, J. (1979). *Human cognition*. Belmont, CA: Wadsworth Press.

Carbonell, J. (1978). Intentionality and human conversations. In B. Weber–Nash & R. Schanh (Eds.), *TINLAP-2: Theoretical issues in natural language processing*. Urbana-Champaign, IL: Association for Computational Linguistics.

Clark, H. (1977). Inferences in comprehension. In D. Laberge & S. Samuels (Eds.), *Basic processes in reading: Perception and comprehension*. Hillsdale, NJ: Erlbaum.

Clark, H. (1984). Language use and language users. In G. Lindsay & E. Aronson (Eds.), *Handbook of social psychology* (3rd ed.). New York: Random House.

Clark, H., & Carlson, T. (1981). Context for comprehension. In J. Long & A. Baddeley (Eds.), *Attention and performance XI*. Hillsdale, NJ: Erlbaum.

Clark, H., & Marshall, C. (1981). Definite reference and mutual knowledge. In A. Joshi, I. Sag., & B. Webber (Eds.), *Elements of discourse understanding*. Cambridge, England: Cambridge University Press.

Clark, H., Schreuder, R., & Buttrick, S. (1983). Common ground and the understanding of demonstrative reference. *Journal of Verbal Learning and Verbal Behavior, 22*, 245–258.

Cohen, P., & Perrault, C. (1979). Elements of a plan-based theory of speech acts. *Cognitive Science, 3*, 177–212.

Cohen, P., Perrault, C., & Allen, J. (1982). Beyond question answering. In W. Lehnert & M. Ringle (Eds.), *Strategies for natural language processing*. Hillsdale, NJ: Erlbaum.

Gerrig, R. (1986). Process models and pragmatics. In N. Sharkey (Ed.), *Advances in cognitive science*. Chitchester, England: Ellis Horwood.

Gibbs, R. (1981). Your wish is my command: Convention and context in interpreting indirect requests. *Journal of Verbal Learning and Verbal Behavior, 20*, 431–444.

Gibbs, R. (1982). A critical examination of the contribution of literal meaning to understanding nonliteral discourse. *Text, 2*, 9–27.

Gibbs, R. (1984). Literal meaning and psychological theory. *Cognitive Science, 9*, 275–304.

Gibbs, R. (1986). What makes some indirect speech acts conventional? *Journal of Memory and Language, 25*, 181–196.

Grice, H. P. (1968). Utterer's meaning, sentence-meaning, and word-meaning. *Foundations of language, 4*, 225–242.

Grice, H. P. (1975). Logic and conversation. In P. Cole & J. Morgan (Eds.), *Syntax and semantics 3: Speech acts*. New York: Academic Press.

Grice, H. P. (1978). Some further notes on logic and conversation. In P. Cole (Ed.), *Syntax and semantics 9: Pragmatics*. New York: Academic Press.

Harris, G., Begg, I., & Upfold, D. (1980). On the role of the speaker's expectations in interpersonal communication. *Journal of Verbal Learning and Verbal Behavior, 19*, 597–607.

Haviland, S., & Clark, H. (1974). What's new? Acquiring new information as a process in comprehension. *Journal of Verbal Learning and Verbal Behavior, 6*, 685–691.

Hobbs, J., & Evans, D. (1980). Conversation as planned behavior. *Cognitive Science, 4*, 349–378.

Jarvella, R., & Collas, J. (1974). Memory for the intentions of sentences. *Memory & Cognition, 2*, 185–188.

Kempson, R. (1975). *Semantic theory*. Cambridge, England: Cambridge University Press.

Kintsch, W., & van Dijk, T. (1978). Toward a model of text comprehension and production. *Psychological Review, 85*, 363–394.

Krauss, R. M., & Weinheimer, S. (1964). Changes in reference phrases as a function of frequence of usage in social interaction: A preliminary study. *Psychonomic Science, 1*, 113–114.

Krauss, R. M., & Weinheimer, S. (1966). Concurrent feedback, confirmation, and the encoding of referents in verbal communication. *Journal of Personality and Social Psychology, 4*, 343–346.

Lewis, D. (1969). *Convention*. Cambridge, MA: Harvard University Press.

Johnson, M., Bransford, J., & Solomon, S. (1973). Memory for tacits implications of sentences. *Journal of Experimental Psychology, 98*, 203–205.

Olson, D. R. (1970). Language and thought: Aspects of a cognitive theory of semantics. *Psychological Review, 77*, 257–273.

Owens, J., Bower, G., & Black, J. (1978). The "soap opera" effect in story memory. *Memory & Cognition, 5*, 185–191.

Robertson, S., Black, J., & Johnson, P. (1981). Intention and topic in conversation. *Cognition and Brain Theory, 4*. 303–326.

Sacks, H., Schegloff, E., & Jefferson, G. (1974). A simplest systematics for the organization of turn-taking in conversation. *Language, 50*, 696–735.

Schweller, K., Brewer, W., & Dahl, D. (1976). Memory for the illocutionary forces and perlocutionary effects of utterances. *Journal of Verbal Learning and Verbal Behavior, 15*, 325–337.

Sperber, D., & Wilson, D. (1982). Mutual knowledge and relevance in theories of comprehension. In N. Smith (Ed.), *Mutual knowledge*. London: Academic Press.

Stalnaker, R. (1978). Assertion. In P. Cole (Ed.), *Syntax and semantics 9: Pragmatics*. New York: Academic Press.

Thorndyke, P. (1976). The role of inferences in discourse comprehension. *Journal of Verbal Learning and Verbal Behavior, 15*, 437–446.

Wilson, D., & Sperber, D. (1981). On Grice's theory of conversation. In P. Werth (Ed.), *Conversation and discourse*. London: Croon Hale.

Wilson, D., & Sperber, D. (1987). Inference and implicature in utterance interpretation. In T. Myers (Ed.), *Reasoning and discourse processes*. London: Academic Press.

Describing Cooperation: Afterward to Section I

Wayne W. Zachary

CHI Systems Inc.
Spring House, PA

The chapters in this first section are all by cognitive scientists, although all with varied "home" disciplines. Their chapters have each pointed out different aspects of what needs to be added to *standard* cognitive theory to develop a model of cooperative interaction. A *standard* or minimalist cognitive approach to cooperation might begin with two intelligent agents, each with a cognitive mechanism, and some communication capability. Through this communication capability these agents could express a shared goal that could be pursued through joint action. In developing a plan to accomplish these goals, the agents each develop a plan (using their assumed cognitive mechanism, much as they would if pursuing the same goal in isolation). These interacting plans, to borrow the language of Hobbs and Evans (1980), provide an implicit control structure that allows the individual agents to organize their actions and communications through time. The chapters in section one all seem to have begun from this premise, evaluated it against empirical data, and then elaborated and/or modified it in some way.

Three issues in analyzing human cooperation emerged in Section I:

- the importance of the situation in which the cooperation occurs, and the effects of the situation on the pragmatics of the interaction;
- the reliance on and assumption of plan recognition as a basic element of cooperative interaction; and
- the nature and effect of social relationships on problem solving and cooperative processes.

The first issue was addressed by all the chapters except Malone's, with Suchman and Gibbs and Mueller making the most forceful points. The second issue was also the main concern of Suchman and Gibbs and Mueller, with Newman also contributing to the discussion. The third point was discussed most directly by Malone, although all the chapters in Section I deal with interagent relationship in some manner.

CONTEXT, PRAGMATICS, AND COOPERATIVE INTERACTIONS

Suchman pointed out that cooperating agents need not only a shared understanding of goals but also of their plans and actions. She argued, with the support of empirical data, that gross misunderstandings can develop if the agents do not share some understanding of *how* the goals are to be pursued. Even more importantly, Suchman argued that the primary interaction was mediated by the situation or context in which the interaction occurs. In Suchman (1988), she makes this point more strongly and in more detail. Although there are many dimensions to her argument, the main point concerns the element of pragmatics that the situation imposes on a plan for accomplishing a given goal. Many *ad hoc* elements must be introduced into the agents' plans, and each agent must recognize and accommodate its plans to the *ad hoc* elements introduced by the other agent and/or by the environment. Even the simple act of initially inserting a book upside down in a copy machine can introduce a situational variable that requires the person doing the copying to revise his or her plan. Both the initial error and the correction each require *ad hoc* adjustment of the plan of the machine producing the copies and of the agent doing the copying. Failure to make these situational adjustments makes the cooperation at first more difficult and then almost impossible. Suchman's main contribution, in this argument, is to note that:

1. this mutual adjustment of plan and knowledge to the situation is non-trivial and necessary for true cooperation; and
2. that people do this so effortlessly as to obscure its difficulty and importance.

One interpretation of this is that an ability to pragmatically adjust plans and to recognize and adjust to other agents' pragmatic adjustments is an essential enhancement of the minimalist model.

Newman further elaborates on the importance of the situation of the interaction in two ways, in terms of its effect on acquisition of shared expertise, and in terms of applying it. According to Newman, learning arises from interaction that occurs in a particular social and cognitive context. The cognitive context is what Vygotsky calls the "Zone of Proximal Development," a situation where the knowledge state of one agent is sufficiently close to that of another, that appropriation of action by the second agent may be analyzed and understood by the first agent. The social context is one where the two agents have a relationship where they are able to cooperate from the beginning, and where one agent is able to appropriate the actions of the other. For a counterexample, note that situations where the student appropriates the teacher's behavior are often termed disruptive and are not tolerated; this asymmetry is part of the teacher–student relationship. More is said below on the role of social relationship in cooperation.

There is also an implied cognitive mechanism in Newman's model, a cognitive capability to induce the meaning of the appropriated action and incorporate in the student's knowledge system. Learning, in Newman's model, occurs from need and ability to understand appropriation in specific contexts. This makes learning a situational phenomenon, not a matter of interacting plans as might be suggested by the standard cognitive model (or most of educational theory). It is not surprising, therefore, to recall that Suchman also used the metaphor of a gifted "coach" who contextualizes instruction to discuss the situational basis of interaction.

PLAN RECOGNITION AND COOPERATIVE COMMUNICATION

Gibbs and Mueller elaborated on the need for mechanisms for coordinating action and for implying the points of understanding that are shared and the way in which they are used in the communicative process. Suchman raised the need for an ability to recognize and accommodate the plans of other agents; Schmidt, Sridharan, and Goodson (1979) have termed this ability *plan recognition*. But whereas Suchman dealt with a form of plan recognition that simply arises out of accommodation of situational information, Gibbs and Mueller pointed out the need for *intended plan recognition*. Cohen, Perrault, and Allen (1981) differentiate this form of plan recognition, where one agent expects and intends for the other to do it from another case, termed *keyhole plan recognition*, in which one agent is observing but not directly cooperating with the other. Simple conversation, according to Gibbs and Mueller, inherently involves transactions where one agent deliberately creates a situation in which the other must infer and recognize the element of the plan or the piece of shared understanding that is being implied by the other. This suggests that cooperating agents must be able to perform "keyhole" plan recognition as Suchman suggests, but must also be able to instrumentally use plan recognition to tailor and support the communicative process.

RELATIONSHIP AS A CONSTRAINING PRINCIPLE IN INTERACTION

Gibbs and Mueller also noted how communication can be used in several ways simultaneously, most notably as an instrumental act in itself as well as a communicative act. The idea of speech acts is certainly not new (see

Austin, 1962; Searle, 1969), but is particularly important in this context. As used by Gibbs and Mueller, the speech act involves a social relationship between the agents. This social relationship may encompass a broad range of actions, expectations, and assumptions that the agents may apply to one another during the cooperative situation.

One major application of this concept of relationship to the cooperative behavior is in the work of Axelrod (1981, 1984, 1986). He began with the simple two-by-two game of Prisoners' Dilemma, with its four possible outcomes (win–lose, lose–win, lose–lose, and win–win). He wondered how strategies that favor the win–win approach could evolve in purely Darwinian systems, i.e., where the rule is "eat or be eaten." The win–win strategy requires cooperation among the actors, a relationship in which each expects and relies on the other to select the win–win combination. When this expectation is realized, both parties win. When it is not, one or both lose. Axelrod staged a natural selection tournament by computer simulation, in which a number of (simulated) agents could interact, with each interaction being structured as a Prisoners' Dilemma game. The game outcome was calculated as an accumulated score, with each player accruing "utiles"—positive for each winner, negative for each loser. At the end of a round, the agents with the most utiles got to survive until another round, and those with the fewest became extinct. Although many strategies were offered, the simplest strategy won, at least in the sense of being the last to avoid extinction. This simple strategy was called TIT FOR TAT and used a simple relationship as the basis for judgement. TIT FOR TAT was in fact a cooperative strategy: on the first encounter with another agent, attempt cooperation (i.e., opt for the potential win–win); after that, "doing whatever the other player did on the previous move" (Axelrod, 1981, p. 308).

Under TIT FOR TAT, if the other agent initially cooperates and continues to cooperate, so will TIT FOR TAT, in all future encounters as long as the cooperative relationship holds. Once the other agent first goes against cooperation (i.e., moves for unilateral victory), then the TIT FOR TAT will reciprocate at the next encounter. If the other agent persists, so will TIT FOR TAT, but if the other agent reverts to cooperation, so will the TIT FOR TAT player. A major implication of this simulation (which was actually a long series of simulations) was to demonstrate that a strategy based on relationship between agents was able to generate cooperation without violating the premise of optimization of one's own goals. It also demonstrated that cooperation could evolve without communication about goals and without explicitly shared goals.

Malone's chapter elaborates much further on the kinds of relationships underlying cooperation. He identifies three kinds of coordination (i.e., synchronization, convention, and negotiation), and notes that they all depend on establishing and maintaining relationships among the agents involved. This set of relationships is what he calls the system organization. Malone is much more conscious of the issues of organization and relationship than the others in Section I because he is looking at larger scale interactive systems. The kinds of cooperative systems studied by Suchman, Newman, and Gibbs and Mueller are all dyads. As such, there is only one relationship involved, so differences in relationship can not affect the cooperative process. In systems involving more agents, this changes rapidly, as the number of dyadic relationships increases combinatorially with the number of agents. When Malone discusses highly distributed systems that are likened to marketplaces, he is implicitly considering cooperative systems that have hundreds, thousands, or even more nodes. In these systems, the relationships among the agents play a major role in coordination of activity. A major implication about the dependency of cooperation on interagent relationship is that the agents must be able to reason and plan about these relationships, which in turn requires an internal representation of it. This then, represents another extension to the minimalist model of cooperative systems.

FINAL THOUGHTS ON PART I

The chapters in this first section give some important elements of cooperative interaction beyond the minimalist cognitive model—situation, context, sharing of understanding, ability to recognize and interpret plans and actions, control through organization by relationship. In doing so, the chapters in this section have made some points that computer scientists and engineers would perhaps prefer not to hear. First, section one argues that at the lowest level (i.e., the dyad), cooperation relies heavily on pragmatics. The idea that simple interacting plans can cooperate in the natural world, which is where ultimately any systems that are engineered must operate, is directly attacked by Suchman, and indirectly by Gibbs and Mueller and Newman. Later in section two, Woods et al., bring up the issue again and continue the attack.

The second unpleasant point made in section one is the importance of shared expectations and for explicit implication of shared understandings among cooperating agents. These capabilities are, perhaps, not computationally as difficult to engineer as situational pragmatics. The need to design these capabilities into a cooperative agent certainly adds to the information-processing overhead. The third design-oriented point from section one is the need to be explicit about relationships. In defining coordinating mechanisms in terms of organizational structures and interagent relationships, Malone has linked the design of cooperative computer systems with that of designing and explicating relationships. In computational terms, this is virgin territory, and again adds to the complexity of designing cooperative systems. In the next section, each of the chapters addresses this problem in some way.

REFERENCES

Austin, J. L. (1962). *How to do things with words.* London: Oxford University Press.

Axelrod, R. (1981). The emergence of cooperation among egoists. *The American Political Science Review, 75*(2), 316–318.

Axelrod, R. (1984). *The evolution of cooperation.* New York: Basic Books.

Axelrod, R. (1986). An evolutionary approach to norms. *The American Political Science Review, 80*(4).

Cohen, P., Perrault, C., & Allen, J. (1981). *Beyond question-answering* (Report No. 4644). Cambridge, MA: Bolt Beranek and Newman Inc.

Hobbs, J., & Evans, D. (1980). Conversation as planned behavior. *Cognitive Science, 4*(4), 349–377.

Schmidt, C., Sridharan, N., & Goodson, J. (1979). The plan recognition problem: An intersection of artificial intelligence and psychology. *Artificial Intelligence, 10*, 45–83.

Searle, J. R. (1969). *Speech acts: An essay in the philosophy of language.* Cambridge, England: Cambridge University Press.

Suchman, L. (in press). *Plans and situated actions: The problem of human machine communication.* Cambridge, England: Cambridge University Press.

II

Design and Architecture of Cooperative Computing Systems

Scott P. Robertson

Department of Psychology
Rutgers University
New Brunswick, NJ

INTRODUCTION

Section I contains chapters that examine cooperative systems from the outside, as observers of all the participants. In this section, the authors discuss designs for the computational component of cooperative human–computer systems.

Woods, Roth, and Bennett start off by observing the behavior of humans working with a computer system designed to diagnose faults in a device. The system is a more or less classic version of a category of systems designed to support decision making (i.e., "intelligent decision support," IDS, systems). The computer in this case is an expert system that instructs the human users to perform certain actions on a device and then queries them about the behavior of the device. The expert system reaches conclusions about malfunctions in the device which the human users may accept or reject. Woods et al. note that the locus of control in the interaction rests with the computer system, although users can react with varying degrees of passivity to the machine's instructions.

Woods et al.'s analyses of protocols reveals considerable brittleness in this paradigm for human–computer interaction. An active human user was much more successful than passive users in diagnosing malfunctions and avoiding errors. Unanticipated deviations in the interaction often led the machine to pursue unproductive paths with little recourse available to the human user. The strict allocation of functions between human and machine and the inability to share information about problem-solving strategies was a core problem with the observed system. Woods et al. suggest that human–

machine systems be reconceptualized as joint problem solvers, and they make several recommendations about the features of such systems.

Durfee, Lesser, and Corkill describe techniques for coordinating decision making among nodes in a distributed problem-solving architecture. This contribution is the only one which does not involve analysis of human agents, but the requirements for distributed problem-solving nodes are strikingly relevant and familiar. A distributed problem-solving network can be thought of as a multiagent planning system where subtasks are formulated and dynamically allocated by the various nodes. Problems within this paradigm involve constraining decision making to the appropriate nodes and avoiding debilitating redundancies, node interactions, and resource tie ups.

After describing various approaches to distributed architectures, Durfee et al. focus on the role of prediction in coherent cooperative systems. Prediction is the ability of a node to guess about the outcomes of processing at other nodes and even about the overall behavior of a network. Successful prediction comes from providing nodes with knowledge about the capabilities of other nodes and the communication patterns between nodes. Durfee et al. describe techniques for acquiring and updating such shared information during problem solving in a distributed network.

In the final chapter in this section, Goodson and Schmidt attempt a broad description of requirements for cooperative systems and propose a development strategy for the design of cooperative human–machine systems. They emphasize that assignment of roles to various agents in problem solving situations should be made according to the information processing capacities of those agents, with constraints on working memory, I/O channel capacity, and knowledge being constant constraints of the human agents. Like Woods et al., Goodson and Schmidt recognize the central role of the communication links between agents in cooperative systems. They propose "communicative plans" to facilitate exchange of information about the states and conclusions of other agents and discuss various types of information dependencies among concurrent problem-solving units. Interestingly, like Gibbs and Mueller in Section I, Goodson and Schmidt propose that communicative "conventions" must be shared among cooperative agents in order to facilitate problem solving. Their design methodology is examined in practice with the presentation of an example human–machine system for tracking submarines.

6

Explorations in Joint Human–Machine Cognitive Systems

David D. Woods

Department of Industrial and Systems Engineering
The Ohio State University
Columbus, OH

Emilie M. Roth

Department of Engineering and Public Policy
Carnegie-Mellon University
Pittsburgh, PA

Kevin B. Bennett

Psychology Department
Wright State University
Dayton, OH

INTRODUCTION

Advances in computer science and artificial intelligence are providing powerful new computational tools that greatly expand the potential to support cognitive activities in complex problem solving worlds, e.g., monitoring, problem formulation, plan generation and adaptation, and fault management. The question that we continue to face is how we should deploy the power available through new capabilities for tool building to assist human performance, i.e., the problem of how to provide intelligent, or more properly, effective decision support (IDS). The application of these tools creates new challenges about how to couple human intelligence and machine power in a single integrated system that maximizes overall performance.

Tool builders have focused, not improperly, on tool building—how to make better machine problem solvers, where the implicit model is a human expert solving a problem in isolation. But effective tool use involves more than this, and the factors associated with tool utilization can have an impact on the very nature of the tools themselves. Building machines that are good problem solvers in isolation does not guarantee high performance in actual work contexts where the performance of the joint person–machine system is the relevant criterion. The key to the effective application of the powers of computational technology is to conceive, model, design, and evaluate the joint human–machine cognitive system (Hollnagel & Woods, 1983; Woods, 1986). Like Gestalt principles in perception, a cognitive system is not merely the sum of its parts, human and machine. The configuration of the human and machine subsystems is a critical determinant of the performance of the system as a whole (e.g., Sorkin & Woods, 1985). The challenge for cognitive science is to provide models, data, and techniques to help designers build an effective configuration of the human and machine portions of a joint cognitive system. In this chapter we will examine some of the issues in the development of effective joint cognitive systems by contrasting two paradigms for configuring human and machine problem solvers: (a) to deploy machine power in the form of cognitive prostheses for people, that is, remedies for human deficiencies; or (b) to deploy machine power in the form of cognitive instruments wielded by a competent practitioner. To illustrate some of the pitfalls inherent in the machine as prosthesis paradigm for IDS, we will examine the results of an investigation of human troubleshooters interacting with an intelligent machine designed within this paradigm (Roth, Bennett, & Woods, 1987). The findings of this investigation with respect to the properties and performance of the joint cognitive system, in conjunction with results from other studies of human–machine interaction during problem solving, point the way to alternative approaches to deploy machine power to assist human performance.

PARADIGMS FOR DECISION SUPPORT

One metaphor used to guide our thinking about how to configure joint cognitive systems is based on an analogy to human–human cooperative problem solving. Because the machine element can be thought of as a semiautonomous cognitive system in its own right, we can begin to think about how this machine cognitive agent might interact with a human cognitive agent during problem solving by transposing how two people interact during problem solving. Another metaphor is to consider the new machine capabilities as extensions and expansions along a dimension of machine power. In this metaphor, machines are tools; people are tool builders and tool users. Technological development has moved from physical tools (tools that magnify man's capacity for physical work), to perceptual tools (extensions to man's perceptual apparatus such as medical imaging), and now, with the arrival of artificial intelligence technology, knowledge or cognitive tools (although this type of tool has a much longer history, e.g., aide-memoires or decision analysis, AI has certainly increased the interest and capability to provide cognitive tools). In this chapter we will consider the factors that contribute to effective joint person–machine problem solving by starting with the metaphor of cognitive tools. Are cognitive tools qualitatively different from the physical and perceptual tools that we have more experience with? What kinds of cognitive tools are needed? What aspects of the human can be amplified with cognitive tools?

Cognitive Tools as Prostheses

What is a cognitive tool? One approach often adopted implicitly or explicitly is to design IDS systems as *prostheses*—a replacement or remedy for a deficiency. In one sense all tool use implies a human deficiency with respect to some goal—if we had hard hands, we would not need hammers. Most current expert systems are cognitive tools in this sense. In the cognitive-tool-as-prosthesis paradigm, the primary design focus is to apply computational technology to develop a stand-alone machine expert that offers some form of problem solution. The technical performance of this system is judged in terms of whether the solutions offered are "usually" correct (e.g., Yu, Fagan, Wraith, Clancey, Scott, Hannigan, Blum, Buchanan, & Cohen, 1979). This paradigm emphasizes tool building over tool use. Questions about how to interface human to machine are secondary to the primary design task of building a machine that usually produces correct decisions. A typical encounter with an intelligent computer consultant developed in this fashion consists of the following scenario: the user initiates a session, the machine controls data gathering, the machine offers a solution, the user may ask for explanation if some capability exists, and the user accepts (acts on) or overrides the machine's solution.

What is the joint cognitive system architecture implicit in this paradigm? The primary focus is to apply computational technology to develop the machine expert. In practice, putting the machine expert to work requires communication with the environment—data must be gathered and decisions implemented. Rather than automate these activities they are typically left for the human (Figure 1). Thus, questions about support for the human portion of the ensemble become, not so much how to interface the machine to the human, but, rather, how to use the human as an interface between the machine and its environment. As a result, locus of control resides with

Figure 1. The joint cognitive system implicit in the expert-machine-as-prosthesis. This paradigm for decision support focuses primarily on building a machine expert that outputs usually correct solutions; the human's role is to gather data and to filter any poor machine solutions (from Woods, 1986).

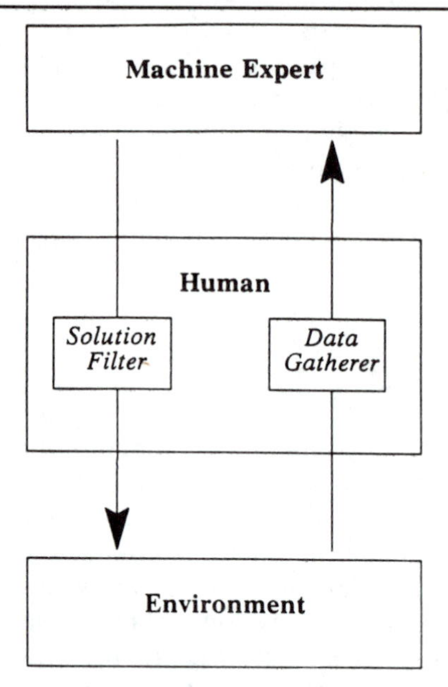

the machine portion of the ensemble, and human–machine interface design focuses on features to aid the person's role as data gatherer (e.g., Mulsant & Servan-Schreiber, 1983) and on features to help the user accept the machine's solution.

In the cognitive-tools-as-prosthesis paradigm, user acceptance and the machine expert's technical performance (again, in the sense of offering problem solutions that are usually correct) are seen as independent issues (e.g., Shortliffe, 1982). "Excellent decision making performance does not guarantee user acceptance" (Langlotz & Shortliffe, 1983, p. 479). Thus, lack of user acceptance is considered to be a problem in the user which must be treated by measures outside of the essential characteristics of the machine expert, rather than a sign that the machine's advice is not useful in the actual problem solving context. One proposed technique is to embed other, useful support capabilities in the same computer system that implements the machine expert, e.g., data management functions such as computerized data entry forms or standard report generation (Langlotz & Shortliffe, 1983). Some designers of machine experts suggest that the interface between the human and machine expert should include features that encourage the user to accept the machine's solution, even if the features that encourage acceptance result in user misperceptions of how the machine expert actually functions. In the case of one medical machine expert, the designers suggested that the interface "provide the physician with the ability to report the facts he considers important (*even if they are not used internally*)," i.e., by the machine expert (Mulsant & Servan-Schreiber, 1983, pp. 11–12; emphasis added). On the other hand, considering the human and machine as parts of a joint cognitive system suggests that problems with user acceptance are very often *symptoms of an underlying deficiency* in the "cognitive coupling" (Fitter & Sime, 1980) between the human and machine subsystems.

The cognitive-tools-as-prosthesis paradigm emphasizes user acceptance of the machine's solution. However, nonroutine events may occur, the problem may be outside the machine's scope of competence, there can be errors in the rule base, or the human may make errors in data gathering. In order to cope with the fact that the machine may not always generate a correct solution, the machine-as-prosthesis paradigm labels proposed solutions as *advice* that the human can always decide to reject or modify if his or her knowledge or circumstances warrant (e.g., solutions are often offered in some form of confidence or likelihood estimate over a set of possible diagnoses). Thus, the human user needs to decide whether a proposed solution is correct or incorrect, given the machine's output and the context of the problem solving episode. How good are people at discriminating correct from incorrect machine solutions, and how does discrimination performance vary with user expertise and with different types of machine explanation? Very little is known about what factors affect human performance at filtering another decision maker's solutions.[1] What level of expertise is needed to recognize erroneous machine output or a situation that is beyond the capabilities of the machine? For example, can people use syntactic cues (this output looks odd for this type of problem) or experience-driven associations (in this situation, the machine often has difficulties) to filter erroneous system output?

Cognitive Tools as Instruments

Cognitive tools can also be seen as instruments, that is, a means for effecting something in the hands of a competent practitioner. People are a gener-

[1] We are referring to the person's ability to filter machine output *after* the machine expert has been developed and deployed. The standard iterative refinement approach for building machine experts depends on human experts/knowledge engineers' ability to detect and correct erroneous machine output.

alized species—we work fairly well in a wide range of environments (almost all man-made today). It is the ability to develop, adapt and utilize tools that supports specialization in any particular environment—if we had hard hands, we might make good carpenters, but we would be poor pianists. From this vantage point, our tool building/using skills are resource utilization in the pursuit of a goal. For example, when repairing a car if someone sees he needs to manipulate something around a corner where his hand cannot reach, he would look for an instrument that affords manipulation around an angle—perhaps a flexible manipulator is nearby or perhaps he can bend an extra pair of needle nose pliers. The point is that tool use involves an active role for the tool wielder in some instrumental context, whereas the cognitive tool as prosthesis view leaves the tool user with a passive role. Anyone who has worked with practitioners in some field has seen the tools that the practitioners fashion to help them carry out their jobs, even tools to help them work around the hindrances often created by the tools that they are given or to make these aids actually useful.[2]

The cognitive-tool-as-instrument paradigm demands a problem-driven, rather than technology-driven, approach to the development of effective joint cognitive systems. In a problem-driven approach knowledge of the requirements and bottlenecks in cognitive task performance is used to provide tools which help the human problem-solver function more expertly (cognitive task analysis, e.g., Woods & Hollnagel, 1987; Rasmussen, 1986; and cf. Mitchell & Saisi, 1987, for one example application). From the cognitive tool as instrument perspective, computational technology should be used, not to make or recommend solutions, but to aid the user in the process of reaching his or her decision.

The instrumental view of tools captures the role of adaptability in intelligent behavior. Research on human performance emphasizes that skill is the capability to adapt behavior to changing circumstances in pursuit of a goal (e.g., Woods, O'Brien, & Hanes, 1987), and some philosophers argue that rationality is intellectual adaptability (Toulmin, 1972). This means that the power or intelligence of an IDS system is related to its ability to increase the human problem solver's adaptability to the kinds of problems that could arise in the pursuit of domain goals. While a large amount of research on intelligent interfaces focuses on building interfaces that adapt to different users independent of a problem world, for example, adapting to different user styles or to different stages of learning, the above suggests

[2] Yet laboratory research into human problem solving continues to examine human performance stripped of any tools; perhaps more could be learned from examining the nature of the tools that people will spontaneously create to work effectively in some problem solving environment.

that research should focus more on increasing the range of problems solvable by the human–machine ensemble.

One critical issue that distinguishes the two approaches is the criterion for judging the effectiveness of a problem-solving system. In the cognitive-tool-as-prosthesis paradigm, the *system* is defined as the machine expert, and *effective* means usually correct machine solutions. In the cognitive-tool-as-instrument paradigm, the system is defined as the combination of human and machine (the human–machine cognitive system) and effective means maximizing joint performance; i.e., performance of the whole should be greater than the performance possible by either element alone (see Hollnagel, 1986, for a treatment of some of the implications of this for the evaluation of intelligent machines).

HUMAN INTERACTION WITH AN EXPERT SYSTEM

We examined the effects of the machine-as-prosthesis paradigm for the development of IDS in a troubleshooting application (cf. Roth, Bennett, & Woods, 1987). We were able to observe human–machine interaction when novel situations arose (problems outside the machine's competence), when adaptation to special conditions was required, in the face of underspecified instructions, and when recovery from human or machine errors was demanded. While the consequences of these situations can be treated in design by minimizing their frequency of occurrence, this cannot guarantee that all such situations have been eliminated both on practical grounds (there are rarely enough resources available to avoid underspecification of instructions) and theoretical grounds (underspecification is an inherent property of language; cf. Gibbs & Mueller, this volume). Unanticipated variability will arise as it did in this application, and the design of IDS systems must take that fact into account. As others have noted, we need to complement design for the prevention of trouble with design for the management of trouble (Ackoff, 1981; Brown & Newman, 1985).

In the study we observed four technicians at a variety of experience levels using the machine expert to troubleshoot malfunctions in an actual device. The troubleshooters were instructed to use the machine expert to diagnose the source of the malfunctions. Two of the four participants, the most experienced technician (D) and the one with the least experience (A), took an active role in initiating observations and measurements and forming judgments beyond those requested by the machine expert (in other words, they tended to treat the machine expert as an instrument they could manipulate to accomplish goals). The other two technicians were more passive in their interactions, following the directives of the expert system as literally as possible and providing minimal extramachine contributions to the problem-solving process (in other words, they tended to act only as data gatherers for the machine).[3]

There were six different problems in the study which were created by introducing faults into an actual operable device. We were able to observe and analyze 18 instances of joint person-machine problem-solving episodes. (Not every technician was presented each problem.) Fourteen of the 18 problem-solving episodes included in the study were eventually solved. Differences in performance on these problems were reflected in the amount of time required to solve the problem and the number and substance of knowledge engineer interventions that were required. Of the remaining four cases, two were solvable problems (both Problem 4) but deviations arose from which the technicians were unable to recover. In the other two cases, the machine expert could not solve the problem (Problem 5). These problems afforded the opportunity to examine the interaction of human and machine under conditions when the problem-solving episode went outside the machine's competence.

The opportunity to observe both passive and active human problem solvers interact with a machine expert in the face of unanticipated variability revealed how the prosthesis approach to decision aiding leads to an impoverished joint cognitive system and its consequences for joint system performance. Contrary to the expectations of the designers of the machine expert, troubleshooters actively and substantially contributed to the diagnostic process. The more the human functioned as a passive data gatherer for the machine, the more joint system performance was degraded. Those who passively followed the directives of the machine expert dwelled on unproductive paths and reached dead-ends more often than participants who took a more active role. Furthermore, because comprehension of instructions depends on active participation, the passive troubleshooters made more errors that contributed to unproductive paths.

In contrast, active human participation in the problem solving process led to more successful and rapid solutions. The active technicians attempted to cope with unexpected conditions, to monitor machine behavior, to recognize unproductive directions, and to redirect the machine to more productive paths (within the limited means available) based on activities and judgments formed outside of the machine's direction. However, the machine-as-prosthesis design of the machine expert not only failed to support an active human role, it actually retarded technicians from taking or carrying out an active role.

---

[3] The passivity demonstrated by at least one of the technicians may have, in part, been due to the perception that the expert system was being introduced to replace technicians or reduce the skill level requirements for technicians.

The Investigation

The machine expert in question was designed to support technicians in troubleshooting a new generation of an electromechanical device.[4] The specific device was newly introduced and based on a shift to digital control technology. The machine expert was designed to be used by technicians who had experience troubleshooting the previous generation of this device, but who were unfamiliar with new generation of the device and with digital technology in general. The expert system was viewed to some extent as a "replacement" for training on how to troubleshoot this new device.

The machine expert was developed along the classic lines recommended in the AI literature for building an expert system for industrial applications (e.g., Hayes-Roth, Waterman, & Lenat, 1983). An expert system development shell that incorporates a mixed forward and backward chaining strategy and confidence value propagation was selected. An expert technician with extensive experience in troubleshooting devices of this general class and knowledge of the functioning of this particular device was selected as the domain expert/knowledge engineer and was assigned the task of building the expert system knowledge base. He had the full time responsibility of writing rules into the expert system development shell. The suppliers of the shell were available to assist him in building the knowledge base and extending the shell's capabilities to fit this application. The standard iterative refinement development strategy was employed. The expert system was built around a small set of cases of malfunctions, and then was modified and extended by the domain expert as problems were discovered in handling additional malfunctions.

At the time of the study the expert system was in the late stages of its iterative refinement development cycle. It had already gone through several evaluation-modification cycles over a period of a year, including some field tests where technicians were observed trying to troubleshoot actual malfunctions using the expert system. The designers of the system and management had judged the system to have reached a sufficient level of maturity to be deployed in the field on a regular basis. The expert system was built on the assumptions that the technician has minimal knowledge about the device, minimal knowledge about how to find faults, and minimal information about the symptoms of the particular device being tested. The assumed scenario is that the technician arrives at the site with the malfunctioning device, receives a brief trouble report, and immediately accesses the machine expert, which directs all troubleshooting activities.

---

[4] There are two machine or systems in this application; to keep terminology clear, we will refer to the system that is to be repaired as the *device* and to the system that attempts to diagnose the device as the *machine* or *machine expert*.

The machine expert uses the conventional expert system question-and-answer style of interaction. The machine expert asks the technician to choose from a list of symptoms the one that best describes the trouble with the device (the machine uses the answer to this type of request to narrow its search to certain kinds of disturbances or to certain areas of the device). The machine also directs the user to make observations about the device's behavior and to take measurements from internal test points. When sufficient evidence accumulates, the expert system draws a conclusion about the possible malfunction and reports it together with a confidence value computer for that hypothesis. The technician has the option of either accepting the hypothesis as the solution and terminating the session, or rejecting the hypothesis (in which case the machine will continue its search for another candidate solution). This process continues until either the correct solution is reached or the machine's search paths are exhausted. In addition to the machine-paced interaction, a technician has available a few commands, for example, a help command to ask for clarification of a question (e.g., location of test points, meaning of terms, description of procedures for performing a test, or a postulated malfunction) and a backup command that allows the technician to step back question by question through the past dialogue to review and/or change past responses.

The machine expert in this application is a representative product of the machine-as-prosthesis paradigm. Locus of control resides with the machine; the design intent is for the human to act as the eyes and hands of the machine. The purpose of the investigation was to observe the interaction of domain practitioners with a machine expert in the course of solving actual domain problems, and not to evaluate the expert system's performance alone or to compare human performance to human plus expert system performance. Given this objective, protocols were collected and analyzed on how each technician interaction with the machine expert during each problem (a total of 18 cases of joint person–machine problem solving were analyzed, because some technicians were not tested on all problems).

A *canonical* solution path was defined by the domain expert for each problem. The canonical solution path specified the series of machine expert requests (e.g., choice, measurement, observation, hypothesis evaluation) and corresponding technician responses required to reach a correct solution. Human–machine interaction was analyzed by charting the flow of judgments (symptom classification, hypothesized malfunctions) and actions (e.g., observations, test measurements) that were made either by the machine or by the person and by comparing the actual path of each problem solving episode with the canonical path. For each problem the technician was given a brief description of problem symptoms (similar to the initial trouble report received in actual troubleshooting situations) and told to use the machine expert to help troubleshoot the device. The machine asked the technician to make choices from lists of symptoms, to observe the behavior of the device, or to take a measurement from test points inside the device. These categories are noted as *choice, observation*, and *measurement*, respectively, in the protocols. The machine also output hypotheses about the suspected fault (noted as *hypothesis* in the protocols), and, very rarely, it output a judgment about the nature of the problem or area of the device it was investigating (noted as *judgment* in the protocols). Machine output and machine-directed activities of the technician are indicated within the shaded regions of the interaction protocols. There were also activities/decisions that the technicians made outside the scope of the machine expert which are noted outside the shaded areas. These activities/decisions took the form of (a) observations and judgments about the device which the technician made before going to the machine expert; (b) observations, measurements, or judgments that were made independent of the machine during the session; (c) judgments about whether or not the machine expert was going down an appropriate diagnostic path, and (d) judgments about whether hypotheses offered by the machine were valid solutions to the device trouble.

The technicians were asked to read aloud the instructions provided by the expert system and to "think aloud" as they interacted with the machine and worked through the problem. The protocols were based on first-hand observation, audio/visual recordings of the sessions, and a computer record of expert system requests and technician responses. Figures 2–4 are examples of the coded protocols produced in the analysis.

The domain expert/knowledge engineer and at least one of the authors were present at all sessions. While the technicians were encouraged to work through the problems without assistance from the observers, they could ask for help in interpreting questions, locating test points, or using the machine expert (e.g., what commands were available and how to execute them). The knowledge engineer or the other observers could intervene if they judged that the technician otherwise could not make progress towards problem solution.[5]

Results on Joint System Performance

Troubleshooting rarely followed the canonical path described by the domain expert/knowledge engineer. Situations unanticipated by the domain expert were the norm rather than the exception. Overall, the problem-

[5] Interventions to bring the technicians back on track were made to ensure the opportunity to observe and evaluate expert system interaction throughout the canonical path (e.g., to identify ambiguous questions, to observe how the technician responded to expert system hypotheses).

Figure 2. Technician C, Problem 6. The technician goes directly from the trouble report and at a choice point selects the more general symptom description. This leads him down an unproductive path. He makes an error in observation which takes him further off track, and the machine expert eventually drops out. He restarts, and based on the knowledge gained, selects the more specific choice item on the second pass. The technician never deviates from the canonical solution path from this point (64 minutes).

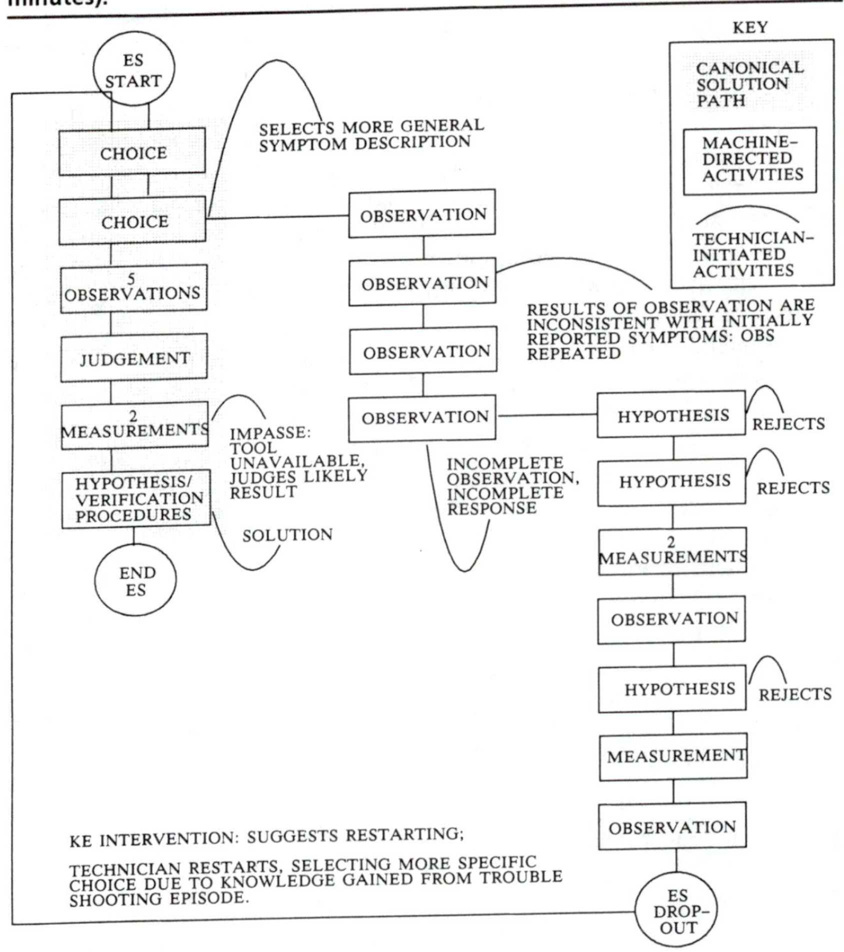

solving episodes diverged form the canonical path in 14 of the 18 cases. In four cases the correct problem solution was never reached. Substantial deviations from the canonical path arose even when the problem was solved (in 10 of the 14 successfully solved cases). We will focus here on aspects of the human–machine expert interaction related to how these deviations

Figure 3. Technician D, Problem 6. The technician performs preliminary observations that lead him to recognize that there are two valid answers at a choice point. He selects the more general symptom description, reasoning that that is likely to be what the machine expert expects. When he realizes it leads to an unproductive path, he aborts the system and restarts, selecting the more specific choice item. The technician never deviates from the canonical solution path from this point (18 minutes).

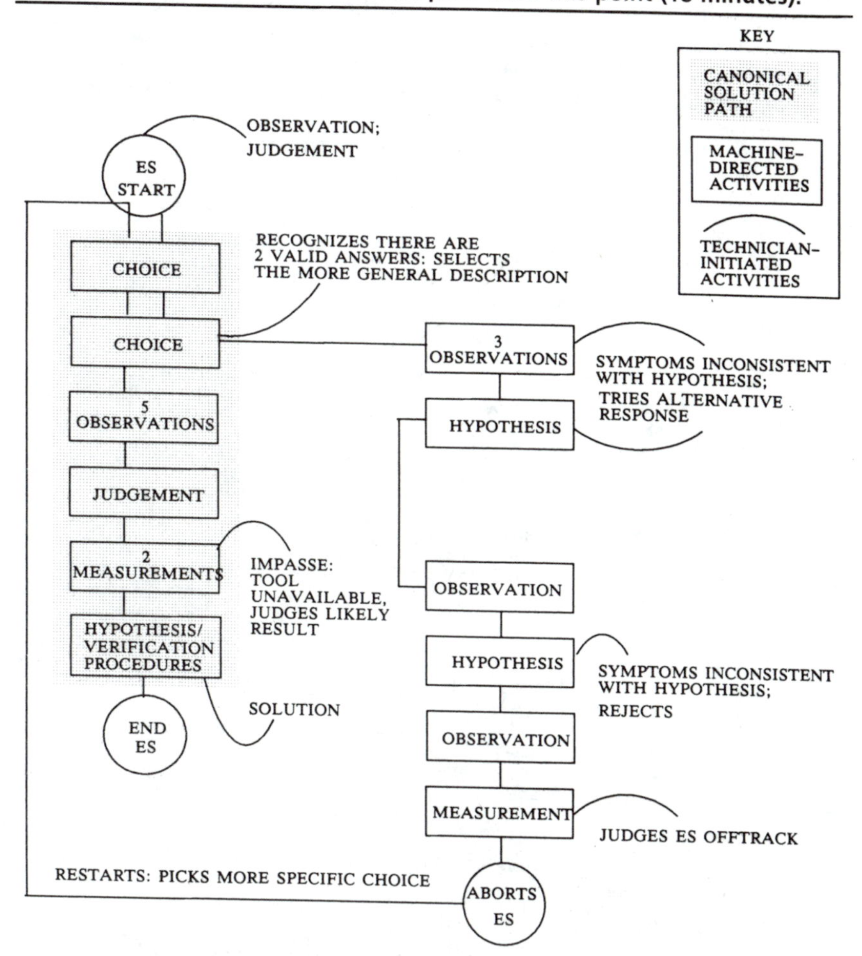

arose and what happened once a deviation occurred (cf. Roth, Bennett, & Woods, 1987, for a complete description of the results). Three protocols for one of the difficult problems are used to illustrate the most salient findings:

- Contrary to the assumption of the machine-as-prosthesis paradigm, the technicians did not and could not function solely as passive data gatherers.

Figure 4. Technician A, Problem 6. The technician was given the Problem 2 trouble report but performs preliminary observations and detects device symptoms associated with Problem 6, which he works on first. An impasse arises when the machine expert directs him to perform a test that is not possible to do given the current state of the device; the technician resolves the impasse by taking actions to change the device state. A second impasse arises when the machine expert directs him to perform a measurement that is not possible because the tool is not available. The technician resolves the impasse by providing a "best guess" response. The machine expert provided a partially incorrect hypothesis and the correct solution was identified by the technician during the verification process (33 minutes).

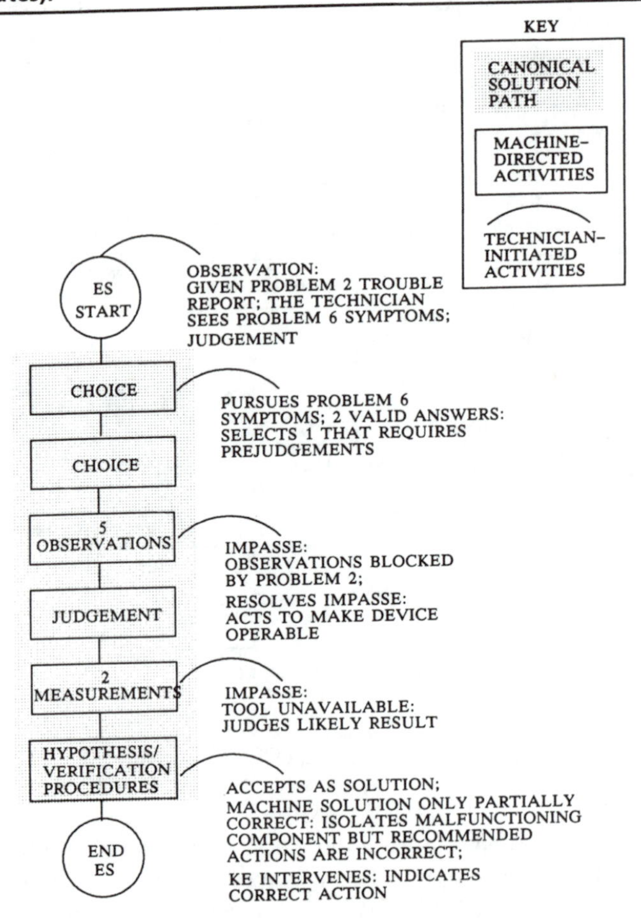

- Successful performance depended on the ability of the technician to apply knowledge of the structure and function of the device and sensible troubleshooting approaches. This turned out to be necessary to follow underspecified instructions, to infer machine intentions, to resolve impasses, and to recover from errors (person or machine) that led the machine expert off-track.
- Once the machine expert was off-track, it could not recover by itself (the machine remained "fixated"). The burden to detect and recover from deviations (direct the machine expert back toward a productive path) fell on the human. However, the machine-as-prosthesis design provided virtually no support for the person to carry out this role.

Deviations from the canonical path occurred during all phases of the joint problem-solving episode. They could occur at the start of the interaction, when a technician chose an initial symptom description that differed from the choice anticipated by the designer of the machine expert; during the course of the question–answer dialogue, and at the point where the machine expert offered hypothesized solutions. Deviations from the canonical path arose from a variety of sources. These included mismatches between the technician's state of knowledge about the device and the state of knowledge assumed by the machine expert; technician entry errors due to "slips," mismeasurements, misobservations, or misinterpretations; technician installation errors; technician errors due to inability to assess the intentions of the machine expert; inherent variability of actual devices; the existence of multiple simultaneous faults; unavailability of test equipment; and "bugs" in the machine expert's knowledge base. In the next section we chronicle the flow of the problem solving episodes to illustrate the kinds of deviations that arose at different points, the opportunities for recovery, and how differences in technician response affected solution outcome.

Sources of deviations. In designing the expert system, the knowledge engineer assumed that the technician would turn to the machine expert immediately upon receiving a trouble report. It was assumed that all troubleshooting activities would be performed under the direction of the machine expert. In actual practice, the technicians often performed preliminary observations and formed tentative judgments about the nature of the problem before ever going to the machine expert. During these observations the technician would begin to make judgments about the kinds of disturbances present and the general area of the device where the fault might be located. These initial observations and judgments (or their absence in some cases) could have a profound effect on the subsequent interaction with the machine.

For example, Figures 2 through 4 contain the protocols for three of the technicians for the same problem (Problem 6). Early in the interaction the machine asks the technician to choose from a list of symptom descriptions the one that most closely matches the observed problem. One of the choices is consistent with the symptoms mentioned in the trouble report; however, a second item on the list provides a more specific description that matches additional symptoms of the device that were not mentioned in the trouble report. The selection at that choice point therefore depends on the state of knowledge of the technician. Technician C, who went directly from the malfunction report to the machine expert, selected the more general symptom description. In contrast, Technician D made preliminary observations of the behavior of the device and judgments before turning to the machine expert allowing him to detect additional symptoms. Consequently, given his perspective and state of knowledge, there were two choices on the list that were consistent with the observed symptoms of the device. He commented on the fact that two choices were possible and decided to enter the item that assumed less knowledge (the more general category). He reasoned that the machine assumed minimal knowledge on the part of the troubleshooter, since the purpose of the machine was to reduce the experience and knowledge needed by technicians. As it happened in this particular problem, the path taken when the technician responds with minimal knowledge (which was the path intended by the knowledge engineer for this fault) led to a dead end for the machine problem solver (this occurred on the first pass in the case of both Technician C and D). The line of questioning and hypothesis generated by the machine expert gradually appeared less and less reasonable, given the technician's state of knowledge about the device. Technician D reacted by aborting the session. He restarted and this time selected the more specific choice, which led to the correct solution. Technician C, who was more passive, persisted down the dead-end path till the machine expert ran out of hypotheses and self-aborted. At this point, he restarted, and having gained more knowledge about the state of the device during the diagnostic process, he too selected the more specific choice item and solved the problem.

The protocol of Technician A provides an interesting twist on Problem 6. In his case, there were inadvertently two simultaneous malfunctions in the device. One malfunction (Problem 2) was deliberately introduced into the device. However, during the test and test recovery for the previous subject, a second malfunction (Problem 6) was inadvertently created. The technician was presented the trouble report for Problem 2. Rather than relying solely on the trouble report, the technician began by making detailed observations of device behavior before going to the expert system. In so doing, he uncovered symptoms that were not in the trouble report (those produced by the fault for Problem 6). He used these observations and judgments about the nature of the disturbance to answer questions posed by the machine which led the interaction down the Problem 6 solution path. Had he gone to the expert system with the original trouble report symptom, as expected by the expert system designers, the Problem 6 malfunction might not have been uncovered.[6] In the case described above, the technician's preliminary observations and judgments enhanced joint system performance. In fact the problem could not be solved without it. There were other cases where joint system performance degraded because the technician knew more about the locus of malfunction than the machine expert assumed he would know. For one class of trouble (Problem 2) the machine "expected" the technician to respond with one symptom description (the most general symptom—"device doesn't work"). However, one of the technicians, in the course of preliminary observations, noticed that the device did operate under certain conditions. As a result, he selected a different symptom description which led to a dead-end path.

Once an initial choice of symptom classification was made, the machine expert directed the technician to make a series of observations and measurements. There were a variety of opportunities for deviating from the canonical path in this part of the interaction that arose due to input errors (i.e., mismeasurements or misentries) and misinterpretations of machine instructions. Despite the fact that the machine expert was specifically targeted at technicians with minimal familiarity with the device, a significant amount of knowledge about the device was required to understand and follow the machine's instructions. This included simple knowledge such as the location of parts and test points or terminology associated with the new technology, and more complex knowledge such as the structure, function, and behavior of the device. The knowledge engineer was required to frequently intervene to clarify terminology, locate test points, and disambiguate expert system statements in order for problem solving sessions to advance.

The more interesting causes of deviation were misinterpretations where the difficulty was not merely a lack of familiarity with terminology or knowledge of test point locations which could be handled by more careful wording of machine statements or layout diagrams. Rather, the difficulty had to do with an inability to assess the intentions of the expert system. In these cases the technician needed to understand the intended purpose behind machine expert requests for observations or measurements. The more active technicians were able to fill in the missing knowledge, infer machine intentions, and avoid deviations more frequently. In contrast, the more passive human partners followed the machine's directives literally, which led to more deviations from the canonical path.

---

[6] No multiple-fault cases were deliberately included in the problem set defined by the domain expert, although this is an important source of information about the capabilities of the machine and of the human–machine ensemble.

For example, in Problem 6 one of the more passive technicians (Technician C) responded incorrectly to a question, which led the machine into nonproductive paths because of an incomplete observation (see Figure 2). The question posed to the technician was whether the symptoms are observed at only one setting of the device. The intention of the machine expert was to test a particular hypothesis—that there was a malfunction in the device that caused a symptom to appear at the two extreme settings of the device. However, the technician, not knowing the hypothesis being tested, failed to appreciate the special status of the two end-point settings. He tested the device at several settings, including the lowest, but failed to check the highest setting. He observed the problem at only the lowest setting, and consequently reported that it occurred only at one setting, when in fact it occurred at the highest setting as well.

A second example of a deviation resulting from a failure to appreciate the intention of the machine expert (or, from another vantage point, a failure of the machine to communicate its intention) arose when the machine expert wanted a measurement to be taken that required a part of the device to be disconnected (Problem 3, Technician C). The machine expert directed the technician to disconnect the part and then take a measurement. When the result was entered, it then requested a second measurement without ever indicating that the technician should first reconnect the part (which the technician failed to do). The machine assumed that the technician would restore the device to its original state before continuing the diagnosis. In the absence of specific instructions, the technician could know whether or not the part should be reconnected at this point only if he understood the role that the requested measurements played in the diagnostic process; for example, there could have been a series of measurements to be taken that were meaningful only if the part remained disconnected.

The machine expert worked by classifying data patterns into preenumerated categories (Clancey, 1985). A source of deviations for this type of machine problem solver is related to the degree of variability of instances within the categories that the machine expert is designed around. For example, in this case, departures from the canonical path occurred because the knowledge engineer did not always take the inherent variability of actual devices into account; e.g., "normal" behavior or measurement readings for one particular device can deviate a great deal from the ideal or prototypical signatures. An example arose in the protocol of Technician A for Problem 6 (Figure 4). The machine expert successfully identified the part of the device that was malfunctioning, but provided the wrong advice about how to adjust the component to correct the problem. The machine expert assumed a nominal range of measurement readings for that class of components, and the current reading was higher. Consequently, the machine expert responded that the component was out of tolerance high and directed the technician to adjust the component to reduce the reading. In fact, for this particular unit, the functional range was higher than the nominal range stored in the machine expert, so that the component was actually out of tolerance low and should have been adjusted to increase the reading. The technician accepted the advice and was about to act on it when the domain expert intervened to clarify the situation.

In some cases, unanticipated conditions arose which did not result in deviations, but rather blocked progress along the canonical path—what we will call an *impasse*—where the machine made requests that could not be fulfilled due to failed preconditions. When an impasse arose, the technician had to supply knowledge and act outside of the scope and direction of the machine expert. Two examples of impasses arose in Problem 6. In one case the machine expert asked for a measurement to be taken that required using a tool that was not available (Technicians A and D).[7] Both technicians resolved this impasse by entering a response based on *their* assessment of the nature of the problem and what the measurement would be expected to be under those conditions. The second example arose with Technician A on Problem 6. His protocol was unique in that, inadvertently, there were two simultaneous faults in the device. In following up one of the malfunctions, the machine expert requested that an observation be made that depended on the device operating, albeit not completely properly; however, the second malfunction made the device inoperable. Again, activity outside of the machine expert was required to resolve the impasse. In this case, the technician initiated a set of activities on his own that made the unit operable (without locating or eliminating the root cause that had made the device inoperable). Having fulfilled the precondition for the machine expert's request (i.e., an operable device), he reverted to working with the machine expert.

Technician responses to deviations. Due to a variety of factors the machine expert sometimes departed from the canonical problem solution path. We were thus able to examine what happens when the problem-solving episode goes beyond the capability of the machine expert. In general, the technicians' ability to detect that the machine's problem solving was off-track and to redirect machine problem solving into more fruitful channels depended on how knowledgeable each technician was and on how actively he chose to participate in the diagnosis.

When the machine offered an hypothesis, the technician had to judge whether the potential solution was in fact the source of the device trouble.

---

[7] This is very realistic since technicians can forget a measurement tool, or only realize what they should have brought once they understand the nature of the problem, or measurement tools may fail.

In general this presented various difficulties for the technicians. They could do this by implementing the appropriate repair and then checking to see if the trouble had disappeared. Instead, the technician frequently simply judged whether or not the proposed solution would resolve the trouble. If he decided it could not, one response was to continue the interaction with the machine expert to see what other solutions the machine might propose (e.g., subject C in problem 6).

The technician could decide that the machine's hypotheses (and behavior in general) were not simply incorrect but implausible given his perception of the situation. In some cases, the technician was able to detect that the machine was off-track only when there was extremely strong evidence—the machine continued along a nonproductive path, suggesting a series of hypothesized solutions that were judged to be less and less plausible and rejected as incorrect until it reached a dead end and terminated (see Figure 2). This machine behavior was spontaneously characterized by one subject who experienced it as "being taken down the garden path." Off-track machine problem solving was also signaled when it made an assertion or hypothesized solution that seemed blatantly inconsistant with the state of the device that the technician could immediately observe (e.g., because of bugs in the rule base). More active technicians used information about the particular symptoms, their judgments about the nature or location of disturbances, and their knowledge of the device to decide that the machine expert's behavior was off-track (Figure 3).

As evidence built that the machine was off-track, the technician had to decide what the source of the problem-solving difficulty was and what could be done to redirect the machine expert. First, an input error or misinterpretation could have occurred that led the machine expert astray. A common response was to back up and review inputs. Failing to detect an input problem, a second response was to try to redirect the machine. For example, in Figure 3, when the technician first suspected that the machine's hypotheses were implausible, he went to the immediately prior entry and reversed it to see if that would produce more plausible output. Another response was to restart the diagnostic process from the beginning. In the above example, when the machine's output continued to contradict the technician's knowledge/expectations, he restarted the entire problem solving process. When they suspected that the machine might be off-track, the technicians were forced to rely on their own assessments of the problem state to determine where the machine's difficulty or misconception arose. Thus, the limited access to the machine expert's view of the problem forced the technicians to perform their own diagnostic activities in parallel with the machine—a redundant rather than diverse joint cognitive system architecture.

The most experienced (and an active) technician (D) tended to decide that a path was unproductive and abort it more readily than the less experienced and more passive technicians (compare Figure 3 to Figure 2). Twice, one of the technicians with a passive style of interaction (Technician C in Problems 5 and 6) persisted in going down the path directed by the machine expert in the face of clear evidence that the machine was off-track. In contrast, in similar situations (Problems 5 and 6) Technician D aborted the diagnostic path soon after detecting an inconsistency. In Problem 6, he restarted joint problem solving from a different starting point; in the other case, he aborted the session and completed the troubleshooting on his own.

TOWARDS EFFECTIVE JOINT COGNITIVE SYSTEMS

The results of the investigation summarized above illustrate many of the issues that arise when machine power is deployed in actual work contexts. The capability to build more powerful machines does not in itself guarantee effective performance. The conditions under which the machine will be exercised and the human's role in problem solving affect the quality of performance. This means that factors related to tool usage can and should affect the very nature of the tools to be used. These results are not new. The breakdowns observed here in human interaction with an intelligent machine have been observed repeatedly with support systems, in a variety of media and technologies, built within the machine-as-prosthesis paradigm. In the following sections, we will use the results of this and other studies to mark the ubiquity of the deficiencies of the cognitive-tool-as-prosthesis approach to deploy machine power, and to identify some ways that exploit machine intelligence to provide cognitive instruments to be wielded by competent practitioners.

Following Instructions

From the instrumental perspective the tool user is in control; he or she wields the tool. The contrast between joint cognitive system performance with passive versus active technicians illustrates again the negative effects on human and total system performance that arise when control of the interaction resides with the machine (e.g., Smith, Cohen, Stammerjohn, & Happ, 1981; Hoogovens Report, 1976). These kinds of breakdowns are observed whenever the human is treated as a passive agent who will blindly follow directions. Suchman (this volume) observed similar cases of performance impasses with users attempting to follow dynamically generated instructions for using a copying machine. She documents cases of impasses that arose because of mismatches between the user's and the machine's assumptions about current world state and objectives.

Similar breakdowns in performance have been observed with other forms of human performance aids. For example, experience with Fully Proceduralized Job Performance Aids (FPJA) and Automatic Test Equipment (ATE) parallel the performance problems observed in the present study (Smillie & Clelland, 1984; Coppola, 1985; Chenzoff, Joyce, & Nauta, 1985) It has been found repeatedly that a machine locus of control for test sequencing prevents the systems from capitalizing on the insights of technicians in guiding troubleshooting. For example, cases have been known where technicians had to wait hours for ATE programs to reach the test for the part of the module that they suspected as the source of the problem (deKleer, 1985). As a result the systems can take long unproductive detours before turning to the correct solution path or can "fixate" in one path, never reaching the solution (Pople, 1985).

These issues are not unique to a particular delivery medium; they have been observed with paper-based instructions as well as with computer-based systems (machine experts such as the one investigated earlier are just computerized versions of paper procedures with a more elaborate control strategy). Instructions, however elaborate, are inherently "brittle" when followed rotely, in that it is difficult to build in mechanisms that cope with novel situations, adapt to special conditions or contexts, or recover from human errors in following the instructions or bugs in the instructions (e.g., Brown, Moran, & Williams, 1982; Woods et al., 1987; Herry, 1986). For example, Woods (1984) found that, when power plant operators responding to simulated accident conditions with paper-based instructions made an error, the errors fell into two categories: one where rote rule following persisted in the face of changing circumstances that demanded adaptation, and another where the human correctly recognized that responses needed to be adapted but failed to adapt correctly due to a lack of knowledge. Following instructions requires actively filling in gaps based on an understanding of the goals to be achieved and the structural and functional relationships of objects referred to in the instructions. For example, Smith and Goodman (1984) found that more errors arose in assembling an electrical circuit when the instructions consisted exclusively of a linear sequence of steps to be executed, than when explanatory material that related the instruction steps to the structure and function of the device was provided. Successful problem solving requires more than rote instruction following; it requires understanding how the various instructions work together to produce intended effects in the evolving problem-solving context (cf. Britton & Black, 1985).

While some of the problems that resulted from misunderstandings of the machine expert's intentions can be eliminated by more carefully worded, detailed, explicit descriptions of requests, this approach has limitations. Even if in principle it were possible to identify all sources of ambiguity and craft detailed wording to avoid them, in practice the resources required for such extensive fine tuning are rarely available. Furthermore, the kinds of literal elaborate statements that would need to be developed to deal with exceptional situations are likely to obstruct the comprehension and execution of instructions in the more typical and straightforward cases.

Fundamentally attempts to eliminate all sources of ambiguity are misguided. Examination of language use in human–human communication reveals that language is inherently underspecified and requires the listener (or reader) to fill in gaps based on world knowledge (Schank & Abelson, 1977) and to assess and act on the speaker's (writer's) intended goals rather than his or her literal requests (Gibbs & Mueller, this volume; Cohen, Perrault, & Allen, 1982; Searle, 1969). In most human–human cooperative activity, metastatements about the motivation for requests are not constantly needed, because the agents share common knowledge about the state of the world and what are meaningful activities in the current context (Clark & Marshall, 1981). Ironically, while there has been significant attention to the issue of how to get intelligent machines to assess the goals and intentions of humans without requiring explicit statements (e.g., Suchman, this volume; Quinn & Russell, 1986; Allen & Perrault, 1980), less attention has been given to how to signal a machine's perceptions, goals and intentions to people (Woods & Roth, 1988).

Human Control of Machine Resources

In order to wield a tool, the user must know its boundaries and instrumental characteristics (effectivities)—what it can do and what are its limits, side effects, preconditions, and postconditions. The opacity of the machine expert hindered the human from effectively utilizing its capabilities to meet domain goals. The machine expert provided few cues about its intentions in pursuing a line of diagnosis, and little possibilities for redirecting its resources. This inhibited the human problem solver in his role as a manager of a semiautonomous resource—to monitor the performance of the subordinate agent in order to detect when it is off-track and to redirect it (again, the same result has occurred with the introduction of other forms of automation when the human's supervisory role has been ignored, e.g., the Hoogovens Report, 1976).

When the technician reached a point where the line of questioning seemed to be focused in an unproductive direction or the system generated a hypothesis that was suspect, the technician had to decide whether the machine expert was still on track or not. The technician did not have access to the machine's perceptions of current world state nor to its troubleshooting strategy. Consequently, he had no way of knowing whether the system was systematically working through hypotheses and would eventually reach the correct one; whether the machine expert misperceived the state of the device, possibly because of an input error on the technician's part; or whether the system was beyond its boundary of competence and responding erratically. Failing to find an input error, the only basis for evaluating whether the system was on track was to judge the plausibility of the line of reasoning based on his own perceptions of device state and his assessment of plausible lines for diagnosis and tenable hypotheses under those conditions. Thus, he was forced to work through the diagnosis process independently and in parallel rather than build on top of the information processing work carried out by the machine.

As Finegold (1984, p. 115) has noted of machine experts as prostheses in the area of integrated circuit design:

> one of the big problems is the tendency for the machine to dominate the human . . . consequently an experienced integrated circuit designer is forced to make an unfortunate choice: let the machine do all the work or do all the work himself. If he lets the machine do it, the machine will tell him to keep out of things, that it is doing the whole job. But when the machine ends up with five wires undone, the engineer is supposed to fix it. He does not know why the program placed what it did or why the remainder could not be handled. He must rethink the entire problem from the beginning.

Furthermore, the question-and-answer dialogue interface is an extremely poor representation of the state of the device and the problem-solving process for the human attempting to solve the problem in parallel with the machine. The burden to remember the state of the device (e.g., a part is disconnected), the history of observations and measurements, and the history of hypotheses tested is placed on the person without any support whatsoever (cf., Woods & Roth, 1988, for an alternative approval).

If the human troubleshooter decided the machine was off-track, there were limited mechanisms available to redirect it. Because of the system design philosophy, the machine expert is in complete control; virtually no features were provided for redirecting the line of diagnosis. There was a backup feature for changing input errors, which was included because the possibility of human misentries had been recognized, but there were no explicit features provided for the user to redirect the system in cases where the source of error lay in misperceptions or wrong assumptions made by the machine. Other than changing their inputs, the only option available to users for redirecting the diagnosis was to use one available command as an abort command and start problem solving over again. Even though this was a crude and time-wasteful mechanism for control and this command had not been designed for that purpose, it did provide a mechanism for manipulating the machine expert as an instrument, and technicians spontaneously and readily adapted it to this purpose.

There are several simple steps that can be taken to convert the power of the machine expert into a more instrumental form. One is to make the machine's knowledge about the state of the device, viable hypotheses, and diagnostic directions available to the human (Roth, Butterworth, & Loftus, 1985; Fitter & Sime, 1980). This means that these aspects of machine knowledge must be represented explicitly (e.g., the shift from mycin to neomycin, cf. Clancey, 1983; cf. also Gruber & Cohen, 1987). Another step is to provide more capabilities for human control of the machine's reasoning. This includes mechanisms for the human problem solver to add to or change the information or knowledge that the machine is using about the state of the device and regions where the fault may lie (Roth, Elias, Mauldin, & Ramage, 1985). More ambitious use of machine power can provide the human with facilities to explicitly manipulate the attention of the machine expert. For example, Pople (1985) is currently developing mechanisms to allow users to direct the diagnostic path pursued by the Caduceus expert system for internal medicine.

Extramachine Knowledge and Domain Variability

One salient result of our investigation of human–intelligent machine interaction was the amount of extramachine knowledge and activity that occurred. The prosthesis approach to IDS attempts to design the human's role as data gatherer for the machine, yet the amount of extramachine human contributions indicates that the human's actual role is to amplify the machine's ability to cope with the unanticipated variety in the world or in the problem-solving process.

Psychologists are fond of discovering biases in human decision making. One judgmental bias is the overconfidence bias where people at all levels of expertise overestimate how much they know (e.g., Wagenaar, 1986). However, we sometimes forget that these biases can apply to the designers of machines as well as to the users of machines. This means that the designer of an IDS system is likely to overestimate his or her ability to capture all relevant aspects of the actual problem-solving situation in the behavior of the machine expert. This result has often happened with other forms of automation and support systems—the designer of the system fails to appreciate all of the variability and complexities of the operational setting, so that the system fails to support or even hinders the operator in achieving his or her goals. For example, operators quickly learn the signatures of automatic controllers—when they are so unreliable that the task must be performed manually, when they are highly sensitive to disruptions and must be very closely monitored, or when they are adequate to the task and require little supervision (e.g., Roth, Woods, & Gallagher, 1986). The result, in accordance with Ashby's law of requisite variety, is that, as automation increases, the human's role is increasingly to amplify the machine's adaptability by dealing with the unexpected conditions,[8] especially failures that arise because of interactions across normal system boundaries (cf., e.g., NUREG-1154 for one real-world example from nuclear power plant emergencies; the Hoogovens Report, for a case in the automatization of a rolling mill; Hirschhorn, 1984, for examples from manufacturing technologies; and Noble, 1984, for examples from numerical control).

When the goal is to completely remove the human from the problem-solving loop via automation, the problem of unanticipated variability arises. Automated systems inevitably solve the simple cases and straightforward portions of the problem, but fall down in the more complex cases (novel events, multiple faults). For example, one problem that has been experienced with ATE is that automated fault isolation often only produces a long list of suspect parts (Coppola, 1985). For the more difficult discriminations the human is left to his or her own resources without any tools but those that he or she can craft personally. Thus, attempts to take the user out of the problem-solving loop can inadvertently increase the user's burden by asking him or her to handle the difficult cases without the benefit of experience and the practice of solving the simpler cases (Woods, 1986; Finegold, 1984). This result may be acceptable in relatively risk free applications, but the costs can be high in risky domains where performance on the rare but catastrophic incident is important (Wiener, 1985; Norman, 1986). These lessons from past increases in automation emphasize the need to conceive of IDS as a tool in the hands of the human problem solver.

What is Good Advice?

Development of effective decision support in an application is essentially an exercise in deciding what good advice is for the domain in question. The passive human and machine-as-prosthesis interaction that we observed parallel descriptions of unsatisfactory advisory encounters found in empirical studies of human to human interactions. Alty and Coombs (1980) and Coombs and Alty (1980) found that unsatisfactory human–human advisory encounters (in a center for advice on using the local computer system) were strongly controlled by the advisor. The advisor asked the

---

[8] Note the irony that increased automation is often justified on grounds of human incompetence, yet in practice it is often that same person who must now help the machine to cope with disturbances beyond its design range.

user to supply some specific information, mulled over the situation, and offered a solution with little feedback about how the problem was solved. While a problem was usually solved, it was often some proximal form of the user's real problem (i.e., the advisor was guilty of a form of solving the wrong problem: solving a proximal case of the user's fundamental or distal problem). The advisor provided little help in problem definition.

On the other hand, Coombs and Alty found that, in more successful advisory interactions, two partial experts (experienced computer user with a domain task to be accomplished, and a specialist in the local computer system) cooperated in the problem-solving process. Control of the interaction was shared in the process of identifying the important facts and using them to better define the problem. In this process each participant stepped outside of his or her own domain to help develop a better understanding of the problem and, as a consequence, appropriate solution methods.

The study of human–intelligent machine interaction summarized above, and these studies of human–human advisory encounters (cf. also Pollack, Hirschberg, & Webber, 1982; McKendree & Carroll, 1986), reveal that good advice is more than recommending a solution; it helps the user develop or debug a plan of action to achieve his or her goals (Jackson & Lefrere, 1984, p. 63). Good advisory interactions aid problem formulation, plan generation (especially with regard to obstacles, side effects, interactions and tradeoffs), and help determine the right questions to ask, and how to look for or evaluate possible answers. This means good advice must be more than a solution plus justification; it must be structured around the problem solving process to help the user answer questions like (e.g., Woods & Roth, 1986; Woods & Hollnagel, 1987): what would happen if *x*, are there side effects to *x*, how do *x* and *y* interact, what produces *x*, how to prevent *x*, what are the preconditions (requirements) and postconditions for *x* (given *x*, what consequences must be handled).

Studies of advisory interactions reveal another important characteristic of joint cognitive systems: the relationship between the *kinds* of skills and knowledge represented in the human and in the machine (Hawkins, 1983; Coombs, 1986). The human user of decision tools is rarely incompetent in the problem domain. As this study showed, even when the design intent is to reduce human skill and knowledge requirements, significant amounts of domain knowledge are required for minimal capabilities to understand and execute machine instructions. The relation of human to machine is not one of novice to expert; instead, the human and machine elements contain partial and overlapping expertise that, if integrated, can result in better joint-system performance than is possible by either element alone. For example, Pople (1985; Gadd & Pople, 1987) models an IDS system on the interactions between multiple medical specialists that occur, often informally, when one of them encounters a difficult problem.

Towards Cognitive Instruments

Fundamentally, the difference between the cognitive-tools-as-prosthesis approach and the cognitive-tools-as-instrument approach to joint cognitive system design is a difference in the answer to the question "what is a consultant." One definition of a consultant is someone (thing) called in to solve a problem for another, on the assumption that the problem was beyond the skill of the original agent. Given this definition, the important issues for building decision aids is to build better automated problem solvers and to get people to call on these automated problem solvers (the acceptance problem). The issues explored in this paper show that questions of tool use cannot be treated as a secondary design problem, to be handled as only an interface issue that is relevant late in the development of an automated problem solver. Rather, the characteristics of the joint person–machine cognitive system (implicit or explicit in the design) have a fundamental impact on the ultimate effectiveness of the new system in the actual work context, and on the definition and architecture of the tools themselves (Coombs, 1986).

The instrumental perspective, on the other hand, defines a consultant as a reference or source of information for the problem solver. The problem solver is in charge; the consultant functions more as a staff member. As a result, this view of joint cognitive systems stresses the need to use computational technology to aid the user in the process of solving his or her problem. The human's role is primarily to achieve total system performance objectives as a manager of knowledge resources that can vary in kind and amount of intelligence or power. In order to put more knowledge in the hands of the human problem solver at the scene, first, raw domain data can be massaged into forms that more directly answer the questions the human problem solver must face as problems unfold. Machine resources can compute such information as when a domain goal is violated, when one alternative is preferred over others, whether there are alternative response strategies, what data are potentially relevant, what the post-conditions are that follow from a strategy, what the preconditions are that must be met before a strategy can be implemented (e.g., Woods & Hollnagel, 1987). This means that knowledge about what the computations carried out by the machine signify about the domain (i.e., the domain's semantic structure) must be used explicitly or implicitly to structure machine power for human utilization. Second, the human problem solver must have capabilities to utilize and direct these computational resources. To build human–machine cognitive systems of this type, there is a need for techniques and concepts to determine what knowledge resources are needed by a domain problem solver, and how to integrate and communicate these results as appropriate during the problem-solving encounter.

How can we avoid the deficiencies of the prosthesis approach? Consider what kinds of power cognitive tools could provide. Physical tools magnify the user's capacity for physical work; perceptual tools extend the user's perceptual range. For knowledge environments, tools can provide calculation power (the original use of computer technology), search and deductive inference power (support for more complete utilization of available evidence)[9] through symbolic processing technology, economy of representation through frame-based representation technology (e.g., Walker & Stevens, 1986), and an extended ability to conceive of both the problem to be solved and possible paths to solutions, i.e., *conceptualization power*.

In a sense both the designer of IDS systems (really, any machine) and the target users of IDS systems face a complex, ill-defined problem-solving task: how to cope with unanticipated variability and avoid brittleness? The best help we can provide to people engaged in these tasks may be conceptualization power through (a) enhancing their ability to experiment with possible worlds or possible strategies; (b) enhancing their ability to visualize, or make concrete, the abstract or uninspectable, or to see the implications of a concept or change (analogous to perceptual tools); (c) to enhance error tolerance by providing better feedback about the effects/results of actions (not that errors are not made, but that error detection and correction is enhanced so that errors do not propagate, e.g., Rizzo, Bagnar, & Visciola, 1987). The most important contribution of AI technology to decision support in the long run may well be in cognitive tools that amplify human powers of conceptualization.

The critical importance of conceptualization power (which could be amplified by appropriate tools) in effective problem-solving performance is often overlooked, because the part of the problem-solving process that it most crucially effects, problem formulation, is often left out of studies of problem solving. We study diagnosis independent of evidence gathering (e.g., Patel & Groen, 1986), or plan generation independent of plan monitoring and adaptation, or diagnosis only when possible categories have been completely enumerated (converting the problem to one of classification; e.g., Clancey, 1985; Nickles, 1980). The problem-formulation aspect can be neatly finessed in the design world by establishing bounding constraints on the problem that reduce it to manageable dimensions. However, this strategy merely displaces the complexity to the person in the operational world, rather than providing support to cope with the true complexity of the actual problem-solving context. For example, the de-

[9]But there are other forms of reasoning that people may and should be using, e.g., Cheng, Holyoak, Nisbett, & Oliver, 1986; as Mill commented, "Logic neither observes, nor invents, nor discovers."

signer of a machine problem solver may assume that only one failure is possible in order to completely enumerate possible solutions and, therefore, use classification problem-solving techniques. However, the actual problem solver must cope with possibility of multiple failures, misleading signals, interacting disturbances (e.g., Pople, 1985; Woods & Roth, 1986). In this example, note that the difficulty arises, in part, because the designer defines the design problem to be solved in terms of the tools available to him or her. The cost of underemphasizing the role of problem formulation and reformulation is particularly high if more adaptive behavior is required to improve performance, for example how to adapt a plan given nonsuccess on an element of the original plan (Alterman, 1986) or how to revise one's situation assessment in a dynamic, event-driven world (Woods & Roth, 1986).

Conceptualization power as an aid to problem formulation and as a means to cope with unanticipated variability suggests several kinds of IDS systems. One approach is representation aids that use direct manipulation and graphic techniques to help the human problem solver find the relevant data in a dynamic environment, visualize the semantics of the domain (that is, make concrete the abstract), and restructure his or her view of the problem (e.g., Hutchins, Hollan, & Norman, 1985; Zachary, 1986). Another path is to deploy machine power in the form of an agent who comments on or critiques human problem solving (e.g., Coombs & Alty, 1984; Langlotz & Shortliffe, 1983; Miller, 1983). IDS systems built in this mold consist of a core automated problem solver plus other modules that use knowledge about the automated problem solver's solution and solution path, knowledge of the state of the user–computer interaction, and knowledge of the user's plans/goals in order to warn the user, to remind him or her of potentially relevant data, and to suggest alternatives. A similar approach is to help the human problem solver construct and evaluate alternative world views that could explain the available data (Pople, 1985). The problem becomes better understood and a solution emerges as the user directs the machine's attention to different subsets of the data, different domain issues, and different hypothesis, much as one person with a difficult problem will bounce data, interpretations, and issues off colleagues to help refine his or her understanding of the problem. Coombs (1986) and Coombs and Hartley (1987) are working to develop model generative reasoning techniques for machine problem solving where the machine itself builds alternative models based on available evidence and then manipulates and evaluates the models to generate possible solutions. In all of these cases, which tools to build are constrained by characteristics of the situations in which the tools are to be used, namely, how to avoid brittleness in the face of unanticipated variability.

Reprise: Tool Building Versus Tool Use

Failures to consider the joint cognitive system in the deployment of new tools for cognitive task performance have contributed to the history of so-called aids that turned out under actual working conditions to fail to support or even hinder people from carrying out their job (Duncan, 1986). However, it is clear that support systems will be built, whatever our understanding of what is effective decision support. The challenge or problem for the person in some work environment will continue to be how to make use of a new "support" device—depend on it as a crutch, work around the hindrances it creates, or use it as an instrument to enhance his or her capabilities in the work context, or a mixture of these.

The problem of providing effective decision support will hinge on how the designer decides what will be useful in a particular application. Can researchers provide designers with concepts and techniques to determine what will be useful, or are we condemned to simply build what can practically be built and wait for the judgment of experience? Is principle-driven design of IDSs possible?

There is always tension between top-down design of what kind of tool is needed and the bottom-up constraints imposed by what can be built practically. It is rarely clear which element drives the development of new support systems—a new tool is created and people look for and adapt it to various uses (e.g., the latest technology always becomes the model for how the mind works), or there is a new need or pressure from problems to be solved that leads to new tools (e.g., war). In the long run, one benefit of AI technologies may be that cognitive tools created via this technology are more likely to contain the characteristics that a top-down analysis would show to be important for effective decision support in that application. Thus, even if the designer fails to understand (or misunderstands) what would be effective support, he or she will come close to that target because of the constraints imposed by the tool-building technology; e.g., using an object-oriented programming environment may automatically lead the designer away from the deficiencies of certain types of problem representations for human problem solvers.

While our ability to build more powerful machine cognitive systems has grown and proliferated rapidly, our ability to understand how to use these capabilities has not kept pace. With increased power we can amplify the impact of both successful and unsuccessful support systems. Today we can describe cognitive tools in terms of the tool-building technology. If we are to achieve principle-driven design of support systems, the key is to balance this technological description with a *cognitive* description of the interaction of domain problem-solving demands, problem-solver characteristics, and characteristics of the available tools. The joint cognitive system perspective (Hollnagel & Woods, 1983; Woods, 1986) is one of many efforts that have begun to establish that cognitive language (e.g., Coombs & Hartley, 1987). Until we are capable of this cognitive description, independent of the language of technologies, effective decision support will be a hit or miss affair, and the checkered past of attempts at decision support will be a statement of the future.

REFERENCES

Ackoff, R. L. (1981). The art and science of mess management. *Interfaces, 11*, 20–26.

Allen, J., & Perrault, C. (1980). Analyzing intention in utterances. *Artificial Intelligence, 15*, 143–178.

Alterman, R. (1986). An adaptive planner. In *Proceedings of AAAI*. Philadelphia, PA: AAAI.

Alty, J. L., & Coombs, M. J. (1980). Face-to-face guidance of university computer users—I: A study of advisory services. *International Journal of Man-Machine Studies, 12*, 390–406.

Britton, B. K., & Black, J. B. (1985). *Understanding expository text*. Hillsdale, NJ: Erlbaum.

Brown, J. S., & Newman, S. E. (1985). Issues in cognitive and social ergonomics: From our house to bauhaus. *Human–Computer Interaction, 1*, 359–391.

Brown, J. S., Moran, T. P., & Williams, M. D. (1982). *The semantics of procedures* (Tech. Rep.). Palo Alto, CA: Xerox Palo Alto Research Center.

Cheng, P. W., Holyoak, K., Nisbett, R., & Oliver, L. (1986). Pragmatic versus syntactic approaches to training deductive reasoning. *Cognitive psychology, 18*, 293–328.

Chenzoff, A. P., Joyce, R. P., & Nauta, F. (1985, July). *Maintenance job aids in the U.S. Navy: Present and future directions* (Tech. Rep. MDA903-81-C-0166-2). Naval Training Equipment Center.

Clancey, W. J. (1983). The epistemology of a rule-based expert system—A framework for explanation. *Artificial Intelligence, 20*, 215–251.

Clancey, W. J. (1985). Heuristic classification. *Artificial Intelligence, 27*(3), 289–350.

Clark, H., & Marshall, C. (1981). Definite reference and mutual knowledge. In A. Joshi, I. Sag, & B. Webber (Eds.), *Elements of discourse understanding*. Cambridge, England: Cambridge University Press.

Cohen, P., Perrault, C., & Allen, J. (1982). Beyond question answering. In W. Lehnert & M. Ringle (Eds.), *Strategies for natural language processing*. Hillsdale, NJ: Erlbaum.

Coombs, M. J. (1986). Artificial intelligence and cognitive technology: Foundations and perspectives. In E. Hollnagel, G. Mancini, & D. D. Woods (Eds.), *Intelligent decision support in process environments*. New York: Springer-Verlag.

Coombs, M. J., & Alty, J. L. (1980). Face-to-face guidance of university computer users—II: Characterising advisory interactions. *International Journal of Man-Machine Studies, 12*, 407–429.

Coombs, M. J., & Alty, J. L. (1984). Expert systems: An alternative paradigm. *International Journal of Man-Machine Studies, 20*, 21–43.

Coombs, M. J., & Hartley, R. T. (1987). The MGR algorithm and its application to the generation of explanations for novel events. *International Journal of Man-Machine Studies, 27*, 679–708. (Also in G. Mancini, D. Woods, & E. Hollnagel (Eds.), *Cognitive engineering in dynamic worlds*. London: Academic Press.)

Coppola, A. (1985). Artificial intelligence applications to maintenance. In J. Richardson (Ed.), *Artificial intelligence in maintenance*. Park Ridge, NJ: Noyes Publications.

De Keyser, V. (1986). Technical assistance to the operator in accidents. In E. Hollnagel, G. Mancini, & D. D. Woods (Eds.), *Intelligent decision support*. New York: Springer-Verlag.

deKleer, J. (1985). AI Approaches to troubleshooting. In J. Richardson (Ed.), *Artificial intelligence in maintenance*. Park Ridge, NJ: Noyes Publications.

Duncan, K. (1986). Panel discussion on cognitive engineering. In E. Hollnagel, G. Mancini, & D. D. Woods (Eds.), *Intelligent decision support in process environments*. New York: Springer-Verlag.

Finegold, A. (1984). The engineer's apprentice. In P. H. Winston & K. A. Prendergast (Eds.), *The AI business: The commercial uses of artificial intelligence*. Cambridge, MA: MIT Press.

Fitter, M. J., & Sime, M. E. (1980). Responsibility and shared decision making. In H. T. Smith & T. R. G. Green (Eds.), *Human interaction with computers*. London: Academic Press.

Gadd, C. S., & Pople, H. E. (1987). An interpretation synthesis model of medical teaching rounds discourse: Implications for expert system interaction. *International Journal of Educational Research, 1*.

Gruber, T., & Cohen, P. (1987). Design for acquisition: Principles of knowledge system design to facilitate knowledge acquisition. *International Journal of Man-Machine Studies, 26*, 143–159. (Special Issue on Knowledge Acquisition for Knowledge Based Systems).

Hawkins, D. (1983). An analysis of expert thinking. *International Journal of Man-Machine Studies, 18*, 1–47.

Hayes-Roth, F., Waterman, D. A., & Lenat, D. B. (1983). *Teknowledge series in knowledge engineering Vol. 1: Building expert systems*. Reading, MA: Addison-Wesley.

Herry, N. (1987). Errors in the execution of prescribed instructions. In J. Rasmussen, K. Duncan, & J. Leplat (Eds.), *New technology and human error*. Chichester: Wiley.

Hirschhorn, L. (1984). *Beyond mechanization: Work and technology in a postindustrial age*. Cambridge, MA: MIT Press.

Hollnagel, E. (1986). Cognitive system performance analysis. In E. Hollnagel, G. Mancini, & D. D. Woods (Eds.), *Intelligent decision support in process environments*. New York: Springer-Verlag.

Hollnagel, E., & Woods, D. D. (1983). Cognitive systems engineering: New wine in new bottles. *International Journal of Man-Machine Studies, 18*, 583–600.

Hoogovens Report. (1976). *Human factors evaluation: Hoogovens No. 2 hot strip mill* (Tech. Rep. FR251). London: British Steel Corporation/Hoogovens.

Hutchins, E., Hollan, J., & Norman, D. A. (1985). Direct manipulation interfaces. *Human-Computer Interaction, 1*, 311–338.

Jackson, P., & Lefrere, P. (1984). On the application of rule- based techniques to the design of advice giving systems. *International Journal of Man-Machine Studies, 20*, 63–86.

Langlotz, C. P., & Shortliffe, E. H. (1983). Adapting a consultation system to critique user plans. *International Journal of Man-Machine Studies, 19*, 479–496.

McKendree, J., & Carroll, J. M. ()1986). Advising roles of a computer consultant. In M. Mantei & P. Orbeton (Eds.), *Human factors in computing systems: CHI'86 conference proceedings* (pp. 35–40). Boston, MA: ACM/SIGCHI.

Miller, P. L. (1983). ATTENDING: Critiquing a physician's management plan. *IEEE Transactions on Pattern Analysis and Machine Intelligence, PAMI-5*, 449–461.

Mitchell, C., & Saisi, D. (1987). Use of model-based qualitative icons and adaptive windows in workstations for supervisory control systems. *IEEE Transactions on Systems, Man, and Cybernetics, SMC-17*, 573–593.

Mulsant, B., & Servan-Schreiber, D. (1983). *Knowledge engineering: A daily activity on a hospital ward* (Tech. Rep. STAN-CS-82-998). Palo Alto, CA: Stanford University Press.

Nickles, T. (1980). Scientific discovery and the future of philosophy of science. In T. Nickles (Ed.), *Scientific discovery, logic, and rationality*. Dordrecht, Holland: D. Reidel.

Noble, D. F. (1984). *Forces of production: A social history of industrial automation*. New York: Alfred A. Knopf.

Norman, D. A. (1986). New views of human information processing: Implications for intelligent decision support. In E. Hollnagel, G. Mancini, & D. D. Woods (Eds.), *Intelligent decision support in process environments*. New York: Springer-Verlag, 1986.

Patel, V. L., & Groen, G. J. (1986). Knowledge based solution strategies in medical reasoning. *Cognitive Science, 10*, 91–116.

Pollack, M. E., Hirschberg, J., & Webber, B. (1982). User participation in the reasoning processes of expert systems. In *Proceedings of the National Conference on Artificial Intelligence*. The American Association for Artificial Intelligence, Pittsburgh, PA.

Pople, H., Jr. (1985). Evolution of an Expert System: from Internist to Caduceus. In I. De Lotto & M. Stefanelli (Ed.), *Artificial intelligence in medicine*. New York: Elsevier Science Publishers B. V. (North-Holland).

Quinn, L., & Russell, D. M. (1986). Intelligent interfaces: user models and planners. In M. Mantei & P. Oberton (Eds.), *Human factors in computing systems: Chi'86 conference proceedings* (pp. 314–320). Boston, MA: ACM/SIGCHI.

Rasmussen, J. (1986). A framework for cognitive task analysis. In E. Hollnagel,

G. Mancini, & D. D. Woods (Eds.), *Intelligent decision support in process environments*. New York: Springer-Verlag.

Rizzo, A., Bagnara, S., & Visciola, M. (1987). Human error detection processes. *International Journal of Man-Machine Studies, 27*, 555–570. (Also in G. Mancini, D. Woods, & E. Hollnagel (Eds.), *Cognitive engineering in dynamic worlds*. London: Academic Press.)

Roth, E. M., Butterworth, G., III, & Loftus, M. J. (1985). The problem of explanation: placing computer generated answers in context. In *Proceedings of the Human Factors Society* (Vol II, pp. 861–865). Baltimore, MD: Human Factors Society.

Roth, E. M., Elias, G. S., Mauldin, M. L., & Ramage, W. W. (1985). Toward joint person-machine cognitive systems: A prototype expert system for electronics troubleshooting. In *Proceedings of the Human Factors Society* (Vol. I, pp. 358–361). Baltimore, MD: Human Factors Society.

Roth, E., Woods, D., & Gallagher, J. (1986). Analysis of expertise in a dynamic control task. In *Proceedings of the Human Factors Society* (pp. 179–181). Dayton, OH: The Human Factors Society.

Roth, E., Bennett, K., & Woods, D. D. (1987). Human interaction with an "intelligent" machine. *International Journal of Man-Machine Studies, 27*, 479–525. (Also in G. Mancini, D. Woods, & E. Hollnagel (Eds.), *Cognitive engineering in dynamic worlds*. London: Academic Press.)

Schank, R. C., & Abelson, R. P. (1977). *Scripts, plans, goals, and understanding*. Hillsdale, NJ: Erlbaum.

Schum, D. A. (1980). Current developments in research on cascaded inference. In T. S. Wallstein (Ed.), *Cognitive processes in decision and choice behavior*. Hillsdale, NJ: Erlbaum.

Searle, J. R. (1969). *Speech acts*. London: Cambridge University Press.

Shortliffe, E. H. (1982). The computer and medical decision making: Good advice is not enough. *IEEE Engineering in Medicine and Biology Magazine, 1*, 16–18.

Smillie, R. J., & Clelland, I. (1984). Acceptance and use of job performance aids within an integrated personnel systems approach. In *Proceedings of the Human Factors Society* (pp. 220–224). San Antonio, TX: Human Factors Society.

Smith, E. E., & Goodman, L. (1984). Understanding written instructions: The role of an explanatory schema. *Cognition and Instruction, 1*(4), 359–396.

Smith, M. J., Cohen, G. B., Stammerjohn, L. W., & Happ, A. (1981). An investigation of health complaints and job stress in video display operations. *Human Factors, 23*, 387–400.

Sorkin, R. D., & Woods, D. D. (1985). Systems with human monitors: A signal detection analysis. *Human–Computer Interaction, 1*, 49–75.

Toulmin, S. (1972). *Human understanding*. Princeton, NJ: Princeton University Press.

U.S. Nuclear Regulatory Commission. (1985). *Loss of main and auxillary feedwater at the Davis-Besse plant on June 9, 1985*. Springfield, VA: National Technical Information Service. (NUREG-1154).

Wagenaar, W. A. (1986). Does the expert know? The reliability of predictions and confidence ratings of experts. In E. Hollnagel, G. Mancini, & D. D. Woods (Eds.), *Intelligent decision support in process environments*. New York: Springer-Verlag.

Walker, E., & Stevens, A. (1986). Human and machine knowledge in intelligent systems. In E. Hollnagel, G. Mancini, & D. D. Woods (Eds.), *Intelligent decision support in process environments*. New York: Springer-Verlag.

Wiener, E. (1985). Beyond the sterile cockpit. *Human Factors, 27*, 75–90.

Woods, D. D. (1984). Some results on operator performance in emergency events. In D. Whitfield (Ed.), *Ergonomic problems in process operations*. Institute of Chemical Engineering Symposium Services 90.

Woods, D. D. (1986). Cognitive technologies: The design of joint human–machine cognitive systems. *AI Magazine, 6*, 86–92.

Woods, D. D. & Hollnagel, E. (1987). Mapping cognitive demands in complex problem solving worlds. *International Journal of Man-Machine Studies, 26*, 257–275. (Special issue on knowledge acquisition for knowledge-based systems.)

Woods, D. D., & Roth, E. (1986). *Models of cognitive behavior in nuclear power plant personnel*. Washington, DC: U.S. Nuclear Regulatory Commission (NUREG-CR-4532).

Woods, D. D., O'Brien, J., & Hanes, L. F. (1987). Human factors challenges in process control: The case of nuclear power plants. In G. Salvendy (Ed.), *Handbook of human factors/ergonomics*. New York: Wiley.

Woods, D. D., & Roth, E. (1988). Aiding human performance: II. From cognitive analysis to support systems. *Le Travail Humain, 51*(2), 139–171.

Yu, V. L., Fagan, L. M., Wraith, S. M., Clancey, W. J., Scott, A. C., Hannigan, J. F., Blum, R. L., Buchanan, B. G., & Cohen, S. N. (1979). Antimicrobial selection by a computer: A blinded evaluation by infectious disease experts. *Journal of the American Medical Association, 242*, 1279–1282.

Zachary, W. (1986). A cognitively based functional taxonomy of decision support techniques. *Human–Computer Interaction, 2*, 25–63.

7

Cooperation Through Communication in a Distributed Problem-Solving Network

Edmund H. Durfee, Victor R. Lesser, & Daniel D. Corkill

Department of Computer and Information Science
University of Massachusetts
Amherst, MA

INTRODUCTION

There is no mystery to why people cooperate. Cooperation occurs when each person believes that he or she will benefit more by cooperating than by acting in some other way. Similarly, groups of people will cooperate for their mutual benefit: businesses cooperate to increase profits, and nations cooperate in part to improve security (increase the survival probability of their people). In such situations, the cooperating *agents*, be they individuals or groups, only cooperate to improve their own self-interests. The prevalence of cooperation in human society, and in the world in general, indicates that there are many ways that selfish agents can interact for their mutual benefit.

Despite its prevalence, however, cooperation is still a poorly understood phenomenon. The advent of computers as agents in human environments has led to many interesting problems that stress issues in cooperation. For example, cooperation among humans is facilitated because people have an understanding of each other—each can predict the others' actions, because they have so much in common (their humanity). Human–computer interactions are often fraught with frustration and misunderstanding because neither agent has an adequate view of why the other is behaving as it is (Oberquelle, Kupka, & Maass, 1983). Furthermore, computer–computer interactions have been similarly problematic, because computers have primitive, if any, abilities to understand other computers. Unlike natural systems where such understanding evolved with the species, artificial systems must explicitly be given such understanding.

The research described in this chapter is a step toward this end. The artificial systems studied use artificial intelligence techniques to solve problems. A network of these problem solvers perform *distributed problem solving* by cooperating as a team to solve a single problem. The principal focus of this research is on developing mechanisms that allow each problem solver to understand what other problem solvers are doing. With this understanding, the problem solvers can act as a more coherent team.

After we more fully introduce our view of cooperation in distributed problem solving networks, we provide an overview of distributed problem-solving techniques and outline our experimental testbed for studying distributed problem solving. In the remainder of the chapter are described mechanisms that improve cooperation by allowing each problem solver to better understand overall network problem solving. First we discuss how more intelligent decisions about the communication of partial solutions can enhance network coherence. These decisions are further improved through mechanisms that enable a problem solver to better understand its own past, present, and intended future actions, and these mechanisms are subsequently described. We then outline how the communication interface between problem solvers is augmented so that the problem solvers can not only exchange *domain-level* information about partial solutions, but can also exchange *metalevel* information that is specifically intended to improve coordination between problem-solving activities. Finally, we summarize our current ideas about distributed problem solving and describe the future directions of our research.

COOPERATION

Since problem solvers can cooperate in many different ways, distributed problem solving provides a rich environment to study issues in cooperation among artificial systems. The actions of each problem solver depend on the desired outcomes of the interactions between problem solvers: The problem solvers base their actions and interactions on one or more goals of cooperation. The possible goals of cooperation include:

- To improve performance (form an overall solution faster) by working in parallel.
- To increase the variety of solutions by allowing agents to form local solutions without being overly influenced by other agents.
- To increase the confidence of a (sub)solution by having agents rederive (verify) each other's results, possibly using different problem-solving expertise and data.
- To increase the probability that a solution will be found, despite agent failures, by assigning important tasks to multiple agents.

- To reduce the amount of unnecessary duplication of effort by letting agents recognize and avoid useless redundant activities.
- To improve the overall problem solving by permitting agents to exchange predictive information.
- To reduce communication by being more selective about what messages are exchanged.
- To improve the use of computing resources by allowing agents to exchange tasks to better balance the computational load.
- To improve the use of individual agent expertise by allowing agents to exchange tasks so that a task is performed by the most capable agent(s).
- To minimize the time agents must wait for results from each other by coordinating activity.

Because the problem-solving agents cannot achieve all of these often conflicting goals simultaneously, they must cooperate differently depending on the particular problem-solving situation. For example, if a solution must be found quickly, the agents should not spend time verifying each other's results or developing a wide variety of solutions. Because of the diversity in the forms that cooperation can take in a distributed problem-solving system, distributed problem solving is an appropriate context in which to study issues in cooperation.

Self-Interested Problem Solvers

A distributed problem-solving network can be viewed in two ways. When viewed as a single entity, the network is a system for decomposing a problem and assigning subproblems to its various subprocessors. Alternatively, the network can be viewed as a collection of independent problem solvers that can communicate. While the first view of distributed problem solving stresses intelligent *network control* (to decompose the problem and assign subproblems appropriately), the second view emphasizes intelligent *local control* of each individual problem solver (to decide which local tasks to pursue in order to best contribute to network performance).

We view a distributed problem-solving network as a set of independent communicating agents, and thus our research has concentrated on developing problem solvers that have sophisticated local control. Each problem solver has the knowledge and intelligence needed to make its own decisions about subproblems to solve and about subproblem solutions to communicate. The emphasis of this work has therefore been on developing mechanisms that allow problem solvers to individually make decisions that contribute to achieving the goals of the overall network.

Since the individual problem solvers make local decisions about what actions to take, each problem solver essentially pursues activities that appear interesting from its local viewpoint. Each problem solver is thus *self-interested*, because it attempts to maximize its local rewards (to achieve its local goals, to follow its local heuristics). Local viewpoints can vary from one problem solver to another, and the compatibility of the individual viewpoints determines whether problem solvers will cooperate, compete, or merely co-exist.

Cooperation Through Self-Interest

Cooperation can evolve in a population due to self-interest. For example, Dawkins (1976) has shown that genes, when viewed as self-interested agents, can cooperate to mutually improve their chances of reproducing. Similarly, Axelrod (1984) has identified attributes of self-interested agents that successfully cooperate when confronted with the prisoner's dilemma. A more extensive discussion about cooperation among various types of self-interested agents is presented elsewhere (Durfee, 1986).

To cooperate, self-interested problem solvers must recognize that cooperation is in their self-interest. That is, they must have local knowledge that guides them into cooperating when cooperation is potentially to their mutual benefit. This knowledge can take many forms. For example, when problem solvers use goals to guide their decisions, cooperation can be promoted by allowing the problem solvers to share high-level goals. Since each self-interested problem solver is attempting to achieve the shared high-level goals, the problem solvers will tend to cooperate. Note, however, that, if the agents have differing views on how to achieve these goals or different interpretations of these goals, then they may perform competitive actions. Examples of this phenomenon occur in human problem solving: Just because the president and the congress share the goal of lowering the deficit, their different perspectives might lead them toward competing solutions.

Rosenschein and Genesereth (1985) misleadingly call agents that share goals *benevolent* agents. Actually, these agents are completely self-interested, since each performs actions only to satisfy its own local interpretations of these goals. Benevolence is neither assumed nor needed for the agents to cooperate.[1] What *is* necessary, however, is some degree of com-

---

[1]Attributes such as benevolence, malevolence, and altruism are conferred on an agent by other agents when the other agents cannot otherwise rationalize the agent's decisions. For example, an intelligent agent may perceive a parent's self-sacrifice as being altruistic, when in fact the parent may have had no choice (its actions are genetically ingrained and it cannot even recognize alternative actions). Altruism is attributed to the parent, because the intelligent agent views the decisions differently. Similarly, an agent that appears to be benevolent (or malevolent) will only seem that way because other agents assign different rewards to its decisions than it does itself.

mon knowledge: Intentional cooperation among self-interested agents requires some amount of common knowledge, so that agents can comprehend and anticipate each other's actions. The approach recommended by Rosenschein and Genesereth, for example, is to allow agents to share decision matrices rather than goals. But since both goals and decision matrices guide agents' decisions, these approaches are essentially the same—they use slightly different forms of common knowledge to lead self-interested nodes into cooperating.

Even if problem solvers want to cooperate due to their common knowledge, effective cooperation in a distributed problem-solving network can still be difficult to achieve because of the changing characteristics of the problem-solving situation. The common knowledge inherent in shared high-level goals or decision matrices might not help agents make useful coordination decisions as problem solving progresses (if new subproblems appear at one or more problem solvers, new subproblem solutions are formed, problem solvers or communication links fail, etc.). The principal focus of this chapter is to describe new mechanisms that improve the dynamic view that each problem solver has of network activity. With this improved view, each problem solver can make more intelligent decisions about what local actions will most contribute to network activity, and the entire network thus behaves more coherently.

DISTRIBUTED PROBLEM SOLVING

Achieving coherent cooperation in a distributed problem-solving network is a difficult task (Davis & Smith, 1983; Lesser & Corkill, 1981). In the type of distributed problem-solving network studied in this chapter, each agent is a semiautonomous problem solving *node* that can communicate with other nodes. Nodes work together to solve a single problem by individually solving interacting subproblems and integrating their subproblem solutions into an overall solution. These networks are typically used in applications such as distributed sensor networks (Lesser & Erman, 1980; Smith, 1980), distributed air traffic control (Cammarata, McArthur, & Steeb, 1983), and distributed robot systems, (Fehling & Erman, 1983) where there is a natural spatial distribution of information but where each node has insufficient local information to completely and accurately solve its subproblems.

Approaches to Distributed Problem Solving

Three important approaches have been taken to improve coordination among cooperating nodes. These approaches, which are by no means mutually exclusive (Smith & Davis, 1981), are multiagent planning, negotiation, and the functionally accurate, cooperative approaches. In the multiagent planning approach, the nodes typically choose a node from among themselves (perhaps through negotiation) to solve their planning problem, and send this node all pertinent information. The planning node forms a multiagent plan that specifies the actions each node should take, and the planning node distributes the plan among the nodes. Since the multiagent plan is based on a global view of the problem, all important interactions between agents can be predicted. If it can be effectively implemented, the approach seems suitable in domains such as air traffic control, where it is imperative that node interactions with dire consequences (such as vehicle collisions) are assured of being detected and avoided (Cammarata et al., 1983). Unfortunately, achieving a global view of the problem might be extremely costly both in communication resources and in time, and the performance of the entire network depends on the planning node and would be compromised if that node fails. The approach may thus be unfeasible in many realistic situations.

In most domains, fortunately, the consequences of unexpected node interactions are not so dire; usually these interactions merely degrade performance. In the negotiation approach (Davis & Smith, 1983), a node will decompose a problem task into some set of subtasks and will assign these subtasks to other nodes (for parallel execution) based on a bidding protocol (Smith, 1980). Since nodes may have different capabilities, the bidding protocol allows a subtask to be assigned to the most appropriate available node (nodes that are already working on subtasks are not available to bid until they have finished their tasks). A node that is awarded one subtask may thus be unavailable to perform a subsequently formed subtask despite being the best node for that subtask. If the node had been able to predict that a more suitable subtask might soon be formed, the node would not have bid on the earlier subtask, so that it would be available later. The inability of nodes to make such predictions can therefore cause incoherence in the problem-solving network: the nodes could make a more coherent team and improve their overall performance if they could assign subtasks to nodes better.

In the *functionally accurate, cooperative* (FA/C) approach to distributed problem solving (Lesser & Corkill, 1981), nodes cooperate by generating and exchanging tentative, partial solutions based on their limited local views of the network problem. By iteratively exchanging their potentially incomplete, inaccurate, and inconsistent partial solutions, the nodes eventually converge on an overall network solution. To cooperate coherently, the nodes would need to predict what partial solutions would be exchanged in the future and when, so that they could modify their problem-solving activities to form compatible partial solutions. To make these predictions, each node needs to understand its own plans and the plans of the other nodes. Without this understanding, nodes may require much more time to converge on a solution, since they may work at cross-purposes.

Prediction is therefore crucial for coherent cooperation. While multi-agent planning requires accurate predictions before it can form acceptable plans, the negotiation and FA/C approaches can perform despite a lack of adequate predictions, but incoherence can degrade their performance. Better predictions in both of these approaches have been achieved through *organization*: by providing nodes with organizational information (the general capabilities and responsibilities of other nodes, the communication patterns between nodes), the nodes have a general understanding of each other and can therefore make better predictions. In the negotiation approach, this allows nodes to use focused addressing techniques in making better subtask-to-node assignments, while, in the FA/C approach, it allows nodes to better decide which of their potential problem-solving tasks is likely to improve network problem solving. However, an organization can only make limited improvements to coherence, since it helps a node predict what other nodes are generally likely to do, but provides little help in predicting when important interactions are likely to take place.

The mechanisms described in this chapter are a step toward providing nodes with the ability to predict their own actions and the actions of other nodes based on exchanged information. Before describing these mechanisms in detail, we first provide an overview of the experimental domain and the testbed in which the mechanisms have been implemented.

The Distributed Vehicle Monitoring Testbed

A vehicle-monitoring problem solver generates a dynamic map of vehicles moving through an area monitored by acoustic sensors. The vehicles' characteristic acoustic signals are detected at discrete time intervals, and these signals indicate the types of vehicles passing through the area and their approximate locations at each sensed time. An acoustic sensor's range and accuracy are limited, and the raw data it generates can be errorful, causing nonexistent (ghost) vehicles to be identified and actual vehicles to be located incorrectly, misidentified, or missed completely. A vehicle-monitoring node applies signal-processing knowledge to correlate the data, attempting to recognize and eliminate incorrect noisy sensor data as it integrates the correct data into an answer map.

In a network of vehicle monitoring nodes, where each is responsible for a portion of the sensed area, the nodes develop partial maps in parallel and exchange these to converge on a complete map. Distributed problem solving is advantageous in this domain for various reasons. Since the nodes work in parallel, the time needed to generate an entire map is reduced.

The nodes can be spatially distributed to reduce the distances over which large quantities of raw sensory data must be transmitted, thereby lowering communication costs. Finally, the system is more robust, since a node failure only disrupts the monitoring of a portion of the area. However, distributed vehicle monitoring must overcome the difficulties of coordination and cooperation common to distributed problem-solving networks.

By simulating a network of vehicle monitoring nodes, the Distributed Vehicle Monitoring Testbed (DVMT) provides a framework where general approaches for distributed problem solving can be developed and evaluated (Lesser et al., 1982; Lesser & Corkill, 1983). By varying parameters in the DVMT that specify the accuracy and range of the acoustic sensors, the acoustic signals that are to be grouped together to form patterns of vehicles, the power and distribution of knowledge among the nodes in the network, and the node and communication topology, a wide variety of cooperative distributed problem-solving situations can be modeled.

Each problem-solving node has a Hearsay-II, blackboard-based architecture (Erman, Hayes-Roth, Lesser, & Reddy, 1980), with knowledge sources and blackboard levels of abstraction appropriate for vehicle monitoring. A *knowledge source* (KS) performs the basic problem-solving tasks of extending and refining *hypotheses* (partial solutions), where each hypothesis tentatively indicates where a certain type of vehicle was at one or more discrete sensed times. The basic Hearsay-II architecture has been augmented to include more sophisticated local control and the capability of communicating hypotheses and goals among nodes (Corkill & Lesser, 1981; Corkill, Lesser, & Hudlicka, 1982). In particular, a goal blackboard, a goal-processing module, and communication knowledge sources have been added (Figure 1).

Goals are created on the goal blackboard to indicate the node's intention to refine and extend hypotheses on the data blackboard. Through goal processing, a node forms *knowledge source instantiations* (KSIs) that represent potential KS applications on specific hypotheses to satisfy certain goals. The scheduler ranks a KSI based both on the estimated beliefs of the hypothesis it may produce and on the ratings of the goals it is expected to satisfy. Appropriate goal processing to modify goal ratings can therefore alter KSI rankings to improve local control decisions. The network organizational structure, for example, is specified as a set of node *interest areas* that affect goal-processing decisions: a node will increase the ratings of goals that fall within its areas of network responsibility while decreasing the ratings of other goals. Since goal ratings affect KSI rankings, the interest areas can influence node activity; but because there are other factors in ranking KSIs (such as the expected beliefs of the output hypotheses), a node still preserves a certain level of flexibility in its local control decisions. The organizational structure thus provides guidance without dictating local

Figure 1. DVMT Node Architecture

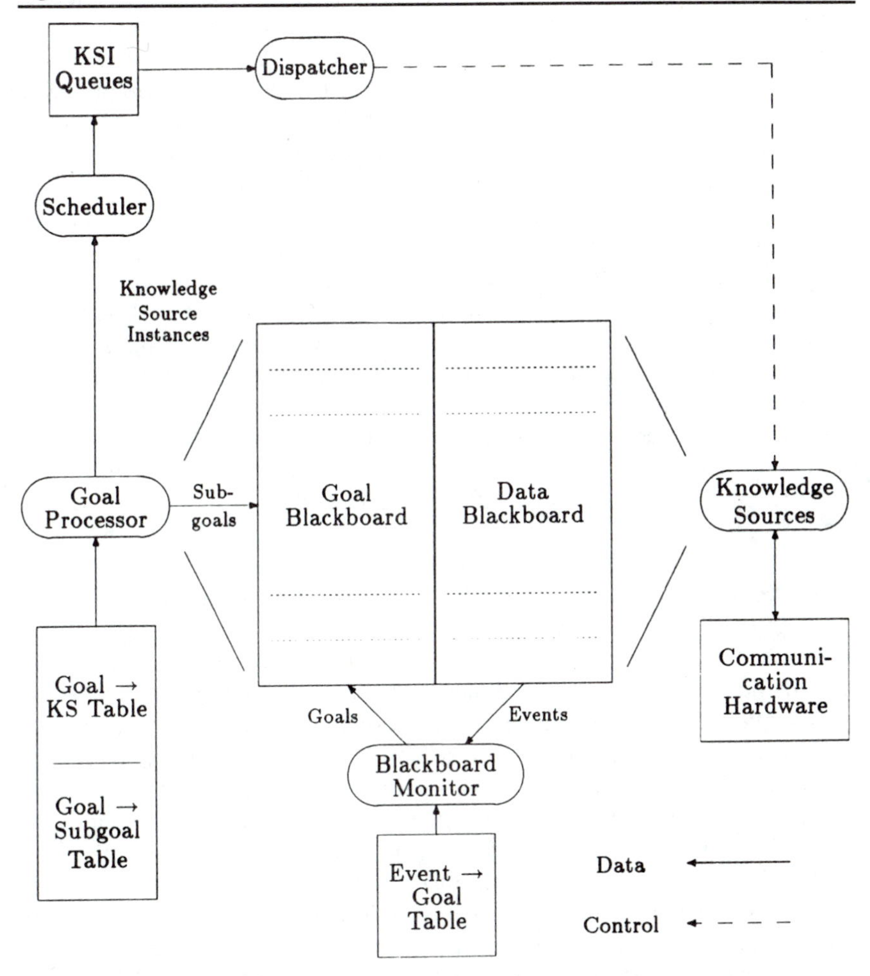

The principal components of the problem-solving architecture of a DVMT node are illustrated.

decisions, and can be used to control the amount of overlap and problem-solving redundancy among nodes, the problem-solving roles of the nodes (such as *integrator*, *specialist*, and *middle manager*), the authority relations between nodes (whether nodes are biased to prefer work based on received data or on locally generated data), and the potential problem-solving paths in the network (Corkill & Lesser, 1983).

Put simply, a node repeatedly performs a problem-solving cycle: it executes a KS to post hypotheses on the blackboard, the new hypotheses trigger the creation of suitable goals for refining and extending them, the goal processing forms and rates KSIs to satisfy these goals, and the scheduler chooses the KSI to invoke next and triggers the appropriate KS to execute. Concurrently, a transmission processor and a reception processor each maintains a queue of communication KSIs (ranked by the rating of the information to be communicated and by the interest areas), and each executes KSs to transmit or receive messages (Durfee, Corkill, & Lesser, 1984). The cyclic problem-solving activity begins when signal hypotheses are posted based on sensor data, and a node then processes its most promising signal data to form tracks. Through goal processing, the goals to extend these tracks boost the importance of KSIs that may lead to the desired extensions. This form of problem solving is called *island-driven*, because it uses islands of high belief to guide further processing. This problem solving is also *opportunistic*, because nodes can react to highly rated new data (perhaps received from another node) to form alternative islands of high belief. Finally, the cyclic problem-solving activity terminates when one or more hypotheses matching a predefined solution are formed.[2]

INCREASING COHERENCE THROUGH COMMUNICATION

There are three major characteristics of the information communicated among nodes that affects coherence: relevance, timeliness, and completeness. *Relevance* measures, for a given message, the amount of information that is consistent with the solution derived by the network. Irrelevant messages may distract the receiving node into wasting its processing resources on attempts to integrate inconsistent information, so higher relevance of communicated information can result in more global coherence (since this information stimulates work along the solution path). *Timeliness* measures the extent to which a transmitted message will influence the current activity of the receiving node. Since timeliness depends not only on the content of the message but also on the state of the nodes, a message's timeliness can vary as node activity progresses. If the transmitted information will have no effect on the node's current activity, there is no point in sending it; however, if the transmitted information will distract the receiving node to work in a more promising area, or if the node needs the

[2] The solution is thus specified before problem solving begins, but a node cannot use this information to guide its processing. Without the availability of such an oracle, termination of problem solving is much more difficult, requiring nodes to determine whether further problem solving is likely to improve upon a possible solution which it has already generated or will cause it to generate another, potentially better, solution. In effect, the termination decision depends on the criteria for deciding whether network goals have been satisfied (Corkill, 1982).

information to continue developing a promising partial solution, then it is important that the information be sent promptly. Finally, *completeness* of a message measures the fraction of a complete solution that the message represents. Completeness affects coherence by reducing the number of partially or fully redundant messages communicated between nodes—messages which negatively distract nodes into performing redundant activity. Furthermore, as the completeness of received messages increases, the number of ways that the messages can be combined with local partial results decreases due to their larger context. Finally, achieving completeness is important to minimize communication requirements in our loosely coupled distributed system.

It is important to note that these three characteristics of communicated information are interdependent (as will be seen in empirical results to be presented later in the chapter). For example, higher completeness leads to higher relevance but also a potential decrease in timeliness. Thus, communication policies often involve tradeoffs among the three characteristics.

These characteristics of communicated information are affected by both the local control in a node (which generates the potential information that can be transmitted), and the communication policy (which decides what information should be sent, to what nodes, and when). Increased global coherence could be achieved by improving only local node control, by improving only communication policies, or by a combination of both. In each of these cases, more intelligent decisions result if a node has a better understanding of what it has already done, what it is likely to do, and the activities and intentions of other nodes.

Implementing More Sophisticated Communication Policies

Our first approach to improving the basic communication policy of sending all candidate hypotheses (as defined in the node's organizational structure) was to develop a simple version of a *locally complete* communication policy. This policy, originally developed by Lesser and Erman (1980), required significant modification to handle the DVMT's more complex task processing structure.

The basic communication policy, which we call *send-all*, often results in a node transmitting a small or irrelevant hypothesis, even though the subsequent internal processing of the sending node will immediately improve upon this data. Because a node might produce a large number of these small hypotheses in the process of generating a more complete hypothesis, the transmission of each of these less-complete hypotheses would be deleterious in terms of the minimizing channel usage and maximizing completeness. Furthermore, because these smaller hypotheses are transmitted immediately (before any activity to support or refute them is done), they are more likely to be irrelevant. Finally, these incomplete and incorrect hypotheses could distract the recipient node into working in inappropriate areas. If nothing else, the recipient node is likely to combine these smaller hypotheses together into a larger hypothesis—a redundant activity, since the sending node both generates and transmits this larger hypothesis as well.

The locally complete communication policy is based on a node transmitting only those hypotheses which it cannot improve upon. This policy minimizes the number of transmissions while maximizing the completeness and relevance of each message. The implementation of this policy in the testbed illustrates how the node's awareness of its activities (in this case future ones) is necessary for intelligent communication. To estimate whether a hypothesis is locally complete, the queue of pending local KSIs must be examined. If there exists a KSI to improve upon the hypothesis, either by increasing its length or belief, then the hypothesis is not yet locally complete. However, the existence of such a KSI does not necessarily mean that an improved hypothesis will be created. The rating of the KSI might be so low that it will never be executed. Furthermore, even if executed, the resulting hypothesis from the KSI might not be an improvement, because the KSI might fall short of its intentions.

Thus, due to the uncertainty as to whether a more locally complete hypothesis might ever be formed, such a policy will not be effective in terms of timeliness—a locally incomplete hypothesis might be useful to another node, and we should not hold it back too long waiting for something better.

To partially rectify this problem, we introduce a time-out mechanism that heuristically balances the desire for completeness with that of timeliness. As in the less sophisticated send-all policy, any hypothesis that meets the criteria for communication stimulates the creation of a communication KSI to perform the transmission. However, we now associate with this transmission KSI an invocation delay. If the KSI is to transmit a locally complete hypothesis, this delay is set to zero; otherwise the delay is set to a *user-specified* value (we discuss this issue later). The delay acts as a timeout value—if sufficient time elapses without the creation of a more complete version of the hypothesis, the transmission KSI is invoked to transmit the locally incomplete version. However, if the improved (but still possibly locally incomplete) hypothesis is created, a KSI to send it is generated (with a new invocation delay), and the KSI of the now inferior hypothesis is deleted from the pending transmission KSI queue.

By providing a node with a chance to improve upon a hypothesis before that hypothesis is transmitted, the locally complete policy will reduce the number of small and irrelevant hypotheses transmitted while allowing the timely transmission of better hypotheses. This can result in improved network performance due to reduction of redundant work and distraction. However, because transmitted hypotheses are used both to focus activity in the receiving node and as data to be combined with the local hypotheses, delaying the transmission of some locally incomplete hypotheses might adversely affect performance. What is needed is a policy which will send locally incomplete hypotheses for predictive purposes as well as locally complete hypotheses for integration.

The communication policy we have developed for this situation has been dubbed *first-and-last*. It is essentially the locally complete policy with the added stipulation that, if a hypothesis that is eligible for transmission does not incorporate any previously communicated hypotheses, then transmit it without delay. Hence, the first partial hypothesis will be transmitted for predictive information, and the last locally complete version will be sent for integration information, but any intervening locally incomplete versions will not be transmitted.

An equivalent set of communication policies was also implemented for goal communication. A goal is locally complete when the node has created as much context for it as it can. Stated another way, a goal is locally complete if the hypotheses that stimulated its creation are all locally complete. Furthermore, the first-and-last policy is considered to be very important for goals, because the primary purpose of a goal is to focus activity at the receiving node. As we investigate more complex environments, we expect the first-and-last policy to become increasingly important.

Experiments with the Communication Policies

We illustrate the communication policies using three vehicle monitoring environments (Figure 2). The *normal* environment consists of four nodes, each attached to a different sensor, where the sensors overlap slightly (Figure 2a). Data is sensed over eight time intervals. The acoustic sensors attach more credence to more strongly sensed (louder) signals than to weaker (fainter) signals; data from the vehicle track has strongly and weakly sensed regions. A moderately sensed ghost track, perhaps caused by echoes from the vehicle or by sensor errors, parallels the vehicle track. In the *overlap* environment (Figure 2b), the overlap between the sensor regions is increased over the normal environment. Since each node has a more global view, the potential for finding the solution rapidly is increased. However, because there is more overlap, there is also more potential for redundant problem solving. Finally, the *twice-data* environment (Figure 2c) is similar to the normal environment, except that there are 15 time intervals, forming additional data points between each of those in the

Figure 2. Sensor and Data Configurations for Three Environments.

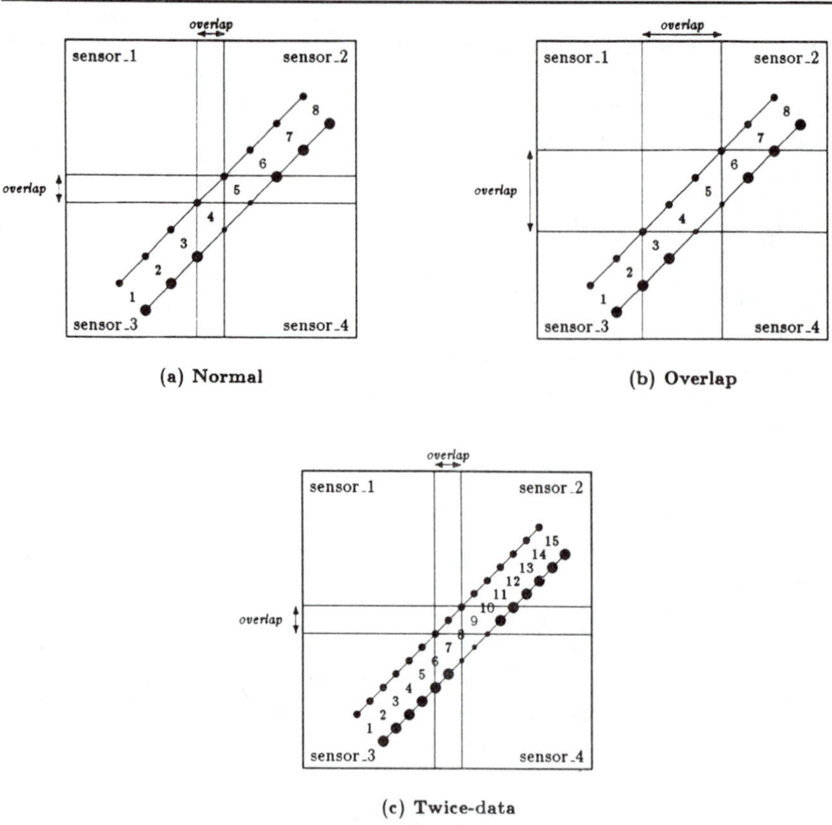

Three experimental environments are illustrated. The normal environment (a) has eight data points on both the ghost track (upper) and the vehicle track (lower). The overlap environment (b) has the same data but more sensor overlap. The twice-data environment (c) has the same overlap as (a) but twice as much data. The size of a data point indicates how strongly it is sensed.

normal environment. Because this effectively doubles the number of locally incomplete hypotheses a node will produce, the locally complete and first-and-last policies should be even more effective on this environment. To simplify our experiments, *all* of the data were available at the start of the problem solving. The invocation delay used by the locally complete and the first-and-last policies was 10 time units, which is usually sufficient to allow a more locally complete version of a hypothesis to be formed.

The experimental results provide evidence supporting our expectations (Table 1). In all cases, the number of hypotheses transmitted under the locally complete policy is less than the number transmitted by the send-all

Table 1. A Summary of Communication Performance.

Environment	Communication Policy	Solution Time	Total Transmitted Hyps
Normal	Send-all	49	25
	Loc-comp	45	10
	F-and-L	49	20
Overlap	Send-all	52	16
	Loc-comp	39	11
	F-and-L	46	13
Twice-data	Send-all	107	74
	Loc-comp	70	17
	F-and-L	99	29

Legend

Environment: The simulated environment
Solution Time: Earliest time at which a solution was found
Total Transmitted Hyps: The total number of hypotheses transmitted

Optimal solution time for the normal environment is 23 regardless of the communication policy and assuming that the nodes perform the most appropriate actions at all times.

policy, with the first-and-last policy falling between. The improved performance in the locally complete policy can be attributed to a reduction in distracting, irrelevant, and incomplete information passed between the nodes. The first-and-last policy is less effective, because it communicates more of this detrimental information.

The degree to which these policies affect performance depends on the environment. In the normal environment, many of the additional communicated hypotheses do not affect the recipient nodes: Either the received hypothesis does not trigger activity to improve upon it or else the triggered activity coincides with the actions the node was already taking. In the overlap environment, however, each node has data from a larger portion of the sensed area. Because a node is more likely to have data to combine with a received hypothesis, the small number of additional messages permitted by some policies cause the nodes to waste more time on unnecessary activities. In the twice-data environment, the abundance of data means that any received hypothesis that does trigger activity will trigger substantial amounts of activity. The first-and-last policy sends the small hypotheses that the locally complete policy withholds, and these hypotheses degrade performance by stimulating large amounts of unimportant work. Although the send-all policy sends many more hypotheses than the first-and-last policy, further performance degradation is modest: most of the additional (larger) transmitted hypotheses only trigger the same activities as earlier (smaller) hypotheses. These communication policies have been used in numerous other environments, including some larger 10- to 15-node simulations, with similar results.

INCREASING THE SOPHISTICATION OF A PROBLEM SOLVING NODE

The user specified invocation delay for locally incomplete hypotheses is inflexible to specific problem-solving situations that nodes may encounter. Instead, we would like a node to determine an appropriate delay for transmitting a locally incomplete hypothesis (or goal) based on predictions about its future activities. Such predictions are unfeasible in the basic problem-solving architecture, which represents future activities simply as a priority queue of KSIs. Because both local activity and received information can stimulate the creation of new, highly rated KSIs, estimating the future activity of a node (or even just attempting to estimate when a particular KSI will be invoked) is a complex and highly uncertain task. A node with a better perspective on its past, present, and potential future activities can make better communication decisions. Such a node can also make improved decisions about its local problem-solving actions and can better understand how its activities fit into its organizational responsibilities.

Although the priority queue of pending KSIs allows for rampant opportunism in a node, nodes often methodically perform sequences of related actions. For example, given a highly rated hypothesis, a node typically executes a sequence of KSIs that drive up low-level data to extend the hypothesis. However, the entire sequence of KSIs is never on the queue at once. We have therefore developed a structure, called a *plan*, to explicitly represent a KSI sequence.

Blackboards, Plans, and Node Activities

Each *plan* represents a desire to achieve a high-level goal by performing a sequence of activities. To identify plans, the node needs to recognize these high-level goals. Inferring high-level goals based on pending KSIs is an inappropriate strategy; it is like attempting to guess a chess opponent's strategy by observing isolated moves. Furthermore, the hypothesis and goal blackboards provide information at too detailed a level to infer these high-level goals. What is required is a structure similar to the blackboards that groups related hypotheses and goals together. We have developed a preliminary version of this structure which we call the *abstracted blackboard*, a structure reminiscent of the focus-of-control database first used in the Hearsay-II speech understanding system (Erman et al., 1980). Our implementation of the abstracted blackboard is incomplete, because it does not adequately incorporate the information from the goal blackboard. However, for the type of processing done in the DVMT, hypothesis abstraction is usually effective.

Related hypotheses are grouped together on the abstracted blackboard based on their blackboard level, their time, and their region characteristics. This clustering allows us to differentiate between areas of the solution space. In our preliminary implementation, the abstracted blackboard takes the form of a two-dimensional array, with level and time indices. When a hypothesis is created, it is incorporated into this structure by stepping through the sensed times of the hypothesis and modifying the appropriate level-time entry in the abstracted blackboard. Each level-time entry contains some number of regions, and, if the location associated with the sensed time can be included in one of these regions (perhaps by enlarging the region within certain bounds), the hypothesis is associated with that region. Otherwise, a new region is formed for the hypothesis.

Each level-time region of the abstracted blackboard is summarized into a set of values that are derived from the associated hypotheses. These values include the maximum belief of the hypotheses in the level-time region, the number of highly believed hypotheses, the number of KSIs stimulated by these hypotheses that have yet to be invoked, the total number of hypotheses in the level-time region and how many uninvoked KSIs are associated with them, and an indication as to the other level-time regions that share at least one of the hypotheses. This information allows the *situation recognizer* to develop a higher-level view of the problem solving. For example, low maximum belief indicates the problem-solving approach in that area should be reevaluated, a large number of equally rated hypotheses could imply that there is uncertainty that should be resolved, and a large number of pending KSIs indicates the need for making an informed and judicious choice of which action to take next. Based on this higher-level view, we can begin to form higher-level goals. A goal might be to merge hypotheses in adjacent clusters, to improve the belief of an established hypothesis, or to extend a highly believed hypothesis into a new region.

The detection of these goals, and the subsequent generation and ranking of their respective plans, is in itself a complex problem solving task. Our current implementation is a first pass toward this end, in which we only consider very simple but important plans. Given the abstracted blackboard, our planner scans down it, looking for regions of high belief. Having found such a *stimulus region*, the planner determines whether there is any indication that the data in this region can be improved (this is done by determining whether any corresponding lower-level regions have higher belief than the upper level regions), and, if so indicated, a plan is formed to achieve this improvement. Otherwise, a plan is generated to extend this highly rated region, either by merging a hypothesis in this region with a hypothesis in an adjacent cluster on the same level (if any), or by driving lower-level data in an adjacent area up to a level at which it can be incorporated. If none of these plans can be formed, then a plan to synthesize the hypotheses in this highly rated region up to a higher blackboard level may be formed.

Plans in our current implementation are not yet fully developed, because a plan should not only involve the specification of an eventual goal, but also of a sequence of actions needed to achieve this goal. Our implementation currently does not represent the entire sequence, but only the next potential step(s) in achieving the desired result. To this end, a queue of KSIs is associated with a plan, comprising the current set of potential tasks that are related to the plan. These are found through the hypotheses in the associated regions of the abstracted blackboard. The queue of KSIs is priority rated—the next activity carried out by a plan is always the highest rated KSI. In turn, the node maintains a queue of plans, ordered based on their respective ratings. Currently, a plan rating is based on a number of factors, including the belief of its stimulus region, the level of its stimulus region, the ratings of its KSIs, and whether the stimulus region represents hypotheses generated locally or it represents received hypotheses (to reason more fully about potentially distracting information received from outside). Therefore, in choosing its next activity, a node will invoke the highest rated KSI in the highest rated plan.

We have therefore made important modifications to the control structure of a node (Figure 3). In a KSI-based node (a node without plans), the highest-rated KSI is always invoked next. The rating of the KSI is based on the ratings of the goals it is intended to satisfy and the anticipated beliefs of the hypotheses it will produce. In a plan-based node, the primary factor in determining the next activity is choosing the top plan. As the figure indicates, the creation and ranking of plans requires the planner to integrate the influences of the long-term strategy of the organizational structure, the medium-term higher-level view of the current situation, and the short-term KSI input indicating actions that can be achieved immediately. Hence, KSI ratings only affect the choice of the next plan to a small extent, being more important in the choice of specific activity once the plan is chosen.

It is important to recognize that our use of plans has not changed the basic opportunistic problem-solving strategy of the node. Because we are working on the best plan at any given time, we are being opportunistic on the plan level instead of the KSI level. Unlike KSIs, however, plans can be interrupted or aborted if something more promising comes along. Therefore, a node using plans need not be any less opportunistic than a node without plans. If an area outside the current plan looks most promising, a plan to work in this area will be formed with a higher rating. This plan will supplant the current plan at the top of the plan queue, and our attention will turn to invoking the activities of this better plan. Note that we have

Figure 3. The Modified Problem-Solving Architecture of a Node.

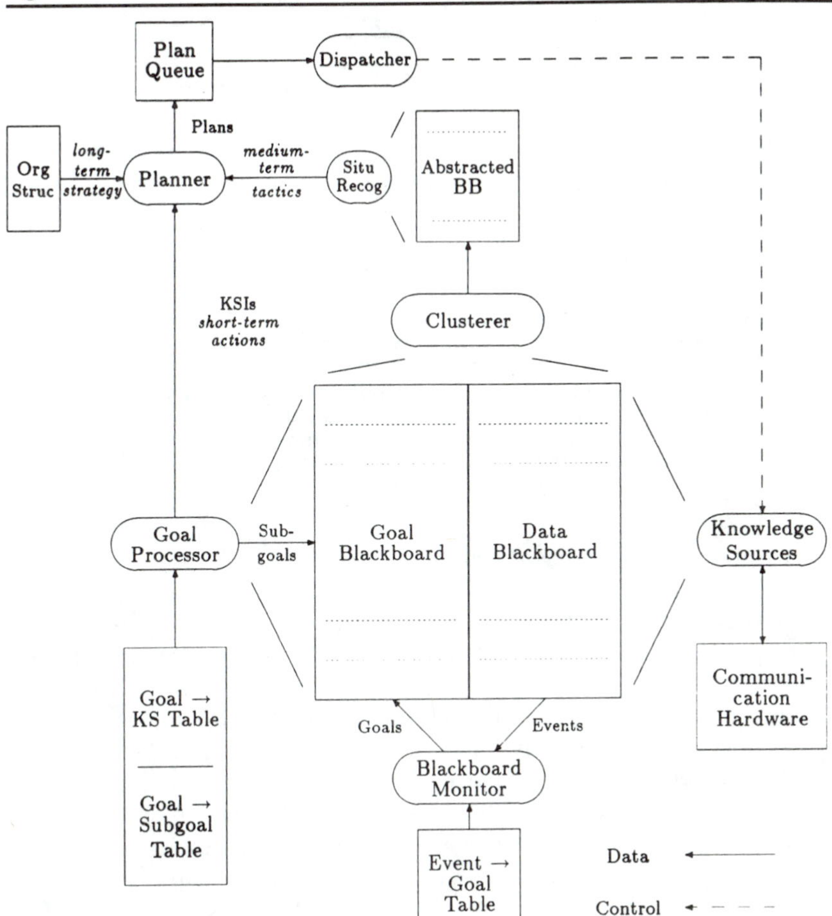

The principal components of the modified problem-solving architecture of a DVMT node are illustrated.

essentially interrupted one plan to work on a better one—once the better plan is fulfilled (or aborted), we can return to the interrupted activities. Finally, by representing sets of tasks as plans, we can make more informed decisions about switching to new tasks. For example, without plans, a promising new KSI can distract the node, despite that node being nearly done with an important sequence of tasks. When plans are introduced, a node can reason about the time invested in a particular area and about whether it really should leave this area for another.

Table 2. A Comparison of Plan-based and KSI-based Nodes.

Environment	Type of nodes	Time
One-node	KSI-based	213
	Plan-based	58
Four-node (ghost above)	KSI-based	49
	Plan-based (internal)	39
	Plan-based (all)	36
Four-node (ghost below)	KSI-based	34
	Plan-based (internal)	31
	Plan-based (all)	26

Legend

Time: Earliest time at which a solution was found

A comparison of plan-based node networks versus KSI-based node networks. Two types of plan-based node are presented, one which includes only internally generated hypotheses in the abstracted blackboard and one which includes all hypotheses in this blackboard.

Experiments with Plan-based Nodes

To illustrate the effectiveness of plan-based problem solving, we present data from three environments in Table 2. The first environment consisted of a single node receiving data from all four sensors shown in Figure 2a. When the node based its activities simply on the KSI queue, solution generation took significantly longer than if the node used plans. This is due to the role that plans play in focusing the problem-solving activity toward extending areas of high certainty, even in regions of lowly rated data. The second environment has four nodes, each receiving the information from a different sensor in Figure 2a. Again, the plan-based control works much better. Furthermore, the importance of using predictive information from other nodes is illustrated. When we incorporate only internally generated hypotheses into the abstracted blackboard, the problem solving is less efficient than if all hypotheses (internally generated and received) are abstracted. Finally, the third environment is identical to the second except that the positions of the vehicle track and ghost track are reversed. Once again, the plan-based control with all hypotheses abstracted is superior.

Numerous other environments have been simulated using plan-based nodes, and all have produced similar results. This success is due primarily to the fact that the mechanisms for recognizing higher-level goals (outlined above) are appropriate in all of the environments we have developed to date. However, as environments which emphasize different situations in distributed problem solving are investigated, it is quite likely that our mechanisms will no longer prove useful. For this reason, we anticipate a need for improving the ability to recognize various situations that the node may face and developing plans which are more appropriate for the particular situation.

Communication Policies and Plan-based Nodes

Because plans represent sequences of activities, our communication policies based on locally complete hypotheses become that much simpler. Currently, we have eliminated associating a delay with a transmission KSI. Instead, in deciding whether to invoke a transmission KSI, we merely compare the transmittable hypothesis with the current plan. If that plan does not involve the hypothesis, we assume that the hypothesis will not be improved upon in the near future, and thus invoke the transmission KSI without delay. On the other hand, if the plan represents an intention to improve upon the hypothesis, then we do not invoke the transmission KSI (but it remains on the queue). If the plan completes successfully, then a transmission KSI for the improved hypothesis is formed and the original transmission KSI is removed. If the plan is aborted or unsuccessful (fails to create a better hypothesis), or if the plan is interrupted by a plan not involving the hypothesis, then the original transmission KSI is invoked. As our understanding of problem-solving activities increases and our representations of plans improves, we may further embellish upon this implementation so that it can reason about the tradeoffs involved in waiting if the plan to improve the hypothesis is near the top of the plan queue. Note that this problem is similar to the problem of making estimations about when a KSI will be invoked in a KSI-based node. However, because the effects of plans are more predictable than those of KSIs, it may be feasible to make projections as to when a particular plan is likely to be at the top of the queue.

The first-and-last policy is again very similar to the locally complete policy outlined above, except we communicate the first piece of a hypothesis even if it is being improved upon by the current plan. The results of running these policies using plan-based nodes on the three environments in Figure 2 are summarized in Table 3. In the normal environment, the locally complete policy is slightly worse than the others, because nodes retain locally incomplete hypotheses that could be used by their neighbors. We therefore see the timeliness drawbacks in a locally complete policy. In the overlap environment, the locally complete policy works slightly better in terms of solution time. The first-and-last policy focuses the nodes on redundant activities, and so, the solution takes longer to generate, but fewer hypotheses are communicated due to the duplication among nodes.

Table 3. Communication Performance with Plan-based Nodes.

Environment	Communication Policy	Solution Time	Total Transmitted Hyps
Normal	Send-all	36	15
	Loc-comp	39	13
	F-and-L	38	15
Overlap	Send-all	38	25
	Loc-comp	36	21
	F-and-L	39	14
Twice-data	Send-all	51	51
	Loc-comp	50	19
	F-and-L	48	21

Legend

Environment: The simulated environment
Solution Time: Earliest time at which a solution was found
Total Transmitted Hyps: The total number of hypotheses transmitted

Finally, the twice-data environment indicates that the predictive qualities of the first-and-last policy can be slightly advantageous in terms of solution time at the cost of requiring more communication than the locally complete policy.

Comparing Table 3 with Table 1, we recognize that increasing the sophistication of the node dramatically improves the speed at which the network finds the solution, regardless of the communication policy. Furthermore, because nodes have an enhanced understanding of their local processing, they are less easily distracted by received information, so that the communication policy has much less effect on the rate of solution generation, even though it still has a significant effect on the number of hypotheses transmitted. However, despite the improvement, the problem solving is still far from optimal as reported in Table 1. Even with increased local sophistication of the nodes, the node network falls prey to redundant and unnecessary activity because a node has a very limited view of the activities at other nodes.

INCREASING THE COHERENCE OF THE PROBLEM-SOLVING NETWORK

We have seen how the performance of the node network can be improved by allowing the individual nodes to reason more fully as to the appropriate activities to perform, based on their local state. We now briefly describe the implications of allowing nodes to communicate this state information to let their neighbors make even more sophisticated decisions about local activities. For example, if two nodes have overlapping sensed areas, each might reason about the other's past, present, and future actions in the overlapping area when deciding whether to work there itself.

By transmitting the abstracted blackboards (or portions thereof), nodes can reason about the *past* activities of their neighbors. Furthermore, if a node knows the current plan of its neighbor, it can reason about the *present* actions of its neighbor. Reasoning about the *future* actions of a node, however, is a complex problem. This reasoning involves considering, not only the current plans in the node's queue and making estimations about their durations and effects, but also what further information the node may receive (from another node or from its sensor) that could affect its activities. A plan may have associated with it some estimations of duration and probability of completion, or even more specific information about how its execution could be affected by received information (Durfee, 1986).

Our current implementation assumes that a node can make completely accurate short-term predictions about future activity based solely on the plan queue. We simulate this best-case scenario by allowing a node access to the abstracted blackboard and plan queue of another node. We have yet to develop more realistic scenarios where nodes must transmit this *metalevel* information as they transmit hypotheses, and must therefore reason about relevance, timeliness, and completeness.

Metalevel information helps nodes to achieve their goals of cooperation. For example, a node may use metalevel information to avoid redundancy. In developing a plan, a node can determine if the plan represents redundantly deriving information that another node has either generated (present in the abstracted blackboard) or is in the process of generating (the top plan). By avoiding redundant activity, improvements in solution generation rate can result because less highly rated but potentially useful activities will be invoked earlier (rather than redundant invocation of highly rated activities). Comparing Table 4 with Table 2, we recognize that metalevel information allows us to come very close to the optimal solution time

Table 4. Performance of Plan-based Node Networks with Metalevel Communication.

Environment	Type of nodes	Time
Four-node	Plan-based + mlc (internal)	27
(ghost above)	Plan-based + mlc (all)	26
Four-node	Plan-based + mlc (internal)	28
(ghost below)	Plan-based + mlc (all)	26

Legend

Time: Earliest time at which a solution was found
mlc: Metalevel Communication

Effects of metalevel communication on network problem solving. Two types of plan-based node are presented, one which includes only internally generated hypotheses in the abstracted blackboard and one which includes all hypotheses in this blackboard. Both types of nodes can both send and receive meta-level information.

without reduction of belief. (Note that the one-node environment will be unchanged because there is no communication, and that the metalevel information is not used in KSI-based nodes, and so, will not affect them.)

Communication Policies and Metalevel Communication

The communication policies for plan-based nodes previously outlined can be modified to exploit the additional knowledge provided by the metalevel communication. In determining whether to send a hypothesis to its neighbor, a node can use the metalevel information to estimate the effects of the hypothesis on that neighbor. The node can reason about the utility of the hypothesis in terms of stimulating activity, avoiding redundancy, or generating the solution. Indeed, in an environment where messages may be lost or garbled, communication of metalevel information can allow nodes to recognize when to retransmit a hypothesis because the anticipated effects of that hypothesis were not seen. In the future, separate communication policies may be obviated, since nodes might be capable of accurately predicting the effects of sending a potential message rather than relying on an unadaptive communication policy to make communication decisions.

In the current implementation, the locally complete policy takes the position that it is important to communicate a hypothesis if the hypothesis is locally complete, if the hypothesis provides information that the receiving node will otherwise redundantly derive, or if improvement of the hypothesis will not affect its anticipated utility. The first-and-last policy is the same as the locally complete, except that the first piece of the hypothesis will be sent for predictive purposes. The effects of these policies are summarized in Table 5. We first note that the policy does not affect the time at which the solution is generated—nodes have such complete views of the network activities that information is communicated in a timely fashion and will not cause distraction. Therefore, our communication policies really only affect the number of hypotheses communicated, and in all cases the locally complete policy sends the least and send-all the most. Secondly, we note that solution generation in the overlap environment requires no more time than in the normal environment. This is an indication of how the exchange of metalevel information can be used to direct nodes away from redundant processing. Finally, it is interesting that the metalevel communication does not significantly reduce the number of hypotheses transmitted (compare Tables 3 and 5). There are two reasons for this. First, because the nodes avoid redundancy, a number of different hypotheses are created and transmitted earlier in the experiment. Second, because the nodes are so well coordinated, they tend to generate and exchange the complete solution at the same time.

Table 5. Communication Performance with Plan-based Nodes, Metalevel Communication.

Environment	Communication Policy	Solution Time	Total Transmitted Hyps
Normal	Send-all	26	15
	Loc-comp	26	12
	F-and-L	26	13
Overlap	Send-all	26	16
	Loc-comp	26	13
	F-and-L	26	14
Twice-data	Send-all	41	27
	Loc-comp	41	16
	F-and-L	41	19

Legend

Environment: The simulated environment
Solution Time: Earliest time at which a solution was found
Total Transmitted Hyps: The total number of hypotheses transmitted

CONCLUSIONS AND FUTURE RESEARCH

Self-interest can lead to cooperation if two conditions are met. First, there must be potential benefits for cooperation in the agents' environment. In many cases, such an environment is purposely designed (a decision matrix is built, high-level goals are shared), while at other times cooperation is potentially advantageous by chance (such as in genetic evolution). The second feature is that agents must exist that can achieve these potential benefits. Once again, they can be intentionally designed (like computer software) or created by chance (gene combinations and mutations).

For natural or artificial self-interested agents to *intentionally* cooperate for their mutual benefit, they must have some common knowledge so that they can anticipate each other's actions and plan cooperative interactions. Cooperating problem solvers, for example, can share high-level goals to improve the chances that they will cooperate effectively. To further improve the coherence of cooperation, the problem solvers can communicate information that increases each other's awareness of network activities.

The focus of this chapter has been on describing mechanisms that allow problem-solving nodes to make intelligent use of communication resources to improve network coherence in the distributed vehicle monitoring testbed. By introducing simple communication policies into the nodes, we recognized the effects of communication decisions on network coherence. Communication decisions were improved by providing nodes with the ability to achieve a high-level view of their activities and to use this view when determining what information to send and when planning their future problem solving actions. Finally, coherence was further improved by allowing nodes to communicate metalevel information based on their high-level views and plans. Through these mechanisms, nodes improved each other's awareness of network activities without incurring excessive communication and computation overhead.[3] Since the nodes have more knowledge in common, the efficacy of their cooperation can be dramatically increased.

Our future research plans are to concentrate on the strong connection between distributed planning and distributed problem solving: effective distributed problem solving requires that agents plan their actions and interactions rather than making isolated and short-sighted decisions. To plan their actions and interactions, agents must have sophisticated planners that can: interleave planning and execution, monitor and revise inappropriate plans, make temporal predictions about when actions and events may occur, and use models of other agents to develop plans that will allow the agent to interact favorably with other agents. This last requirement involves using and improving the communication mechanisms that we have described in this chapter. Currently, to create this planner we are improving the representation of the state of a node, enhancing the mechanisms to recognize problem solving situations, and extending the plan structures to incorporate more information (Durfee, 1986). Our preliminary experiments indicate that these developments should significantly improve the performance of distributed problem-solving networks, and may also be useful in blackboard-based problem-solving systems in general.

REFERENCES

Axelrod, R. (1984). *The evolution of cooperation*. New York: Basic Books.

Cammarata, S., McArthur, D., & Steeb, R. (1983, August). Strategies of cooperation in distributed problem solving. In *Proceedings of the Eighth International Joint Conference on Artificial Intelligence* (pp. 767–770). Palo Alto, CA: Morgan Kaufmann.

Corkill, D. D. (1983, February). *A framework for organizational self-design in distributed problem solving networks*. Unpublished doctoral dissertation, University of Massachusetts, Amherst. (Available as Tech. Rep. 82-33, Department of Computer and Information Science, University of Massachusetts, Amherst.)

Corkill, D. D., & Lesser, V. R. (1981, June). *A goal-directed Hearsay-II architecture: Unifying data and goal directed control* (Tech. Rep. 81-15). Amherst, MA: Department of Computer and Information Science, University of Massachusetts.

Corkill, D. D., & Lesser, V. R. (1983, August). The use of meta-level control for

[3] Evaluations of the costs of these mechanisms are discussed elsewhere (Durfee, Lesser, & Corkill, 1985).

coordination in a distributed problem solving network. In *Proceedings of the Eighth International Joint Conference on Artificial Intelligence* (pp. 748–756). Palo Alto, CA: Morgan Kaufmann.

Corkill, D. D., Lesser, V. R., & Hudlicka, E. (1982, August). Unifying data-directed and goal-directed control: an example and experiments. In *Proceedings of the Second National Conference on Artificial Intelligence* (pp. 143–147). Palo Alto, CA: Morgan Kaufmann.

Davis, R., & Smith, R. G. (1983). Negotiation as a metaphor for distributed problem solving. *Artificial Intelligence, 20*, 63–109.

Dawkins, R. (1976). *The selfish gene*. New York: Oxford University Press.

Durfee, E. H. (1986, March). *An approach to cooperation: Planning and communication in a distributed problem solving network* (Tech. Rep. 86-09). Amherst, MA: Department of Computer and Information Science, University of Massachusetts, Amherst.

Durfee, E. H., Corkill, D. D., & Lesser, V. R. (1984, December). Distributing a distributed problem solving network simulator. In *Proceedings of the fifth real-time systems symposium* (pp. 237–246). Silver Spring, MD: IEEE Computer Society Press.

Durfee, E. H., Lesser, V. R., & Corkill, D. D. (1985). *Coherent cooperation among communicating problem solvers* (Tech. Rep. 85-15). Amherst, MA: Department of Computer and Information Science, University of Massachusetts.

Erman, L. D., Hayes-Roth, F., Lesser, V. R., & Reddy, D. R. (1980, June). The Hearsay-II speech understanding system: Integrating knowledge to resolve uncertainty. *Computing Surveys, 12*(2), 213–253.

Fehling, M., & Erman, F. (1983, April). Report on the third annual workshop on distributed artificial intelligence. *SIGART Newsletter, 84*, 3–12.

Lesser, V., Corkill, D. D., Pavlin, J., Lefkowitz, L., Hudlicka, E., Brooks, R., & Reed, S. (1982, October). A high-level simulation testbed for cooperative distributed problem solving. In *Proceedings of the Third International Conference on Distributed Computer Systems* (pp. 341–349). Silver Spring, MD: IEEE Computer Society Press.

Lesser, V. R., & Corkill, D. D. (1983, Fall). The distributed vehicle monitoring testbed: a tool for investigating distributed problem solving networks. *AI Magazine, 4*(3), 15–33.

Lesser, V. R., & Corkill, D. D. (1981, January). Functionally-accurate, cooperative distributed systems. *IEEE Transactions on Systems, Man, and Cybernetics, SMC-11*(1), 81–96.

Lesser, V. R., & Erman, L. D. (1980, December). Distributed interpretation: a model and experiment. *IEEE Transactions on Computers, C-29*(12), 1144–1163.

Oberquelle, H., Kupka, I., & Maass, S. (1983). A view of human-machine communication and co-operation. *International Journal of Man-Machine Studies, 19*, 309–333.

Rosenschein, J. S., & Genesereth, M. R. (1985, August). In *Proceedings of the Ninth International Joint Conference on Artificial Intelligence* (pp. 91–99). Palo Alto, CA: Morgan Kaufmann.

Smith, R. G. (1980, December). The contract-net protocol: high-level communication and control in a distributed problem solver. *IEEE Transactions on Computers, C-29*(12), 1104–1113.

Smith, R. G., & Davis, R. (1981, January). Frameworks for cooperation in distributed problem solving. *IEEE Transactions on Systems, Man, and Cybernetics, SMC-11*(1), 61–70.

8

The Design of Cooperative Person–Machine Problem-Solving Systems: A Methodology and Example*

John L. Goodson

GE Advanced Technology Laboratories
Morristown, NJ

Charles F. Schmidt

Psychology and Computer Science Departments
Rutgers University
New Brunswick, NJ

INTRODUCTION

Not too many years ago the hypothesis that it is possible to create intelligent machines was considered highly implausible by most people outside the field of artificial intelligence (AI). Now, perhaps due as much to the media as to our scientific advances, this hypothesis has been transformed into a rather widely held belief. If intelligent machines are developed, then the question arises of whether it will be possible and desirable to be able to interact with these intelligent machines.

Of course, we have always interacted with computers in a minimal way. The interaction has been minimal in the sense that the person has had to bear the major responsibility for the nature and course of the interaction. The machine has typically exhibited a rather implacable indifference to, and ignorance of, the goals and intents of the human actor. Even user-friendly environments retain this ignorance. Their friendliness is limited to making helpful information more accessible and to casting the operations of the system in familiar terms and concepts.

---

*The authors wish to thank Henry Mendenhall, Jim Stokes, and John Smith for their effort towards the implementation of these ideas. Julie Hopson and Wayne Zachary have provided support and criticism essential to the direction and execution of the work.

The focus of our research and development is not to develop friendly human–computer systems per se. Rather, the goal is to contribute to the development of cooperative human–computer problem-solving systems. Cooperative problem solving, whether between persons or machines, involves several prerequisites:

- The problem-solving task must be decomposable into subproblems.
- A problem-solving method that is congruent with the problem decomposition must be identified or developed.
- The available problem-solving agents must be assigned tasks and roles in the problem-solving method.
- A language must be available that allows the agents to pass and receive problem relevant information.
- Communication paths between the agents, as required by the task assignment, must be established.
- Messages must be capable of being passed and responded to unambiguously and in synchrony with the problem-solving process.

Viewed in this way, person–machine problem-solving systems are simply one example of distributed problem solving systems. However, there are unique features of a human–computer system that require us to view it as a very special exemplar. It is special because one agent, the person, possesses a relatively fixed information-processing architecture. This architecture has definite limits on working memory, on the kind of processing that can be done either serially or in parallel, on the rate at which different kinds of processing can take place. It has particular patterns of long-term memory storage and retrieval, and its performance can be degraded by factors such as fatigue and stress.

Because of these relatively fixed features of the human hardware or wetware, as designers of person–machine systems we cannot reconfigure the human component in the same sense that we might be able to reconfigure the machine components. Further, there will always be a degree of uncertainty and imprecision associated with our knowledge of the human architecture. Our knowledge of its properties is indirect and rests upon the results of empirical study and scientific theory. (Some may claim that we aren't actually that much better off with a complex computer system.)

Despite these fixed properties of the human architecture, we also know that humans can adapt this basic processing power to a wide variety of problem situations in a flexible and robust manner. Experts often acquire a variety of strategies for solving particular problems. And the strategies themselves may change as a result of new knowledge and experience. What we can be sure of is that, whatever the strategy employed by the human expert, it is one that has evolved in response to both the constraints imposed by the problem itself and by the processing architecture of the expert.

It is important to the designer to know the type of problem-solving method employed by the human expert, since any cooperative problem-solving system will require the coupling of machine problem-solving methods with the expert's method. Again, precise knowledge of the human expert's method is difficult to obtain. We can observe the operation of the method, and at times the expert may be able to provide useful information about the strategy that is being used. But, finally, our knowledge of the expert's method will rest on empirical study and scientific theory.

Any particular machine has many limitations that are analogous to those that have been outlined for the human agent. but the machines of today are extensible and fast enough such that, in most situations, we can create from the machine's computing power a "virtual machine" that has whatever properties we can imagine with enough precision to implement on that machine. Consequently, we argue that, at least in the early stages of the design process, we can largely ignore the limitations that the machine places on the design of a cooperative human–machine problem-solving system.

The need to know, at least at some level of approximation, how the person can or does perform some problem-solving task stems from the major design principle that we would like to apply to the creation of person–computer problem-solving systems.

This principle is to assign to the participants, person, and machine information processing roles that are congruent with their information-processing capacities, a "to each according to its ability" principle. In the case of the person, this resolves to a problem of determining the information-processing capacities available for the problem-solving task. In the case of the machine, this involves defining and implementing the desired problem-solving method on the machine.

DESIGN METHODOLOGY

Overview

It is one thing to argue for the desirability of the goal of building cooperative person–machine systems, and quite another to develop a research and development strategy for moving toward this goal. The development of such a methodology is particularly important for two reasons. First, the tasks for which such person–machine systems will be constructed will typically be complex tasks. Second, these tasks will not be will understood. If a task is well defined and understood, then there may be little reason to retain the person in the problem-solving process. Or, if the task is one which the person can do efficiently, safely, and well, then there may be little motivation to design a machine to aid in this task. What this entails is that those optimistic enough to attempt to design such cooperative person–machines system will typically be concerned with problem-solving tasks:

- which are ill-defined;
- where algorithms relevant to the task are unknown;
- where little is explicitly known concerning how the human expert solves the task.

As researchers who were optimistic enough to begin work on such a project, probably the most interesting information that we can convey is an account of the research and development methodology that has emerged from our efforts. We will first sketch the methodology and comment on its rationale. Then, the majority of this chapter will simply illustrate the application of this strategy within a specific problem domain in which we have been working.

The specific problem that we are concerned with involves an existing person–machine system which attempts to locate a vehicle, determine its current speed and direction, and identify its current plan of travel with sufficient accuracy to support a prediction as to its course and location in the very near future. The vehicle involved is a submarine, which is designed and operated in a manner that makes the achievement of this goal rather difficult. For a variety of reasons, we consider only some aspects of this overall task and discuss the task in a somewhat idealized fashion.

Steps

Table 1 provides a summary of our research and development strategy for designing cooperative person-machine systems. Each step represents a subgoal in the overall design methodology.

The first two steps in our research and development strategy are the most problematic. Both steps involve discovery. the first step requires discovery of an appropriate model of problem solving within which to formulate the specific problem of interest. The second involves discovery of the particular problem solving method or methods implicit in the problem solving activities of the expert. These steps are crucial to our ability to carry out the remaining steps shown in Table 1. However, the specific problem of interest will usually be complex. That is, the problem may often involve reasoning in differing but related problem spaces, and a variety of different but integrated problem-solving methods may be utilized in the overall problem solving activity. Further, the specific problem will generally not be explicitly defined, but rather only implicitly defined within the subculture of the experts who carry out this problem-solving activity and train others to become experts.

Step 7, looping back through the first six steps, represents an explicit

Table 1. A Design Methodology for Cooperative Person–Machine Systems.

1. Formulate the general problem.
2. Construct a first-order theory of the human expert's problem-solving method.
3. Identify a subproblem difficult for the human expert.
4. Design and implement a problem-solving process for the identified subproblem.
5. Integrate the implemented process into the overall problem-solving method.
6. Evaluate the impact of the redefined system on overall problem-solving performance.
7. Loop.

recognition that, due to the difficulty of the design task, the research and development strategy is typically one that involves successive refinement of the solutions for each step. The next step or subgoal is attacked when the previous steps have a partial solution that supports further refinement of the solution to the subsequent steps.

The initial subgoal of obtaining a formulation of the problem involves discovering the essential features that characterize the specific problem of interest. Problem formulation is achieved when the problem has been explicitly represented within an appropriate problem-solving model. Applied mathematics, logic, computer science, particularly artificial intelligence, are some of the disciplines concerned with defining and investigating problem-solving models. A problem-solving model can be thought of as consisting of explicit statements which specify:

- the conditions under which a model is applicable to a problem,
- what constitutes a solution to a problem,
- a method for arriving at a solution.

Additionally, in some cases it may be possible to prove certain statements about the model. For example, it may be possible to prove that, if the problem satisfies the applicability conditions, then the method will find the optimal solution.

It will rarely be the case that this subgoal can be fully achieved for complex real world problems. What must be achieved before proceeding to the next subgoal is a partial formulation that specifies:

- the goal of the problem-solving process or a test that can be used to recognize a solution,
- at least one way in which to decompose this goal into an AND tree of subgoals.

There may be many plausible decompositions and the more that can be identified, the better.

Achieving this level of problem formulation is crucial for two reasons. First, a decomposition provides a basis for considering possible ways in which to distribute the problem-solving tasks among the participants—person and machine. Second, it will provide a basis for comparing the logically possible decompositions of the problem with the hypothesized decomposition that is developed in step 2 to describe the problem-solving model of the expert's performance. This comparison significantly contrains the solution to other steps in this strategy. First, generally it is desirable to choose for further development that problem decomposition that is similar to or identical with that used by the expert. This will facilitate finding a solution to step 5; namely, integrating the problem-solving activities of the person and the machine into the overall problem-solving method.

Second, note that step 3 involves identifying a subgoal in the expert's problem-solving method that is difficult for the human expert, and assigning

There is an additional possible useful result of performing this step. Although there may be no single known problem solving model that uniformly applies to the problem of interest, it may be possible to identify specific subgoals for which a known problem-solving model is appropriate. If this model is strong in the sense developed by Newell (1969), then this constitutes a good candidate for assignment to the machine.

To determine the information processing capacities of the person, a first order theory of an expert's problem-solving method is formulated in step 2. In many cases, results from AI and cognitive science may suggest the outline of a theory or model of expert performance which explicitly identifies the problem constraints and information-processing constraints to which the expert's problem-solving method is adapted. Of course, these hypotheses must be supported and often refined, with respect to the problem under study, by empirical test.

A theory of expert performance provides a basis for identifying aspects of the method that may strain or exceed the expert's information-processing capabilities. These capacities or resources fall into three general categories: working memory, input/output channel capacity, and knowledge. Step 3 involves identifying an aspect of the problem-solving method used by the expert that is difficult for the human expert. The model of the expert's problem-solving provides the basis for designing ways to experimentally assess the expert's performance to determine if the hypothesis about a particular resource limitation does negatively affect the expert's problem solving. If the hypothesis is empirically verified, then this subproblem constitutes a candidate for assignment to the machine. Often the empirical results will also provide a basis for specifying under what conditions the expert's performance is affected and for estimating how seriously the expert's performance is degraded.

expert's performance is affected and for estimating how seriously the expert's performance is degraded.

Having chosen a subproblem for assignment to the machine, an appropriate problem solving model must be identified and implemented in Step 4. Here it is necessary to completely satisfy this subgoal. However, the details of the implementation will often be affected by the next major task, step 5.

The fifth step involves specifying one or more ways in which to cooperatively link the problem-solving activities of person and machine. These linkages are required at all points where it is necessary or desirable for one of the agents to have information about the problem solving state or result of the other agent. At these points, a plan must be created that achieves the necessary information exchange. These plans are termed *communicative plans*, because the goal is to transfer or receive information from the other agent. Such plans are required since neither process can be given direct access to the problem-solving state of the other.

There are two dimensions that define the nature of the dependence between two separate processes. The first may be termed informational dependence. Two types of information about one process may influence another process. One type of information is the output or result computed by a process. The other is information about some aspect of the problem solving state of a process. Process A is informationally dependent on process B if information about the state or result of process B influences the computation carried out by process A. Process A is influenced by the information provided by B if either the result computed by A is affected by B or if the computation carried out by A is influenced by the information provided by B. For example, A may computer its result in fewer steps due to the information provided by B. Note also that information dependence is not a symmetrical relation.

Two strengths of informational dependence can be distinguished. Process A is strongly dependent upon process B if A cannot begin its computation until it has information about the state or result of process B. If process A can continue computation without information from process B, then process A is weakly dependent on process B. If both the problem-solving state and result of process A are unaffected by information from some other process B, then process A is informationally independent of process B.

If one process is informationally dependent upon another, then there is a second dimension that further defines the relation between the two processes. This is a temporal dimension. If process A is informationally dependent upon process B, then process A is fully temporally coordinated with process B if the information from B is always available at or before the time when process A requires that information. If A may have to suspend computation until the information is made available by B, then A is not fully temporally coordinated with B. Note further that, if process B provides information to A before it is required by A, then A will require some buffer to hold this information until it is needed.

Whenever there is a dependence in any of the above senses, it is necessary to specify some rules to govern the interaction between the two processes. The pattern of these rules in relation to the problem-solving activity define the nature of the role or roles that a process plays in the overall problem-solving process. We will call these rules *normative rules* to convey the sense that the rules assign responsibilities to the problem-solving agents. A normative rule can be thought of as a three-part rule which provides a declarative representation of the knowledge needed to decide *when, what,* and *how* some information should be communicated to another process. The first part of the rule specifies when a norm is applicable. The applicability conditions will typically refer to the problem solving state of the problem-solving component to which the norm is assigned and usually also to the problem-solving state of the intended recipient of the communication. One process, knowing that it has ruled out a hypothesis and knowing that another process is still actively considering that hypothesis, is an example of a possible conjunctive set of applicability conditions.

The second part of the normative rule specifies the intended effect of a communication of some information to another process. This intended effect may be to simply inform, or explain, or persuade, and so on. The third part of the rule specifies whatever conventions, if any, have been established between communicator and recipient to accomplish the communicative goal specified in part two of the rule. Viewed in another way, a normative rule provides an abstract specification of aspects of the communicative plan. Part 1 and 2 of the rule specifies the conditional goal of the communicative plan. Part 3 specifies aspects of the action sequence to follow in achieving the communicative goal.

Intentional relations such as *believes* and *knows* are used to specify the applicability conditions, and speech acts such as *inform, request, explain* are used to specify the intended goal of the normative rule. This intentional language is used for two reasons. First, it is used as an abstract specification language within which the various patterns of communication and control implied by a set of norms defined over a set of problem solvers can be examined. For example, an obligation to request information from another process under certain conditions requires that a norm be specified that requires the other process to be responsive to the request. Secondly, the use of this intentional language highlights the fact that person and machine are distinct entities neither of which can directly access the problem solving state of the other. Rather, the designer must be concerned with providing the computational strategies and resources required by the problem solvers to meet and carry out the obligations that the role assignment places upon them.

The third part of the rule, the conventions, provides a way to fully or partially specify the means by which the obligation should be fulfilled. Knowledge of conventions must, of course, be shared. Thus, whenever fixed conventions are adopted, both communicator and recipient must share knowledge of these conventions.

Cooperative problem solving involves both a plan for doing the problem and a plan for synchronizing and integrating the efforts of the problem solvers. The specification of norms which are intended to govern the interaction of the problem-solving entities provides a language within which to begin to express the responsibilities that each entity must meet in order to achieve the desired cooperation. Each norm carries with it a responsibility to monitor for the conditions associated with the norm and the responsibility to determine how to achieve the communicative goal associated with that norm. The desired solution is one that avoids additional responsibility for the person where meeting the responsibility would exceed a resource limitation. For example, determining and encoding the proper communication to achieve cooperation with the machine may also place requirements on the same working memory limit that led to the design of a machine agent. The inability of the person to affect the machine within resource and problem constraints (e.g., memory capacity and time criticality) results in a person–machine system that may at best reach the unaided expert's level of performance.

Even if we follow this rational design method, as designers our knowledge is likely to be incomplete; thus, many of the implications of the previous design steps will not be foreseeable. Therefore, in Step 6, the impact of the redefined person–machine system on overall problem-solving performance must be evaluated. The evaluation proceeds from a comparison of expert alone or baseline person–machine system performance with performance of the cooperative person–machine system. Having followed this design method, we have several advantages over the typical application of this classical design step. First, the problem formulation, first-order theory, and role assignment steps provide detailed expectations about how and where performance should be enhanced. Such expectations allow us to focus the evaluation experiments on specific aspects of performance. Second, this theoretical framework provides the basis for systematically varying aspects of the problem and the design to determine the robustness of the performance enhancement. Third, where our expectations are not met, the framework and systematic experimentation should isolate the area of the design that requires further consideration. Finally, the results of experiments conducted to refine and test the first order theory as well as those performed to identify subproblems difficult for the expert may serve as baseline performance estimates.

Steps 1 through 6 represent a rationally guided search along one or a few paths in a larger design space of possible cooperative person–machine systems for the problem solving task. Even though we may have expended considerable effort to identify and select the area of greatest potential enhancement, other subproblems may be difficult for the human expert. By looping back through steps 1 to 6, we may incrementally produce additional enhancements to overall problem-solving effectiveness. We would expect that, on the average, successive iterations would inherit knowledge, expertise, tools, etc., developed in previous passes. Thus the cost or value of applying this design strategy should take into account a design history of some length. The illustration of the application of this strategy described in the following sections corresponds to an initial pass through steps 1 to 5.

PROBLEM FORMULATION

The Interpretation Task

The specific person–machine system on which we have focused is concerned with locating and tracking the movement of an intelligent agent, a submarine, over a spatial domain, some area of the ocean. This task is typical of the genre of person–machine systems that have been developed in response to the ability to distribute remote sensors that can detect the presence of types of objects within a region. The expert who pursues this interpretation task commands an aircraft with a crew of human specialists as well as various machine systems for sensor interpretation, sensor deployment, etc.

In order to systematically study this problem in the laboratory while retaining many of the real-time complexities, we have constructed a real-time simulation of the expert's environment. The *problem testbed* emulates the currently operational tactical workstation as well as the outputs of other system components such as sensor operators. Recording capabilities built into the testbed permit the capturing of simulation events and actions taken by an expert for replay or analysis. The testbed is well documented elsewhere (Deimler et al., 1983) and will not be presented in detail here.

Airborne submarine localization and tracking is a particularly difficult task because only a small fraction of the environment in which the agent moves can be sensed at any one time. Although a variety of sensors are used, and the quality of the information varies with conditions, we focus on an idealized worst-case. That is, a worst-case from the point of view of the human expert's processing limitations. Two properties of this problem contribute to a worst case.

First, a particular sensor can only resolve the location of the vehicle to a set of spatially discontinuous sets of locations. These sets are of two types which we will term *disks* and *rings*. Figure 1 graphically displays these sensed regions. The circular area immediately surrounding a sensor, indicated by the letter *S* and a number, represents a disk. The area concentric to the disk, indicated by solid lines, represents a ring. Each of these sets, the disk or ring, may contain regions, sets of locations which arise from the deployment of multiple sensors. One of the regions of the disk or ring may currently contain the vehicle when one or more sensors are in contact. Thus a signal from a single sensor is consistent with a very large disjunctive set of possible locations. Referring to Figure 1, a signal from sensor S3 is consistent with the vehicle being located within any one of five regions labeled 1 to 5. The circle intersections represent regions that are sensed by more than one sensor. Note that these intersections of sensed regions further subdivide the sets of possible locations from the point of view of sensor resolution. In this example, the disks represent an area whose diameter is 12 miles, the rings are located at a radius of 30 miles from each

Figure 1. Aspects of the Localization and Tracking Problem.

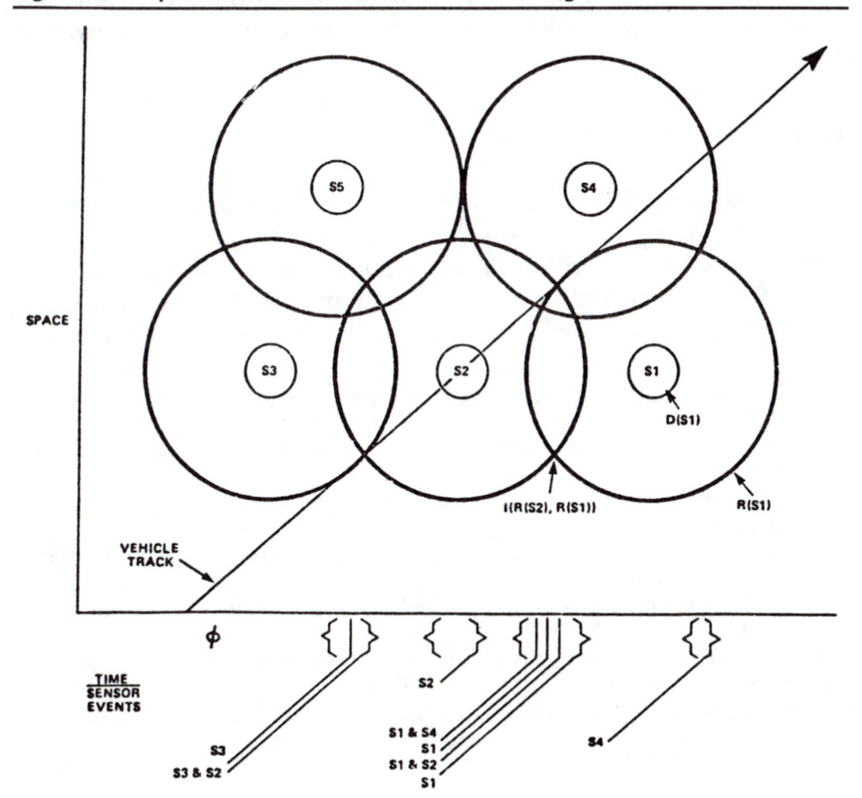

sensor and each ring has a "thickness" of 3 miles. Thus within each "sensed" and "not-sensed" region (e.g., region 7 in Figure 1) there are many possible vehicle locations.

The second difficulty in this task arises due to the fact that, most of the time, no sensor will be in contact with the vehicle. Consequently, even the extremely uncertain information about location that is provided by a sensor contact is typically only available for a small proportion of the total observation time. This is represented in Figure 1 by the large segments of the vehicle track that do not intersect a sensed area. Thus, the event that involves no signal from any sensor may be consistent with a disjunctive set of possible locations, e.g., regions labeled 6 through 9, each of which contains a large number of locations one of which may be occupied by the vehicle.

The expert's task is to locate the vehicle, determine its current speed and direction of movement, and identify its current plan of movement with sufficient accuracy to support a prediction of location and course. The expert has a limited number of sensors to deploy and they may be deployed at any time during the task. The sensors have a limited lifetime and may cease to operate at various points during the task. Most importantly, the sensors typically provide little information about speed or direction of movement.

The Initial Person–machine System

Properties of the initial person–machine system, the emulated workstation, are presented in Table 2.

Sensor location, gain or loss of sensor contact, and aircraft location and direction are automatically and dynamically presented to the expert via updates of a graphic display representing a two dimensional area of ocean

Table 2. Aspects of the Initial Person–Machine System.

1. Sensor location is communicated to the expert via screen display (distinct icons on a 2-D surface).
2. Sensing events are communicated to the expert via screen display (by "highlighting" sensor icons).
3. Expert can retain previous or predicted events by marking the display (with a set of "chalkboard" functions).
4. Sensor deployment is requested of the machine via screen display (by creating and placing icons).
5. Aircraft location and direction of movement is communicated via screen display (by rapidly refreshing the position and orientation of icons).
6. Sensed areas are only implicitly available via screen display (disks, rings, and regions are created by the expert using chalkboard functions).

surface. The expert controls aircraft movement and the deployment of sensors by creating and positioning symbols in the display area. An important feature of the initial system is that the sensed areas are only implicitly represented. That is, the expert can "draw" the disks and rings which define the disjunctive set of possible locations, but they are not provided automatically and the machine has no explicit knowledge of them. Observing expert performance on testbed problems was a key source of information about the task and fed directly into our attempts at problem formulation. In addition, an analysis of the constraints on vehicle movement suggested the ways in which the interpretation task might be decomposed.

Constraints on Vehicle Movement

The four properties of vehicle movement presented in Table 3 were identified as sources of constraint that might be exploited by an interpretation system, either person or machine.

The first three constraints arise from the physics of the situation and the operational characteristics of the vehicle, while the last constraint arises from the intents or tactical goals of the agent controlling the vehicle. The first constraint refers to the macrophysical property that an object can only be at one place at a time. When motion is added, there is continuity in space and time. A vehicle cannot "jump" instantaneously between spatially discontinuous regions. The conservation of momentum restricts the range of possible next locations. The vehicle cannot "backtrack" in the next instant of time, but must gradually turn or stop and reverse. Finally, the tactical plan under execution by the vehicle limits the possible next location. This intentional constraint is somewhat softer than the others since the agent controlling the vehicle can modify or radically alter the plan at any time.

In our example, there are two classes of tactical plan that have quite different implications for patterns of vehicle movement. If the vehicle is pursuing a transit plan, the objective is to get from one place to another,

Table 3. Types of Constraints on Vehicle Movement.

1. A vehicle can occupy only one location at any point in time.
2. A vehicle moves in a spatially continuous fashion.
3. Current speed and direction of the vehicle constrain the possible next locations of the vehicle.
4. The tactical plan pursued by the vehicle constrains the possible next location of the vehicle.

usually with time constraints. In this case, the pattern of movement or next locations should be generally monotonic towards the destination. At least there should be no reversals where the vehicle closes on the point of departure. This "rationality" assumption does not apply to on-station movement plans. Here the object is to remain within a certain region and move in a fashion that makes detection and tracking difficult.

Decomposition

Our problem is an instance of what has been termed, in AI, a plan recognition task (Schmidt, Sridharan, & Goodson, 1978). In this particular case, the indirectly sensed movements of the agent being observed are part of a plan to transit from one location to an other or to remain within a general area. Usually, the goal includes the constraint that detection is to be avoided. The goal of our expert's real-time plan recognition task is to know the location of the vehicle at some time *t*, in the future where of course *t* is bounded. Our previous work on plan recognition problems (Schmidt et al., 1978; Sridharan, Schmidt, & Goodson, 1980; Sridharan, Bresina, & Schmidt, 1983) together with the analysis of constraints on vehicle movement suggested the problem decomposition summarized in Table 4.

The decomposition of the overall goal, to know the location and intended movement of the vehicle, is a goal refinement yielding three related subgoals and their corresponding search spaces. One or more of the constraints discussed in the previous section underly the definition of each logical space. Constraints 1 and 2 from Table 4 together with sensor resolution and deployment pattern, serve to logically define Space 0. This

Table 4. Decomposition of the Plan Recognition Task.

Search Space 0		
	Goal:	Know the general location of vehicle at time *t* and Allocate sensing resources such that the general location can be known at $t + 1$
	Space:	Paths of contiguous regions, defined by sensor deployment and resolution (and consistent with sensor events)
Search Space 1		
	Goal:	Know the general course and speed of vehicle over $t \ldots t + n$
	Space:	Point paths, tracks, with associated speeds (consistent with sensor events)
Search Space 2		
	Goal:	Know accurately the current tactical course plan of the vehicle
	Space:	2a—transit plans (consistent with sensor events)
	Space:	2b—on-station plans (consistent with sensor events)

space consists of alternate region paths which may start in any region and continue with any adjacent region up to an arbitrary length. Clearly, the number of logical possibilities is quite large for typical problems where 10 or more sensors are deployed with some degree of overlap in their coverage. The logical space can be constrained by taking into account observational history, the sequence of sensor events. The initial sensor event defines the permissible start points; namely, those regions that are sensed by the sensor in contact. Subsequent observations define the legal continuations of paths. Even in this reduced space, which we shall refer to as Space 0, the number and length of alternate paths can grow quite large. Also recall that the logical and thus the reduced space can change radically given the dynamic deployment and expiration of sensors.

Let's return to Figure 1 for an example of the reduced space. At the initial sensor event, contact on sensor S3, there are five initial paths, of length one, corresponding to the five regions consistent with an S3 signal. At the next sensor event, contact on S3 and S2, there are four different paths of length two, corresponding to movement from segments of the S3 ring (regions 1, 3, 4, and 5) into adjacent regions defined by the intersection of the S3 and S2 rings. Note that the path originating in the S3 disk is no longer viable, since any continuation of this path would violate the observational constraint.

The logically possible vehicle tracks that comprise Space 1 must fall within the region paths of Space 0. A track is a course and speed refinement of a region path. Consequently, there will be many tracks consistent with each region path. The one-to-many mapping from a region path to tracks suggests that developing tracks must be pursued in a highly selective manner in order to be tractable for either person or machine.

Space 2, as we have formulated it, consists of the logically possible movement plans that the vehicle might be executing. There are two general types of movement plans underlying submarine movement in our problem. When executing a transit plan, the goal of the vehicle is to reach a specific location, usually under time constraints. If we ignore deception, these plans are generally characterized by monotonic movement, away from the origin and towards the destination. While on-station, the vehicle has the goal of remaining within a specified area. In both cases, plans typically include the constraint that detection should be avoided. Consequently, frequent changes in course and speed are typical.

Usually the observer has scant knowledge of the vehicle's goal and plan. Thus Space 2 only weakly contrains the possible tracks or region paths. In addition, the actual plan may not remain constant. The agent controlling the vehicle can modify the plan at any time. Because the input information is so incomplete, a data-driven plan recognition strategy is difficult to follow even in the best of circumstances, especially in Space 1 or 2. Unfortunately, a data-driven control regime is often desirable over long periods of the task. In order to be used effectively, such a strategy requires maintaining a relatively large set of disjunctive hypotheses and an accurate history of a portion of the previous sensor contacts. The size of the memory for previous sensor contacts is not fixed but depends upon the particular sensor distribution and sensor contacts. Both of these requirements can be fulfilled only if a large working memory is available.

FIRST ORDER THEORY

Research in human reasoning in general (Newell & Simon, 1972), and human plan recognition in particular (cf. Schmidt, 1976, 1984; Schmidt et al., 1976), suggest that human performance deteriorates rapidly for any task that requires a large working memory. This suggested that our initial design should focus on how to assign to the machine aspects of the task that require a large working memory. In order to confirm our earlier results in understanding human plan recognition and to refine the theory to this particular task, experiments involving human experts were conducted.

Expected Properties of Human Plan Recognition

Table 5 presents the properties of human plan recognition, based on our first order theory, that were expected to characterize expert performance in the vehicle tracking task.

Assuming that these predictions from the first order theory generalize to this vehicle tracking task, we expected experts to follow the same strategy, termed *hypothesize and revise*, that characterizes human plan recognition in common sense domains. A central property of this process is that a very few, usually one, hypotheses are developed and maintained despite the fact that the observations are consistent with a large number of alternate hypotheses. Current results and theory suggest that this property arises from a fixed and quite limited working memory. Persons appear to have adapted by developing the hypothesis at a level of detail that is consistent

Table 5. Expected Properties of Expert Plan Recognition.

1. The number of alternative hypotheses (plans, associated tracks and region paths) considered are few, usually one or two.
2. Observed events are interpreted using the current hypothesized plans that the observer believes the actor might be pursuing.
3. The plan recognition process is predictive.
4. Predictions invalidated by observations trigger a revision process.

with many possibilities that might be observed next, but that is still specific enough to allow prediction of what should be observed next and thus to allow matching given what is observed. Given a mismatch between prediction and observation, a person attempts to revise their hypothesis to keep it in line with the observations seen so far.

Refining these theoretical expectations in terms of the vehicle tracking task, we expected that experts would develop one or a few hypothesized paths at the level of Space 0 or 1. Since it was not likely that they could exhaustively maintain much less explore these spaces, they could be wrong. That is, their hypotheses could be incorrect relative to the true situation and could even violate the physical constraints operative in the problem. We expected that the experts would make predictions, based on their current hypotheses, about the next sensor event or possible vehicle location. When these expectations were not met, the experts should attempt to revise their hypothesis and bring it in line with the observations or their memory of the observations.

Experiments with Human Experts

We adapted to the vehicle tracking task an experimental paradigm developed in our earlier work on human plan recognition (cf. Schmidt et al., 1976, 1978; Schmidt, 1976). In this case, an observable feature of an actor's behavior is a sensor event, the gain or loss of contact by one or more sensors, through time. The expert is presented, in real-time, a sequence of contacts to be interpreted. At each observation, the expert is asked to communicate the current hypotheses and any rationale for holding the hypotheses. In the vehicle tracking task, at each sensor event (gain or loss of contact) the expert was asked to communicate any current hypotheses about vehicle location, direction of motion, speed, and tactical plan. This repeated request encouraged, if anything, a more exhaustive and analytic approach to the problem than that typically adopted by the experts.

Procedure. The sequence of sensor events was presented via the testbed in the form of a real-time localization and tracking problem. The expert could indicate a hypothesis by placing a reference mark on the dynamic display showing sensor deployment, sensor contact and aircraft position. In addition, the expert could utilize any of the standard drawing and calculation functions available at the workstation (e.g., functions for calculating bearings and distances in the tactical display). At each point, the expert was asked to connect any of the current hypotheses with previous hypotheses to indicate the continuity of hypotheses through history. All of the expert's display manipulation activity was recorded, relative to the state of the problem simulation, using the recording/playback facility provided by the testbed. Since the workstation provides a fairly restricted vocabulary for communicating responses, the expert was asked to verbally annotate display based responses and to provide a running commentary on the problem-solving activity. For the most part, the experts' beliefs about vehicle speed, the relative goodness of hypotheses, and when hypotheses should be revised, were communicated verbally. Verbal responses were recorded using tape recording equipment.

The problem presented to experts is summarized in Figure 2. The sensor field deployed at the initiation of the problem is partially depicted to enhance this presentation. Besides the six sensors shown here, several more extended above sensors 14 and 15, and to the right of sensors 17, 18, and 19 following the same regular spacing and thus intersection pattern. The black dot marks the location of the vehicle at the start of the problem and

Figure 2. Sensor pattern and Vehicle Track-Experiment 1.

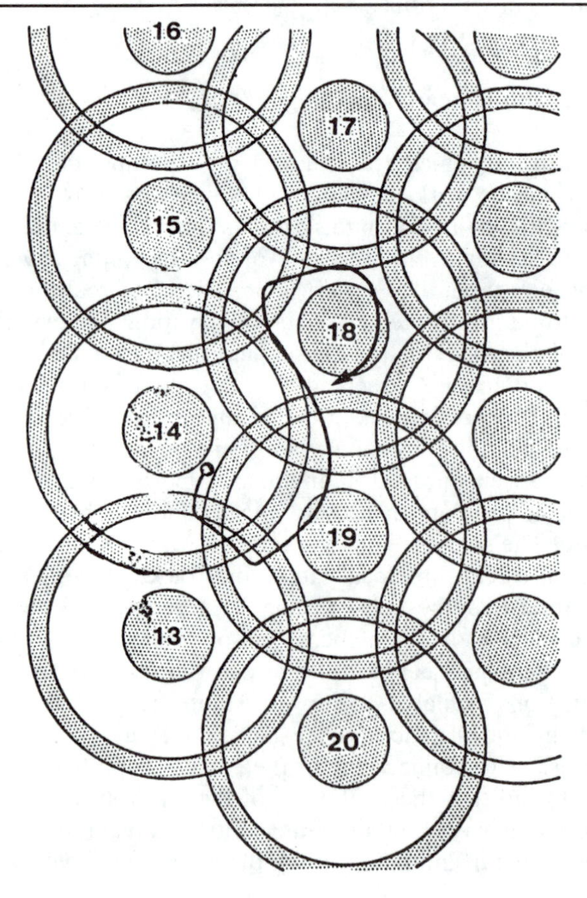

the arrow marks its final location at the termination of the experiment. The vehicle moved along the indicated path at a constant speed of 12 nautical miles per hour. Where the path crosses unshaded area, no sensor was in contact. Where the path crosses a shaded area, one sensor was in contact. Note that ring intersections were avoided in order to maximize the ambiguity of location suggested by any one contact event. The ring-overlaps and *S*-like form of the path further heighten the ambiguity of location and direction of motion. This path generated 24 sensor events of which half were contacts on either sensors 13, 14, 15, 18, or 19, and half of the events were absence of contact. The experiment lasted approximately 2 hours.

Two experts with operational experience in this type of problem participated in the experiment. Their display-mediated and verbal protocols were represented in a variety of ways. Due to the complexity and extent of such problem-solving protocols, only selected aspects of the data will be presented and discussed here. The experts' hypotheses were at times stated as specific regions within which the submarine was thought to be without any resolvable reference to path, speed, or direction of the sub's movements. Most of the time, the expert provided information which indicated that his hypotheses referred to a specific path or paths. For both of these cases, we determined the region or regions that the expert believed might hold the submarine during that event. Figures 3 and 4 graphically depict the results of viewing the data in this way. Figure 3 presents the data of expert 1, and Figure 4 that of expert 2.

In these figures, the x-axis lists the sequence of sensor events. The y-axis represents the number of distinct regions associated with a sensor event. Three different types of association of regions with a sensor event are depicted through the use of different shadings of the bars in these figures. First, there is simply the number of distinct regions consistent with a sensor event considered independently of the requirement that a region must be contiguous to a previously possible region. This is shown to provide an upper bound on what regions could be identified if the event were considered independent of its historical context. Recall that there were 68 regions that could not be heard by any sensor. Consequently, for sensor events which involve absence of contact, these shaded bars all extend to 68 on the y-axis. For sensor events which involve sensor contact, the upper bound varies with the deployment pattern.

The second association of a region with a sensor event involves determining those regions that are consistent with the sensor history—that is, obey the region contiguity constraint. This is the set of regions identified by our Space 0 process. It is, of course, always a proper subset of the regions associated with an event when the contiguity constraint is ignored. The number of these contiguity-constrained regions is indicated by the bars

Figure 3. Regions of possible location and Expert's Location Hypotheses-Experiment 1, Expert 1.

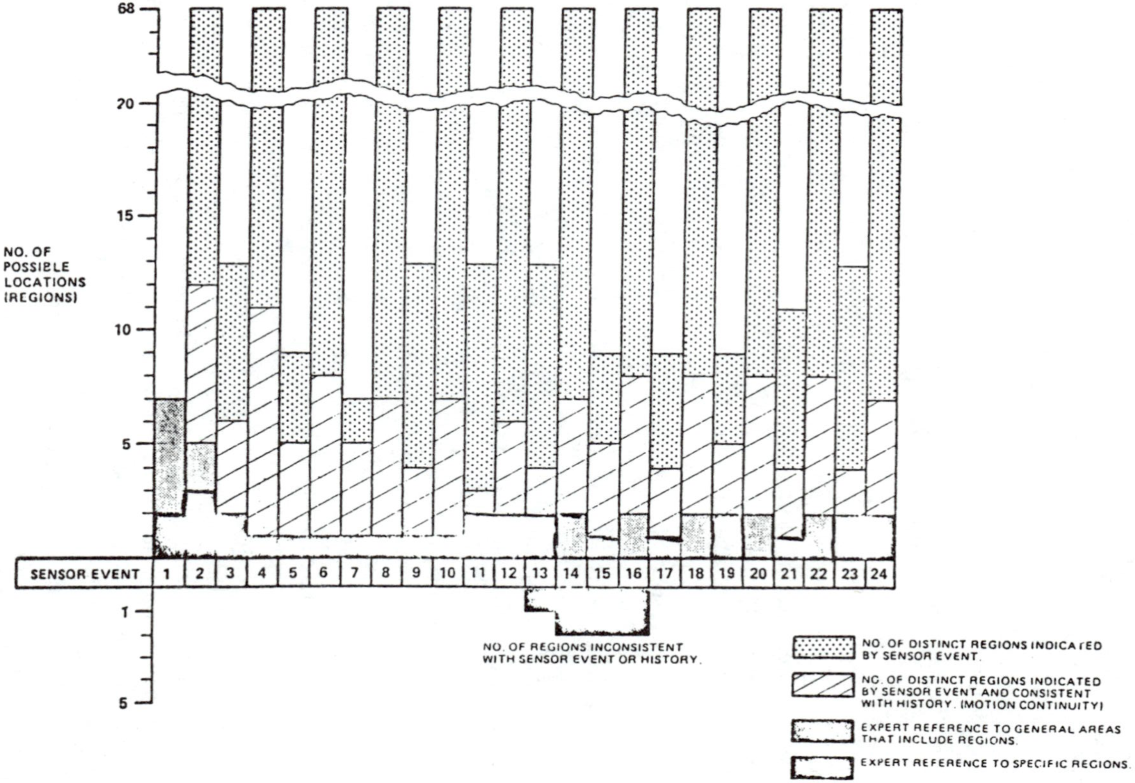

Figure 4. Regions of possible location and Expert's Location Hypotheses-Experiment 1, Expert 2.

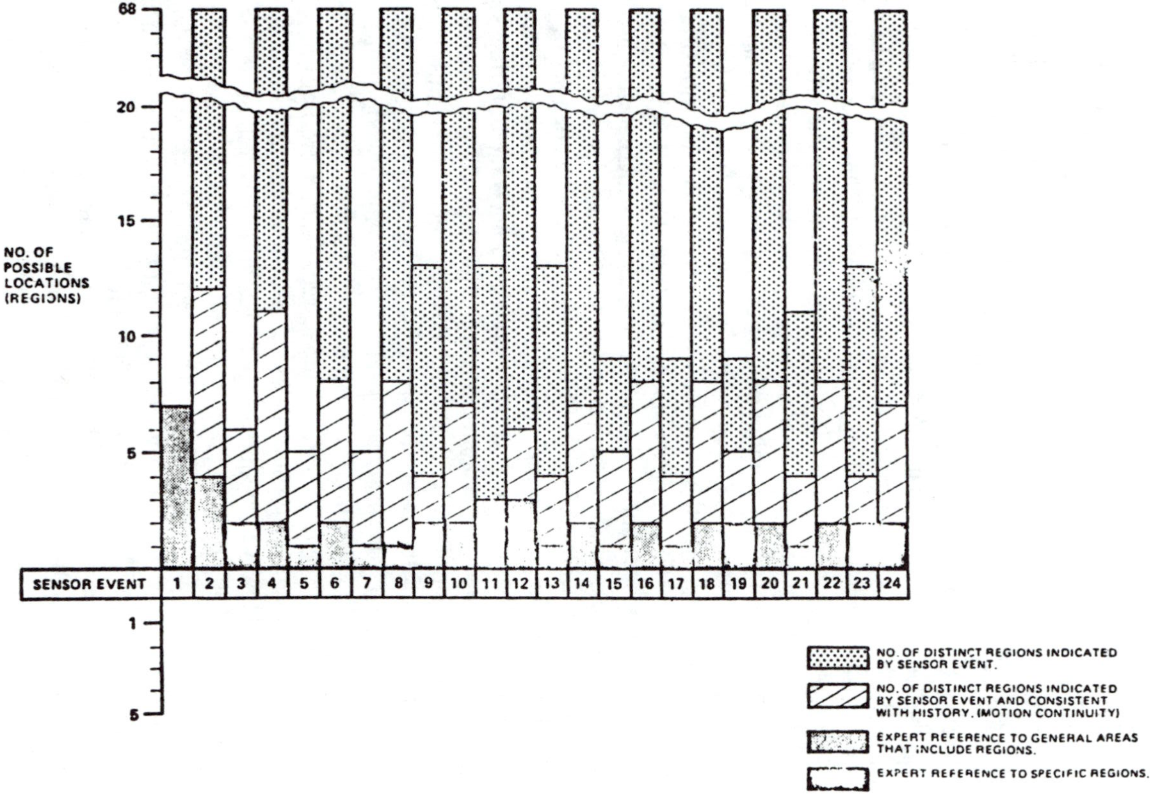

with slightly darker shading. For the first event of the contiguity constraint does not delimit the possible regions and these two sets of regions are identical.

The number of distinct regions considered by the expert during a sensor event is depicted in relation to these bounds. This is the third type of association between a region and a sensor event. If the expert referred to a specific regions by graphically creating reference marks in the display, then this is depicted as a bar with the darkest shading. In his verbal comments, often the expert would refer to a region as being within a ring or disk of a sensor rather than referring more specifically to the distinct region or regions within that ring or disk. Such a reference was counted as referring to all regions within the ring or disk mentioned. This case is depicted by slightly lighter shading than that used to indicate specific reference. Note that this rule for interpreting the expert's reference to a general area yields an upper bound on the number of specific region hypotheses that he might have been entertaining.

If the expert's hypotheses concerning current region location area always consistent with the contiguity constraint, then the regions he refers to will always be a proper subset of these contiguity constrained regions. This is the case for expert 2. However, expert 1 at times explicitly referred to regions which were not part of this contiguity constrained set of regions. This case is depicted by the shaded bars that project below the x-axis.

Results. Although the experts exhibited somewhat differing styles of reasoning, there are certain strong commonalities between them. First, both experts consider the full set of possible region locations of the submarine only in response to the first sensor event. For the remainder of the session they consider only a subset of the contiguity-constrained set of possible region locations of the submarine. As noted previously, expert 1 does entertain hypotheses that refer to regions that are not in this contiguity-constrained region set. This occurs in response to sensor events 13 through 16. However, both experts do a good job overall of staying within the set of possibilities consistent with the sensor event history.

However, they consider only small portions of the regions consistent with this history. In fact, if we examine only those cases where a specific region is mentioned, the darkest shading, then we see that, for one-third or more of the sensor events, both experts are considering only one specific region as the possible location of the submarine. More specifically, expert 1 refers to one distinct region on 9 occasions and two distinct regions on 9 occasions. This represents 18 of the 24, or 75% of the possible occasions. Expert 2 refers on 11 occasions to no specific region, on 8 occasions to a single specific region, and on 5 occasions to two specific regions. This completely accounts for the 24 possible occasions. Thus, even at this coarse level of analysis where we are not explicitly considering the paths that the expert believes the submarine is following, we see that the expert typically considers only a very few of the possible region locations of the submarine. And, in fact, the specific regions indicated by the expert rarely correspond to the actual region locations of the submarine in the simulation. Thus, it is not the case that the expert is employing a strategy that allows him to accurately track the submarine. Neither expert knew the correct location or direction of the submarine at the termination of the session.

A more detailed view of the strategy employed by the experts can be obtained by considering the structure of their hypotheses. Their hypotheses usually do not consist simply of a set of current region locations. The hypotheses usually consist of a path of regions, a direction, and speed that the submarine is believed to have followed. This full structure is rarely stated. It is particularly difficult to ascertain to what degree the expert is using previous hypotheses and events to generate or extend his current path hypotheses. When a mistake is made such as that made by expert 1, then it is clear that the implications of previous sensor events have not been used to fully constrain the current hypotheses.

Despite his lack of total success, the first expert attempted to develop and maintain hypotheses that were specific and constrained by the history of the sensor event sequence. As can be seen in Figure 3, this expert explicitly referred only to specific regions on all but three trials. A rough indication of his strategy can be obtained by considering the relation between the number of regions mentioned in response to adjacent sensor events. If the number of regions, j, mentioned in response to event n is less than the number of regions, k, mentioned in response to event $n + 1$, then this indicates a branching of the expert's hypothesis space. If the number of regions, j, mentioned in response to event n is greater than or equal to the regions, k, mentioned in response to event $n + 1$, then this represents no branching. Rather, it signals either a simple continuation of the previous hypotheses or a pruning of the previous hypotheses. Of 23 such pairs, 6 are branching pairs and 17 are nonbranching pairs. All but one of these six branching pairs involves adding a single disjunctive extension to the previous space of hypotheses.

Figure 5 provides the more detailed analysis of this expert's strategy. Here sensor events are again shown on the x-axis. The y-axis, labeled *path length*, indicates the maximum length path that is consistent with the way this expert developed his hypotheses. Two distinct connected boxes are shown. This corresponds to the two totally distinct paths considered by this expert. One of these distinct paths can be traced historically back to his first hypothesis. The other distinct path was introduced in response to sensor event 7. It was considered concurrently with the other path, starting with sensor event 7, and continued until sensor event 14, when this path was abandoned. Within the major path, branches from one box to the next

Figure 5. Hypothesized Vehicle Tracks-Experiment 1, Expert 1.

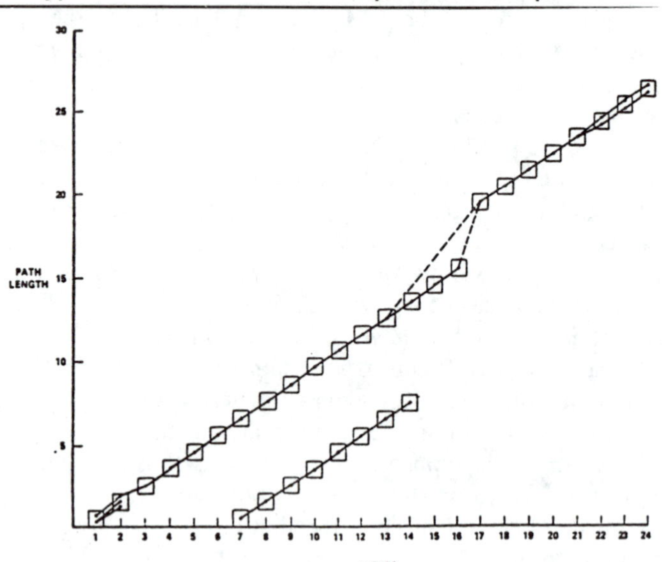

denote an introduction of disjunctive possibilities local to that path. Lines not continued denote a pruning of a disjunct. The dotted lines indicate that the connection of the later part of the main path with the earlier part of that path violates the contiguity constraint. That is, it implies that the submarine was located in regions where it could not possibly have been located.

This examination of the more detailed structure of the hypotheses entertained by expert 1 shows us how few of the possible paths are considered. In the next section, we will present and enumerate the number of possible region paths consistent with the sensor events. Here we simply note that this set of paths starts with 7 of length 1 at sensor event 1, and rapidly increases until there are hundreds of possible paths after event 24.

Expert 2 uses a somewhat different strategy, but the basic result is the same. As can be seen in Figure 4, this expert does not explicitly refer to specific regions on 11 of 24 possible occasions. Rather, when he is uncertain, he often refers to a single spatial region, such as a ring or a quadrant. This represents a way of using a single general spatial term to refer to an implicit set of disjuncts. We have made this implicit set explicit in Figure 4. Thus, his reasoning appears to involve more branching than expert 1. However, even leaving this consideration aside, in 15 of the 23 adjacent pairs of hypotheses, this expert explicitly or implicitly referred to an equal or fewer number of regions after the n + 1st event than after the nth event. In eight cases the converse was true. So again we see that branching in the hypothesis space is less likely to occur than pruning. Further, 5 of the 8 branching cases involve considering an implicit disjunct when a sensor event involving no contact followed one that involved contact with a submarine.

This expert's overall strategy assigned a great deal of weight to the most recent sensor contact. He was less explicitly concerned with using the history of his hypotheses to guide his current reasoning. This more locally dominated strategy, coupled with his extensive use of indirect reference to specific regions, made it difficult to try to reconstruct and display the distinct paths considered by him. Consequently, we do not show a figure for this expert comparable to Figure 5.

Conclusions. The data presented in the previous section, and an analysis of verbal protocols, provide several results and trends that supported our expectations based on the first order theory of plan recognition. First, location hypotheses at the level of Space 0 (regions) are rarely developed. They are considered when the first sensor signal occurs and when the Space 1 (course and speed) and Space 2 (type of movement plan) hypotheses are disconfirmed.

Second, for the most part, a single Space 1 and Space 2 hypothesis is maintained. The experts quickly develop a track hypothesis that assumes constant speed and direction. This combined Space 1 and 2 hypothesis is used to predict what should be observed next, an expectation of the next sensor event, sometimes with an associated time estimate. Note this use of prediction can be viewed as a strong heuristic to limit branching in the space of hypotheses.

Third, the mismatch of observed and predicted events leads to revision of the hypothesis. The revision is based on the previous one or two events. At these revision points, the expert may develop and maintain several alternate Space 1 hypotheses. A major mismatch of prediction and observation occurred for expert 1 at event 17. This necessitated a major revision of his main hypothesis. Figure 6 provides a static reconstruction of the way in which this hypothesis evolved. In this figure the regions actually occupied by the submarine are depicted. Sensed regions through which the submarine passed are denoted by a heavy stippling. Regions with no sensor coverage through which the submarine passed are lightly stippled. White regions are regions never occupied by the submarine. A cumulative representation of the expert's major hypothesis is superimposed on this figure and indicated by the heavy lines. Darkened lines indicate that portion of this hypothesis that was consistent with the sensor history. The dashed lines indicate the points where the hypothesis was inconsistent with the implications of the sensor history that can be derived using the contiguity constraint. Note that the inconsistency occurred at the point where a major

Figure 6. Vehicle Path and Cumulative Hypotheses Experiment 1, Expert 1.

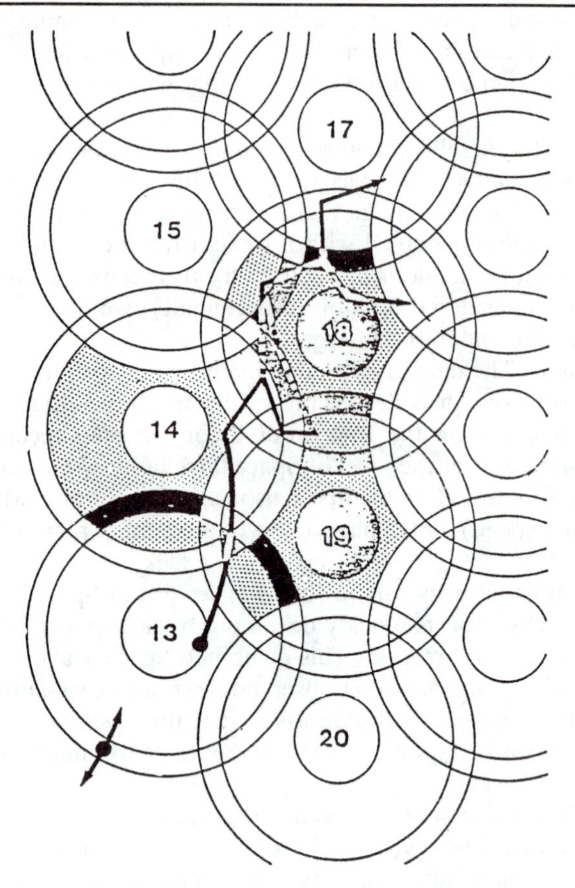

revision was required because of the mismatch between observations and his hypothesis. This graphically shows the difficulty that an expert can encounter using the hypothesize and revise strategy. Such a result is consistent with our assumption that the expert lacks the memory resources required to employ a kind of chronological backtracking revision strategy. Thus, the more or less depth-first strategy for hypothesis development, coupled with the inability to backtrack, can, in this worst case, lead to logically inconsistent revisions.

Reasoning in Space 0, a Difficult Subproblem

The confirmation of our earlier plan recognition results, and the expert's almost exclusive maintenance of a single Space 1 and 2 hypothesis, suggested that reasoning in Space 0 is a difficult subproblem. Several sources contribute to this difficulty. The number of distinct regions is large and only implicitly represented in the current person–machine system. The number of distinct regions changes dynamically with the deployment and expiration of sensors.

A single observed signal is consistent with a disjunctive set of spatially discontinuous hypotheses about the location of the vehicle. The space of location hypotheses can only be pruned if a complete and accurate history of these hypotheses is maintained. In order to reduce the number of location hypotheses by propagating the contiguity constraint, the region possibilities must be constructed and their history maintained.

The importance of reasoning in Space 0 arises from the fact that contiguity is a fundamental constraint. That is, it underlies or holds in the space of course and speed and well as the space of movement plans. Additionally, reasoning in Space 0 is reasoning at the level of resolution of the sensors (at least, in the worst case, this is the only information available). Since the space is large but not infinite in problems likely to be encountered, the human expert has difficulty managing this space because of working-memory limitations. However, a machine might exhaustively enumerate this space in a data-driven fashion that appears to be required by aspects of the problem.

A DATA-DRIVEN SEARCH PROCESS FOR SPACE 0

A data-driven exhaustive search process was designed and implemented for space 0. In AI parlance, the process can be characterized as a data-driven constraint propagation algorithm. The constraint propagated is contiguity of vehicle location and the propagation involves extending region paths, based on a new sensor event, such that the new region added to a path is spatially adjacent to the last region in the path and the new region is consistent with the sensor signal. The process decomposes into two subprocesses, region extraction and region path construction.

Region Extraction

In order to construct region paths, regions must be extracted from a 2-D representation containing sensor location and coverage (the extent of the rings and disks). In our case, we are using top-down connected-component labeling of quadtree representations of sensor coverage (cf. Samet, 1984, 1985, 1981). In addition to the performance gains we have already achieved using quadtree techniques, the quadtree scheme admits readily to parallel implementations that might be required by application scale up and fielding.

Table 6. Process Cycle.

```
1. Accept a new sensor signal
2. Extend OR-graph of region paths:
      For each current region possibility, i,
          Find all regions j such that
              j is adjacent to i
                  and
              j is sensed by the new sensor(s) in
                  contact
          If j is an empty list
              then prune i from the graph and
                  propagate deletion back through
                  the graph
              else extend all paths through i to
                  each j
```

Region Paths

Table 6 presents a sketch of the overall process cycle.

Region paths are represented using an OR-graph representation. This version of the process is strictly data driven, since the construction cycle is driven by sensor events. Referring back to the example shown in Figure 1, the top of the graph would consist of an OR-node with the five region possibilities (labeled 1–5) associated as the disjunctive start points. The node is constructed when the first sensor event occurs.

When the second event occurs (contact on S3 and S2), each region possibility in the graph is extended by associating an OR- node and disjuncts if regions can be found that meet the constraints. In this case, the path starting with the S3 disk cannot be extended, because it is not adjacent to a region covered or sensed by the sensors in contact, S3 and S2. The graph is pruned, deleting the paths that cannot be extended.

The Space 0 process has been implemented on a Symbolics lisp machine. The region extraction and path construction cycle, accepting input and updating the OR-graph of region paths, takes a few seconds for a problem such as the one utilized in the experiments described above. The region path that contains the actual vehicle track used in the example problem is graphically depicted in Figure 6. Each shaded region corresponds to one or more steps in the path. The application of the Space 0 process to the example problem is summarized in Table 7.

The size of the space of region paths over the course of sensor events is presented in the column labeled Space 0. For example, at event 11 (contact on sensor 18) there are 42 paths of length 11 in terms of region units. After processing event 12, no contact on any sensor, all 42 paths are extended, yielding a total of 84 paths. From examination of the se-

Table 7. Number of Region Paths Constructed for Experiment 1 Problem.

		Number of Region Paths		
Sensor Event	Sensor in Contact	Space 0	35-Knot Pruning	15-Knot Pruning
1	13	7	7	7
2	Ø	13	13	13
3	19	12	11	10
4	Ø	23	21	19
5	14	17	13	9
6	Ø	32	24	16
7	13	37	28	19
8	Ø	69	52	35
9	19	62	44	25
10	Ø	116	81	46
11	18	42	33	22
12	Ø	84	66	44
13	19	84	66	44
14	Ø	160	125	84
15	14	194	125	84
16	Ø	354	224	150
17	15	186	92	62
18	Ø	372	184	124
19	15	406	184	124
20	Ø	778	368	248
21	17	338	132	88
22	Ø	676	264	176
23	18	608	264	176
24	Ø	1132	462	308

quence, note that the number of paths approximately doubles when moving from a sensed event to an event denoting presence in a region that is not sensed. In general, regions that are not-sensed are contiguous to a larger number of regions than sensed regions. Consequently, in path construction they have, on average, the effect of enlarging the space of paths whereas, on average, the introduction of sensed regions allows less branching in the space of paths. This property is dependent on the particular pattern in which sensors are deployed and on the particular acoustic conditions which give rise to the size and presence of one or more rings.

The size of Space 0, the disjunctive set of location hypotheses, at any point in the history is less than or equal to the number of location hypotheses that are consistent with the sensor event at that point. An example of the reduction achieved by propagating contiguity is presented in Figure 7. Returning to the example problem, the location hypotheses consistent with contact on sensor 18 (event 11) are graphically represented as regions

Figure 7. Possible Vehicle Locations at Event 11.

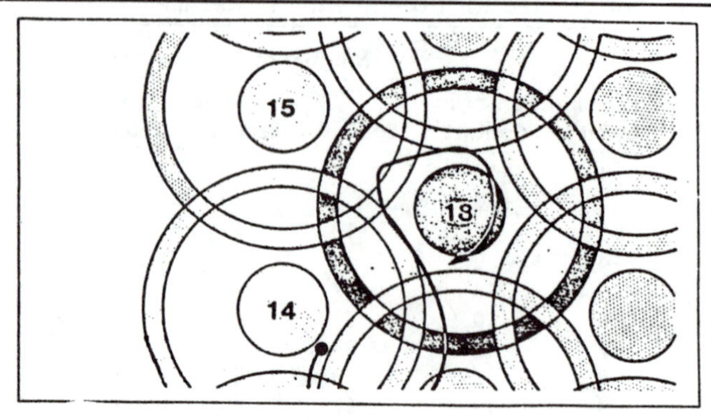

a. Considering Only Event 11.

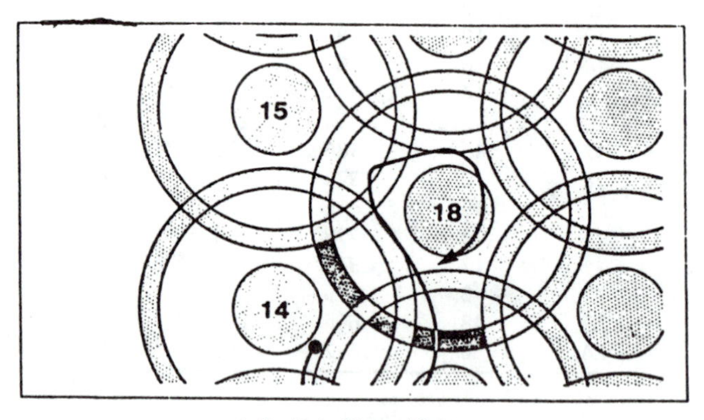

b. Considering History of Events.

with horizontal shading in Figure 7a. Note that there are 13 possibilities. When location hypotheses are generated based on contiguity and event history, this number is reduced to three. The reduced space is depicted in Figure 7b.

The OR-graph of paths, and thus the location hypotheses, can be further reduced if additional information is available. For example, if the maximum straight line speed of the vehicle under observation is known, then the duration of a sensor contact and the maximum straight line speed can be combined to yield the upper bound on distance traveled (assuming constant direction) during the contact. For a given region supported by the sensor contact, at least one straight line crossing (from predecessor region to successor region in a path) must be found which is less than or equal to the upper bound on distance traveled. If not, then the path may be ruled out on the grounds that motion along it is outside the vehicle's motion capability. If all paths in which a given region participates are pruned, the region is pruned from the set of location hypotheses.

Such pruning may not be warranted if the information on which it is based is incomplete, e.g., a default assumption about speed or an estimate of maximum speed. That is, the "true" hypothesis might be eliminated from the space. A variety of strategies can be explored to handle the incompleteness. For example, speed information can be applied to Space 0 in the form of a heuristic evaluation function which ranks the paths. We have explored the use of several different types of information, in addition to speed, that can be used as heuristics in reasoning about Space 0. The application of such information takes the form of either pruning rules or evaluation functions defined over portions of Space 0. These efforts will not be described here except to say that, like the construction of a Space 0 representation, they require considerable computational capacity. Thus, subgoals which involve applying these heuristics are additional candidates for assignment to the machine.

INTEGRATION OF THE EXPERT AND SPACE 0 SEARCH PROCESS

The Space 0 process constitutes a problem solving entity that knows: (a) the history of the sensor observations, (b) the region paths that are currently consistent with this history, (c) the set of regions within which the submarine might currently be located, and (d) the set of regions where the submarine might be located in the immediate future. From our theory and experimentation with the expert, we know that that the expert is aware of only a small subset of this information. And, of course, the expert holds many beliefs about course and direction, overall goals and tactics, and so on which are not known by the specialized Space 0 problem solver. Thus, neither the machine Space 0 problem solver nor the human expert have direct knowledge of the problem-solving state of the other. In fact, their problem-solving methods differ and the language in which their problem solving states can be described, while related, also differ. Nonetheless, we desire to introduce a pattern of information dependence between these two problem solvers that cooperatively links their problem solving activities.

In this section, we propose various ways of coordinating the activities of these two problem solvers. A proposed pattern of information dependence is partially specified by stating a set of normative rules that must be realized in order to coordinate and control the proposed interaction. Recall that a normative rule represents a partially specified communicative plan. The first part of the rule specifies when the interaction should occur. When is not specified in terms of some absolute time metric. Rather, it is specified relative to the problem-solving state of one or both of the problem-solving agents. The second part of the rule specifies the communicative goal of this plan. Taken together, these two parts represent the fact that the execution of communicative plans is typically conditional upon the state of the communicating and receiving agents. The third part of the rule provides parameters that are to be used in planning a way to realize the communicative goal. When communication is between person and machine, achievement of a communicative goal will invariably involve translating from the language of one problem solver into a language understood by the other.

In order to highlight this fact, we have termed the third part of the normative rule, *conventions*. The conventions represent the rules for translating into this shared language.

Our use of a set of normative rules to specify the interaction between the cooperative problem solvers is similar in spirit to Rosenschein's proposal for the use of abstract theories of discourse as a formal specification of programs that converse (Rosenschein, 1981). This partial specification of the communicative plans that must be realized in order to control and coordinate a proposed pattern of interaction is important from a design point of view. Achievement of a proposed pattern of information dependence itself requires computation resources. Determination of when to communicate may require recognizing aspects of the problem-solving activity of the other agent. This is itself a kind of plan recognition problem. Constructing the message to communicate is a kind of planning problem. Thus, any proposed pattern of interaction requires the designer to provide a solution to these attendant plan recognition and generation problems. The solution must leave each agent with the computational resources needed by the main problem-solving activity.

The question that now must be addressed is to propose, implement, and evaluate various ways of coordinating the activities of these two problem solvers. What information known by one problem solver should be shared with the other, when or under what conditions should this information be shared, and how should this information be communicated? Our work on this aspect of the design process is ongoing. Consequently, we will simply sketch various proposals for ways in which to coordinate the activities of the two problem solvers in order to exemplify our method of approach.

Several possible roles for the Space 0 search process have been considered. These can be labeled as:

- Preprocessor of sensor events
- Critic of the expert
- Critic and Advisor to the expert.

Preprocessor of Sensor Events

In the preprocessor role, the Space 0 search process is given complete responsibility for monitoring the sensor events, determining, and then communicating the region paths consistent with these events to the expert. The applicability condition for the norm defining this role is straightforward. Communication to the expert occurs after the graph of possible paths has been updated in response to a new sensor event. Thus, this machine agent only requires knowledge of an aspect of its own problem-solving state to determine when it is obligated to communicate with the expert. Consequently, no computational resources are required to monitor the problem-solving state of the expert.

The action required by this norm is to communicate the current graph of possible region paths to the expert. If this graph is large, the conventions or means of accomplishing this communicative intent may demand considerable computational resources. A convention of communicating these paths graphically in conjunction with the display of the sensor deployment pattern has the advantage of resolving problems of reference to the regions involved in the region paths. The disadvantage, from the point-of-view of the machine, is that its screen size is limited. When the number of paths is large, this full graph of paths may be impossible to display within the screen size available. For example, the particular machine we have been working with for display purposes, a state-of-the-art graphics machine, cannot display all paths, encoding them distinctly, when the space is large and includes many intersecting paths. Of course, a more complicated convention could be adopted which presents the information in a sequential but systematic fashion.

Before devising clever schemes for graphic display, it is advisable to consider the complementary role that the expert must play in this proposed scheme of role assignment. The expert no longer has responsibility for monitoring the sensor events or determining the possible current region paths and locations of the submarine. Rather, the expert's responsibility is to use the information about currently possible paths to restrict the generation and evaluation of hypotheses stated in the language of Space 1 and 2. However, we have already observed that there are typically a very large number of Space 1 hypotheses concerning speed and direction that are consistent with a hypothesis from Space 0 concerning a region path. And theory and experimental evidence suggest that the expert is able to concurrently reason about only a very few hypotheses concerning the vehicle's speed and direction. Consequently, although we can certainly assign this role to the expert, it is also rather clear that the expert will be unable to meet the requirements of this role.

Rather, the expert will probably attempt to do his best, which will involve selecting from the information provided by the machine that information which can be used to evaluate and extend his current Space 1 and 2 hypotheses. That is, the expert will continue to use his hypothesize-and-revise strategy. However, we have added to his processing task the additional task of selecting from the information provided by the machine that subset which is relevant to his problem-solving activities. And, of course, the machine has had to devote considerable computational resources to the task of communicating *logically correct* and *useful* information which is, in fact, *functionally* irrelevant to the expert.

Note that with this role assignment, the expert's problem solving process is dependent in the strong, information sense and the temporal sense defined above. The dependence is assymetric; the expert is unable to influence either when or what information the machine makes available to him. Rather, the expert must attempt to bring his method in conformance with the machine's method. In order to alter this state of affairs, we can next consider a role assignment where the machine problem solver is expected to serve as critic.

Critic

In this role of critic, the machine problem solver is expected to monitor the hypothesis being considered by the expert and to determine whether each of his hypotheses is consistent with the machines' current hypotheses of possible region paths. If an expert's hypothesis is consistent, then it is obligated to inform the expert of this fact. In our language of norm rules this might be expressed as:

Applicability
Conditions: Believe (M, Holds (E,H)) and Knows (M, Implies (O,~H))
Goal: Inform (M, E, Implies (O,~H))
Conventions: E.g., by graphical display of H, of regions consistent with O and of the region that is not consistent with O but of which a portion of H is a refinement

where M refers to the machine
E refers to the expert
H refers to an hypothesis
O refers to an observation

Note that now the machine must monitor the expert in order to determine the expert's current hypotheses. In order to represent the fact that we may need to require the expert to aid the machine component in this task, we might give the person the normative responsibility to inform the machine of his current hypotheses. This norm might be expressed as:

Applicability Conditions: Knows (E, (Holds (E, H.)) and Believes (E,~Believes (M, Holds (E. H.)))
Goal: Inform (E, M. Holds (E, H))
Conventions: E.g., draw track(s) terminating the track "near" where the vehicle is currently thought to be

In order to carry out this norm, the expert must be cognizant of his current set of hypotheses and of whether the machine has already been informed of them, and, if not, then inform the machine of the new hypotheses using whatever conventions we as designers provide.

Alternatively, if we are able to provide the machine with a reasonably efficient and accurate strategy for recognizing the expert's current hypotheses from actions that the expert pursues, such as drawing tracks or placing reference marks on the display, then the machine could initiate a query to the expert concerning his hypotheses only when this recognition strategy failed.

The point is that the assignment of roles will invariably give ruse to demands on the computational resources of person and/or machine to carry out these roles. Part of the design task is to insure that the role assignment does not create demands that exceed the capacities of either agent nor interfere with either agent's problem-solving activity.

Returning to our definitions of information dependence above, we can see that the expert's problem-solving process is weakly dependent on the machine's, and the machine's is strongly dependent on the expert's. In this case the expert will only be informed of information that is relevant to his current problem-solving activities. However, the expert may ignore or discount the information provided. We may wish to impose upon the expert the complementary norm of dutifully listening to and using the criticism provided by the machine. That is, the expert may be required to abandon any hypotheses which the machine component has informed him are inconsistent with the sensor events.

Critic and Advisor

The activities of the two problem-solving agents can be more tightly coupled by giving to the machine an advisory role in addition to its role as critic. In this advisory role, the machine is required to inform the expert of hypotheses which are similar to but are not included in the space of hypotheses that the expert currently holds. This information might be provided unconditionally to the expert or only when the expert's space of hypotheses becomes small or empty. Of course, we must implement a procedure for evaluating how similar one path hypothesis is to another, in order to allow the machine to carry out this role.

These latter two role assignments, critic as well as critic and advisor, are currently being developed and evaluated in a more elaborate form than presented here. Implementation of these roles also requires detailed specification and evaluation of the conventions adopted to enable and facilitate effective communication between the problem-solving entities. We have not considered this aspect of the task here, since the conventions used will often strongly depend upon the particulars of the machine architecture.

What has been presented is an example of the use of the framework provided by representing roles as the assignment of normative rules to the agents involved in the overall problem-solving activity. This framework brings us one step closer to the detailed considerations that must be taken into account in integrating the activities of person and machine. It is apparent from our example that machine and expert possess different limits on their computational resources, may use different knowledge and methods in their problem-solving activities and are typically assigned different tasks in the overall problem-solving activity. Each of these differences can importantly influence the choice of role assignment in attempting to design a cooperative person–machine system that enhances overall problem-solving performance.

The final step in our design cycle involves submitting the design to empirical test and evaluation. Based on the results of such a test, we may return to an earlier step in the cycle to either redesign or refine the design.

Conclusion

The conditions required to achieve cooperative interaction between agents—human or machine—are easier to abstractly specify than to realize within an instantiated organization of intelligent agents working on a common problem. The methodology proposed here represents our current strategy for systematically investigating the important but complex issues that arise in attempting to design a cooperative person–machine system. The methodology represents a blend of theory, from both AI and cognitive science, engineering, and empirical investigation. Our hope is that the use of this methodology may also contribute to each of these areas as well as provide a rational basis for the design of practical systems that will yield cost-effective enhancement of selected human problem-solving activities.

REFERENCES

Deimler, J., Goodson, J., Kavitsky, P., Stokes, J., Weiland, W., & Zachary, W. (1983). *Human factors in distributed intelligence systems design III. Laboratory and simulation testbed implementation* (Tech. Rep. 1788.01.03). Willow Grove, PA: Analytics.

Newell, A. (1969). Heuristic programming: Ill-structured problems. In J. S. Aronotsky (Ed.), *Progress in operations research: Relationships between operations research and the computer* (Vol. III). New York: Wiley.

Newell, A., & Simon, H. (1972). *Human problem solving*. Englewood Cliffs, NJ: Prentice-Hall.

Rosenschein, S. (1981). Abstract theories of discourse and the formal specification of programs that converse. In A. Joshi, B. Webber, & I. Sag (Eds.), *Abstract theories of discourse* (pp. 251–265). New York: Cambridge University Press.

Samet, H. (1981, July). Connected component labeling using Quadtrees. *Journal of the Association for Computer Machinery, 28*(3), 487–501.

Samet, H. (1984, June). The Quadtree and related hierarchical data structures. *Computing Surveys, 16*(2), 187–260.

Samet, H. (1985, January). A top-down quadtree traversal algorithm. *IEEE Transactions on Pattern Analysis and Machine Intelligence, PAMI-7*(1), 94–98.

Schmidt, C. (1976). Understanding human action: Recognizing the plans and motives of other persons. In J. Carroll & J. Payne (Eds.), *Cognition and social behavior*. Hillsdale, NJ: Erlbaum.

Schmidt, C. (1984). Partial provisional planning: Some aspects of commonsense planning. In J. Hobbs (Ed.), *Formal theories of the commonsense world*. Norwood, NJ: Ablex Publishing Corp.

Schmidt, C., Sridharan, N., & Goodson, J. (1978). The plan recognition problem: An intersection of artificial intelligence and psychology. *Artificial Intelligence, 11*(1 & 2), 45–83.

Schmidt, C., Sridharan, N., & Goodson, J. (1976). Recognizing plans and summarizing actions. *Proceedings of the Conference on Artificial Intelligence and the Simulation of Behavior* (pp. 290–306). Edinburgh, Scotland: AISB.

Sridharan, N., Bresina, J., & Schmidt, C. (1983). *Evolution of a plan generation system* (Tech. Rep. CBM-TR-128). New Brunswick, NJ: LCSR, Rutgers University.

Sridharan, N., Schmidt, C., & Goodson, J. (1980). The role of world knowledge in planning. *Proceedings of the Conference on Artificial Intelligence and the Simulation of Behavior* (pp. 1–11). Amsterdam: AISB.

Designing Cooperation: Afterward to Section II

Wayne W. Zachary

CHI Systems Inc.
Spring House, PA

The three chapters in Section II are all by computer scientists and deal with specific computational problems and approaches in building computer systems that behave cooperatively. The chapters varied in their focus and level of abstraction. As in the first section, the emphasis was on dyadic systems and interaction. Both Goodson and Schmidt, as well as Woods et al. considered one-person/one-machine dyads; only Durfee et al. focused on larger scale multiagent systems. Just as the chapters in section one all seemed to begin with an implicit minimalist model of cognition, the chapters in section two all began with an implicit minimalist model of a problem-solving device. This minimalist problem-solving model includes a symbolic processing machine, capable of applying some set of internal knowledge via an inference mechanism to specific data about a problem and problem conditions, in order to achieve some goal within the problem space. The three chapters dealt with three widely studied architectures that capture this minimalist model: Goodson and Schmidt deal with plan recognition architectures; Woods et al. with production rule architectures, and Durfee et al. with blackboard architectures. Each author then elaborated on or modified this model in some way so as to identify or resolve some computational problem in building a dyad or network of cooperating problem solving devices.

Three issues in designing cooperation emerged in Section II:

- Task decomposition and distribution (by what principle, with what performance criteria, into what architecture, etc.)
- Communication (how often, in what form, at what level, etc.)
- Relationship (what kind, how engineered, how formalized, etc.).

The first issue was addressed by all the chapters, although only Goodson and Schmidt were explicit about the fact. The second issue was the main concern of Durfee, Lesser, and Corkill, while the third issue was considered both by Woods et al. and by Goodson and Schmidt.

TASK DECOMPOSITION FOR COOPERATIVE SYSTEMS

Task decomposition, in this context, refers to the problem of dividing a task or problem into component aspects and assigning them to different agents for cooperative solution. Durfee, Lesser, and Corkill analyzed cooperation among sets of generalized problem-solving agents and in particular those cases where the individual agents were machines and were pursuing essentially "local" goals as part of a system-wide problem-solving process. In this regard, their concern was similar to that of Axelrod (whose work was described in the Afterword to Section I). Both Durfee et al. and Axelrod were interested in how cooperation emerges in systems where every agent is acting on the basis of local interests. One of the first points made by the Durfee et al. chapter (although one not pursued to great extent in the chapter) is that the goals of a cooperative system are dependent on the observer's stance—the goals or functions of the overall system can be very different from the local goals or functions of the individual agents that comprise the system. The agents may explicitly share the goals that define the higher-level cooperative system (whether that higher-level system is a dyad or a society), but it is not necessary to do so. In a system composed of multiple agents, the agents do not have to endeavor to cooperate, nor even be aware of the system-wide goals for cooperation to occur. This is certainly the case in marketplace situations, where the participants may well perceive themselves as competing, not cooperating. Yet a marketplace is clearly homeostatic and cooperative *at a system wide level*, as Malone noted in Section 1. Durfee et al. used this analysis only to justify the choice of a specific stance—the agent level—in studying cooperation, choosing to examine how locally-directed agents can be designed so as to exhibit cooperative behavior at the system level.

An implication of the system-level/agent-level distinction, not noted by Durfee et al., is that the decomposability of goals/functions at the system level strongly affects the way (and degree to which) a task is amenable to cooperative solution. Goodson and Schmidt considered some of the ways by which a system goal may be decomposable, and the effects of these ways on cooperative system design. It is an open question whether *all* problems are candidates for distributed solution. It is also a critical question in engineering disciplines, such as Human Factors Engineering and Industrial Engineering explicitly, which are concerned with allocation and distribution of functions among human teams. There is an old saying "too many cooks will spoil the broth" which can be taken to mean that some problems seem inevitably to work best when solved by a single agent. This point has empirical significance here, because Durfee et al. were not able to derive, in any of their variants, a cooperative system that achieved the same efficiency as a single, optimal, problem solver acting along.

Both Durfee et al. and Goodson and Schmidt dealt with decomposition of a problem for cooperative solution in terms of segmentation of the problem-solving process, not of the goals or problem per se. Decomposition on the basis of process requires some explicit consideration for the architecture and control metaphor of the resulting cooperative system. Durfee et al., emphasizing an agent's ability to predict and form expectations about other agents as the basis for coherence, targeted a specific form of architecture which they call functionally accurate and cooperative (FAC). This is a form of cooperation in which the agents communicate only about intermediate or final results, which are expressed in terms of propositions or hypotheses about the state of the environment. Each agent is able to incorporate the communicated results of the other agents into its own problem-solving process. Once incorporated, the agent is able to act on these just as it would its own intermediate conclusions developed internally directly from raw data. If problem solving is pictured, as in Figure 1, the FAC approach can be termed a vertical partitioning (Figure 2). In a vertical partitioning, each agent has a capability at all levels and could in theory solve the problem alone, but has in practice been assigned responsibility for only one subset of the data. It is an example of what Malone has called a marketplace organization.

Kornfeld (1981) has likened this decomposition to the scientific com-

Figure 1. Problem decomposition into functions and stages.

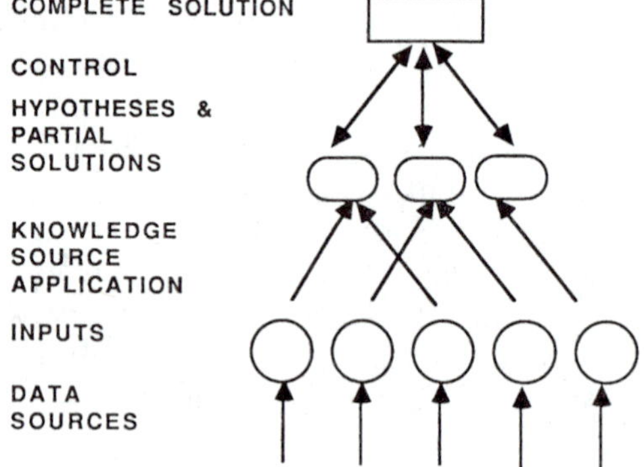

Figure 2. Strictly vertical problem decomposition.

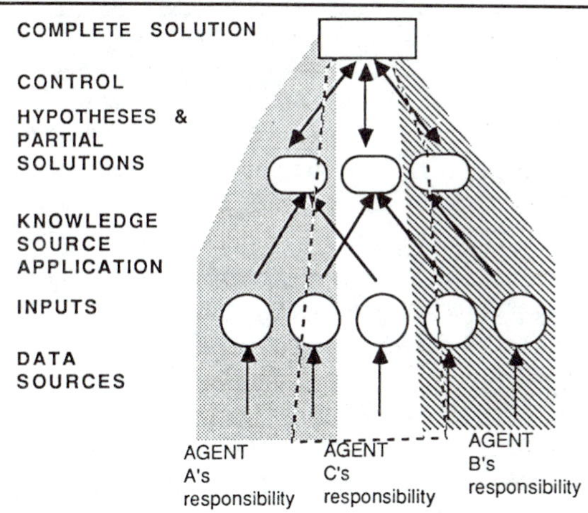

munity, where each scientist works on one small piece of the universe of data and publishes his/her results and hypotheses for verification, refutation, and/or use by other scientists. An alternative decomposition that could be either vertical or horizontal was proposed by Smith (1981), and termed the negotiation model by Durfee et al. In Smith's view of cooperation, agents can have localized expertise that can be used only in one part of the problem-solving process (i.e., at only one layer, see Figure 3). They can also have all levels of expertise, but access to only part of the data (like Durfee's model), or a combination of the two, in which there are multiple local experts at each level, each with access to only one part of the data from the next lower level. A third variant, termed multiagent

Figure 3. Mixed horizontal and vertical decomposition.

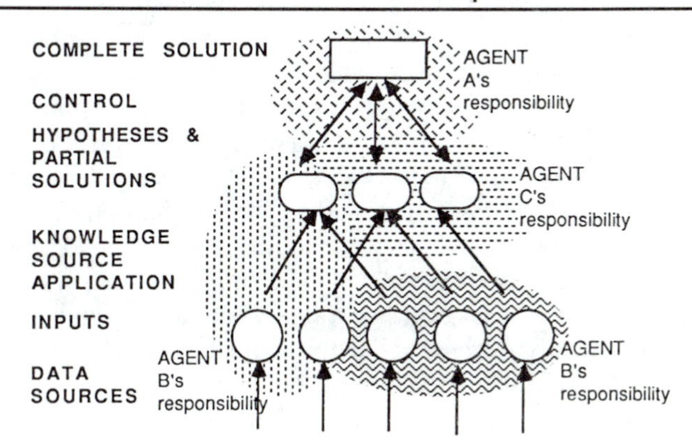

Figure 4. Strictly horizontal problem decomposition.

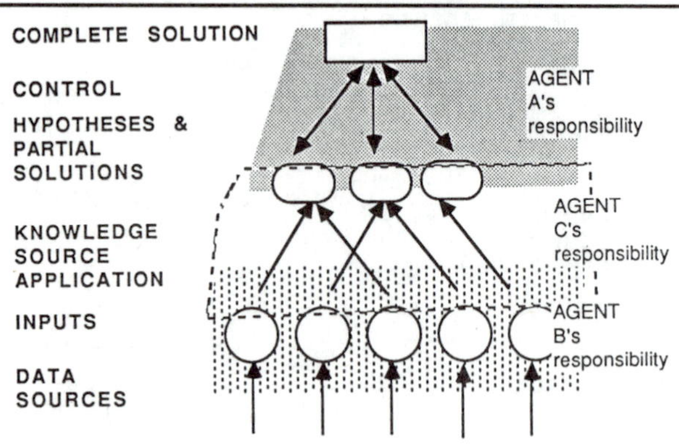

planning by Durfee et al., is strongly horizontal (see Figure 4). As Malone noted, the horizontal and vertical decompositions correspond to the functional and divisional principles (respectively) in organizational theory. It should be noted, however, that in practice virtually all human organizational structures are strongly horizontal, successive layers of management separating the nodes that process the inputs and outputs (i.e., workers) and those that only plan and coordinate (i.e., senior management).

Goodson and Schmidt treated the task decomposition problem as one of replacing a single agent problem solver (i.e., an unaided human) with a joint person–machine system with greater effectiveness. They also took a process decomposition approach to this problem, based on the ideal of near decomposability (Simon, 1981). Their method called for hierarchically decomposing the problem-solving process practiced by the human expert, and then assigning the pieces to the person or machine. In this approach, they sought the decomposition that had the fewest dependencies between the two agents. This criterion was in recognition of the fact that the more interdependent the person's and machine's problem-solving processes are, the more communication the two agents will have to maintain, and the more likely the entire procedure will degenerate into an avalanche of communication. Goodson and Schmidt's solution was to explicitly map out the logical dependencies (both informational and temporal) between processes and subprocesses and then allocate functions so as to minimize the dependencies among the parts assigned to the person and those assigned to the machine. Once the functions were allocated, the dependencies formed constraints on how the two agents must communicate and interact. Thus, they advocated neither a vertical nor a horizontal architecture, but rather one that is empirically derived from the logical structure of the specific problem-solving process to be decomposed.

RELATIONSHIPS AMONG AGENTS IN A COOPERATIVE SYSTEM

A strong conclusion that could be drawn from the similar approaches by Durfee et al. and Goodson and Schmidt to task decomposition is that a decomposition of a problem for cooperative solution seems to lead inevitably to a problem-solving architecture in which the agents' roles in the problem-solving process are well defined. The relationships and communications among these agents must then be designed and fitted into this architecture, at least when one or more of the agents is a computational device with no inherent capability for establishing relationships. Woods et al. and Goodson and Schmidt offered some thoughts on the problem of designing the relationships among problem-solving agents, at least in human–machine dyads.

Woods, Roth, and Bennett began from the perspective of a common problem-solving device, the so-called *expert system*, which is technically better described as a production-rule program. This currently represents the most common class of problem-solving programs in use today. Production-rule programs contain some base of knowledge about one domain expressed as production rules, a mechanism (termed an inference engine) for selecting rules (either by forward or backward chaining) and applying them, and a facility to communicate with some other agent, usually a human, who provides problem-specific data. This last component is particularly important, because the machine is unable to acquire data on its own, lacking eyes, ears, hands, etc. The easily confused Bluebonnet machine, examined by Suchman in Section I, was, in fact, a production-rule program. It had these same limitations, so some of the inherent problems with a blind, deaf, and insensate machine have already been explored. Woods et al., however, considered the fact that it is precisely because the machine has no other way to interrogate its environment, that it must assume control of the dialog with the person and use the person for these purposes. This establishes a relationship between the person, a machine that has the person in a subordinate role, responding to the requests or questions of the machine. Their data showed that humans in this role can easily become passive, and that this passivity leads to substantially degraded problem-solving performance on the part of the person–machine system. This result also seems to generalize to other person–machine contexts where the machine is in control, for example, highly automated aircraft cockpits (Weiner & Curry, 1981).

Although they do not describe it in these terms, Woods et al. were really struggling with the problem of engineering an appropriate relationship in the person–machine dyad. Their proposed solution was to begin the design process by designing the person–machine relationship, and then to define the functional requirements and computational capabilities of the machine from that. Their general suggestion was to place the person in the superordinate role, because the person will always be (or least for the foreseeable future) responsible for integrating information that is outside the boundary of the machine's expertise and for accommodating the problem-solving process to situational factors (including person and/or machine errors). In this relationship, they likened the machine to a staff member, the person to a manager, or someone "in charge," and developed design approaches based on this relationship metaphor.

Goodson and Schmidt were also concerned with the process of engineering the relationship between the person and machine in a joint person–machine cooperative system. Like Woods et al., they considered a dyadic person–machine system designed to perform a single task. In such systems, there is a single high-level goal which is established and enforced by standardization (in Malone's terms). Goodson and Schmidt further complicated the picture by considering an empirical case of a person–machine dyad observing and interpreting the actions of a third agent (a vehicle). In this problem, the goal of the person–machine system is to recognize and infer the plan of the vehicle, making the overall task a plan recognition task. They did not, however, apply plan recognition directly to the person–machine cooperation problem, as have others, such as Croft et al. (1983).

COMMUNICATION BETWEEN AGENTS IN COOPERATIVE SYSTEMS

The reasons for communications between cooperating agents, as well as the content of those communications, depends on the way in which the problem has been decomposed and on the relationships between the agents involved.

In person–machine dyads, Goodson and Schmidt argued for engineering the interaction between agents through the medium of norms that govern the communicative process. These norms, which they formalized as tripartite rules, defined:

- *when* a given communication is appropriate, defined in cognitive terms as X *believes* something, Y *knows* something, etc.;
- *what* should be communicated, defined by the intended effect of the communication on the other agent, and
- *how* the communication should occur, defined by the specific communicative conventions and protocols available in the system.

They also described a methodology for building specific norms based on a given type of relationship that is desired between the machine and the person. From a design perspective, this establishes a (computable) formalism for expressing norms and for defining norms from a specified person–machine relationship. Goodson and Schmidt's method thus picks up where Woods et al.'s relationship design argument leaves off (i.e., at the point where desired relationship is targeted for the person–machine dyad), and provides tools to operationalize the relationship.

Durfee, Lesser, and Corkill were concerned with a slightly different set of communicative issues because none of their agents were people. They could not fall back on the human's communicative competence as a baseline, as did the other authors in Section II. In one sense, this was a difficulty, because it forced them to deal with all levels of communication, not just the higher (i.e., semantic and pragmatic) levels. In another sense, however, it was a simplification because it allowed easy partitioning of their results without regard to the complexities of human interaction. In other words, when they concluded that communication of plans, for example, added a certain increment to cooperation, they did not have to consider such human issues as deceptiveness, communicative errors, pragmatic (e.g., social) dimensions, and so on. Every computational agent behaved exactly as designed—functionally accurate and cooperative.

It is therefore especially interesting that the findings of Durfee et al. provide such clear bridges back to the analyses in Section I. They began with the minimalist model, in this case a network of blackboard-based problem solvers, with an ability to send and receive messages between each other, all working on a vehicle tracking problem in a FAC arrangement. Vehicle tracking is clearly an appropriate problem for this type of agent, as indicated by the success of the HASP system for submarine tracking (see Nii et al., 1982). Given that the problem could be solved alone by this type of problem-solving device, Durfee et al. sought to determine what capabilities needed to be added to the basic blackboard architecture to achieve different levels of functionally accurate, cooperative behavior.

Using a vertical problem decomposition, the blackboard-based agents cooperated simply by communicating all intermediate hypotheses and goals to all other agents, and by processing all communicated hypotheses and goals as their own. After showing that this minimalist approach lead to very suboptimal problem solving at the system level, Durfee et al. examined how different communicative strategies would affect both the global problem-solving process and the internal structure of the blackboard-based agents. A first level of system-wide improvement was made by adding to each agent the ability to differentiate among its various hypotheses and goals and using a communication strategy that transmitted only new hypotheses/goals, and an old one that could no longer be pursued or improved on locally. This capability is somewhat analogous to the observation made by Gibbs and Mueller in Section I that communication is often restricted to key initial and decision points in the problem-solving process under the assumption that the other agent can replicate the intermediate steps, making their communication necessary. If we re-examine the example used both by Gibbs and Mueller and by Suchman:

Customer: Do you have coffee?
Waitress: Cream and Sugar?

we can see Durfee et al.'s first and last communicate strategy in action. Note that the first agent (customer) initially communicates a new goal (which is also a goal that can not be processed further without additional data), and at some later point the second agent (waitress) responds with a hypothesis that she can not process further without information from the first.

The second addition to the minimalist blackboard agent by Durfee provided another, more surprising, connection with the analyses from Section I. Even after the better communicative strategy was implemented, Durfee et al. noticed that each agent was still internally processing as if it were solving the entire problem itself, rather than as a local part of a distributed team. This led them to add a construct by which the agent could plan its activities based on reasoning about the overall problem-solving situation (in the network), as conditioned by its own local goal(s) and the overall (system-wide) goal(s). To accomplish this, they had to introduce both a planning construct in the agents, and more importantly a situation analysis and reasoning component. The implementation of this planning construct achieved still greater gains in efficiency and coherence, particularly when the planning was based on the overall situation in the network. They thus demonstrated the benefits of a situation-based planning and interaction strategy, which is very similar to that proposed by Suchman, Newman, and Gibbs and Mueller as an explanation of human cooperative behavior. Even more importantly, Durfee et al. empirically demonstrated:

- that such a capability does not arise spontaneously from the minimalist problem-solving model;
- what is involved in providing this situation-based communication strategy within a minimalist problem-solving device.

These results provide a further connection between the analysis of human cooperation in Section II and the analysis of machine cooperation in Section II.

FINAL THOUGHTS ON SECTION II

Section I raised some issues that seemed unpleasant for those involved in the design of interactive and/or cooperative computer systems. The chapters in Section II seem to point out how these issues may be dealt with in the design process. The idea of a human–computer dialog that has a true mixed initiative has been around for some time (see Licklider, 1960). The authors in Section I provided some behavioral requirements for the mixed initiative dialogs based on the characteristics of human interaction. Some of these included:

- Interaction that is goal based but sensitive to the pragmatics of the situation or context of the cooperation;
- Organization of the interaction and cooperative process on the basis of relationships between the cooperating agents;
- Communication based on goals, implied facts, and intended plan recognition.

All of these characteristics have been addressed in a constructive or design-oriented way in Section II. Woods et al. provided a theoretical analysis of why mixed initiative dialogs are essential when humans are interacting with machines. In the course of doing this, they also introduced an analysis technique (their diagramming method) that can be used to identify and analyze the initiative in a human–machine dialog. Durfee et al. provided a theory of establishing relationships among agents based on problem decomposition, and Goodson and Schmidt provided a method for doing so. The latter also provided a formalism for building communicative norms based on specified interagent relationships. Durfee et al. demonstrated that features, such as situation-based interaction and intention-based communication, add substantively to the efficiency of systems of cooperating machines. By themselves, each of these results in Section II increase our overall understanding of cooperation. Together, and in conjunction with the analyses of human cooperative behavior in Section I, they suggest that there may, in fact, be some deeper level principles operating in cooperative systems.

The chapters in the third and final Section of the book take a turn away from this general view of cooperation and consider one aspect of one kind of cooperation in detail: human expectations of machines in cooperative human–computer dyads. This is a more practical concern, because of the rapid growth in interactive and personal computing, and the problems experienced by people working with computers interactively. Thus, the last section tries to take some of the more general concepts of cooperation, particularly the role of expectations in defining relationships and establishing communicative norms, and apply them in more concrete terms to the design of interactive computer systems.

REFERENCES

Croft, W. B., Lefkowitz, L. S., Lesser, V. R., & Huff, K. E. (1983). Interpretation and planning as a basis for and intelligent interface. *Proceedings of the conference on artificial intelligence*. Rochester, MN: Oakland University.

Kornfeld, W. A., & Hewitt, C. (1981). The scientific community metaphor. *IEEE Transactions on Systems, Man and Cybernetics, SMC-11*(1), 24–33.

Licklider, J. C. R. (1960). Man-computer symbiosis. *IRE Transactions on Human Factors in Electronics* (pp. 4–11).

Nii, P., Feigenbaum, E., Anton, J. J., & Rockmore, A. J. (1982). Signal-to-symbol transformation: HASP/SIAP case study. *AI Magazine, 3*(2), 23–35.

Simon, H. (1981). *Sciences of the artificial* (2nd ed.). Cambridge, MA: MIT Press.

Smith, R. (1981). *A framework for distributed problem solving*. Ann Arbor, MI: UMI Press.

Weiner, E., & Curry, R. (1981). *Flight deck automation: Promises and problems* (NASA Technical Memorandum NASA TM-81206). Moffett Field, CA: NASA.

III
Studies of Cognition and Cooperation

Scott P. Robertson

Department of Psychology
Rutgers University
New Brunswick, NJ

INTRODUCTION

The final section of this book contains a set of studies of the human agents in human–computer situations. It is clear from all of the contributions thus far that cognition is central to cooperative systems. The agents in cooperative situations must share knowledge, both about communicative conventions and about their various states and conclusions during problem solving. While it is possible to design the computational components of such systems, the knowledge that humans utilize must be investigated empirically. Cognitive psychologists, who comprise the authors in Section III, are most familiar with the methodologies for observing human behavior and inferring cognitive mechanisms from it.

In the first contribution, Graesser and Murray explicitly discuss a methodology for exploring the cognitive representations which underly cooperative interactions between people and intelligent systems. Verbal protocol analysis gives cognitive science researchers access to the goals and plans of people as they pursue a task. Graesser and Murray have adapted a controlled question–answering methodology first used in studies of prose comprehension to the study of human–computer interaction. By combining their question-answering data with a well-explicated theory of information processing in human–computer tasks, they are able to generate *conceptual graphs* that represent human knowledge about computer commands. It is worth noting that directed question asking (either by requests that are propagated in a computer network, hypotheses presented to human users by an expert system, or verbal questions arising in dialogue) is a method used in cooperative situations to gain knowledge about other participants and about shared goals.

Kay and Black utilize the more traditional methodology of multidimensional scaling to build representations of knowledge for computer commands. By looking at computer-naive, -novice, and -expert groups of subjects, the experimenters were able to trace changes in cognitive representations as users gained expertise with a computer system. They propose four distinct phases of learning, distinguished by the nature of conceptual representations utilized by human users to understand the system with which they are working. Any notion of cooperation through a learning period must recognize the changing nature of human representations of knowledge.

In the third chapter in this section, Mack looks at the expectations of new users of a computer system. Kay and Black note in the previous chapter that, in the first phase of learning, people utilize prior expectations about the behavior of a system that may be based on the natural language semantics of the system's command language or other features related to general knowledge. Mack emphasizes this point by looking exclusively at novice users. In a detailed analysis of new users' verbal explanations and predictions, Mack shows both the intrusion of related knowledge (i.e., typewriter knowledge used to understand a computer text editor) during learning and the simple causal structures ("one action, one outcome") into which new users try to fit their expectations about system behavior. Mack goes a step further by describing a prototype training system that does not violate new user expectations. In a sense, such a system is cooperative in that it accommodates the cognitive representations of its users.

In the final chapter in this section, Sebrechts and Marsh combine the concerns of Kay and Black with changes in conceptual representations, and the concerns of Mack with supportive training systems. Sebrechts and Marsh use a set of questions designed to elicit information about users' knowledge to examine the effects of different instructional materials. The instructional materials differ on the degree to which they support *conceptual elaboration* and *syntactic elaboration* of the material to be learned. They trace changes in conceptual models of the computer system in all of these conditions and provide a model, which they call *integrative modeling*, to explain them.

9

A Question-Answering Methodology for Exploring a User's Acquisition and Knowledge of a Computer Environment*

Arthur C. Graesser

Psychology Department
Memphis State University
Memphis, TN

Kelly Murray

Department of Computer and Information Science
University of Massachusetts
Amherst, MA

When intelligent systems communicate successfully, there is a cooperative interaction between them. Each system, whether human or machine, must anticipate what the other system will or can do at each transaction. For example, consider the interaction between people and computer systems. The human anticipates the computer's actions and reactions after the person enters input. If the computer is well designed and cooperative, the computer system "expects" a specific range of user input and presents error messages that help the user to correct erroneous, unexpected input. When people learn how to use a new computer system, they are essentially learning how a communication system operates.

Whenever cooperative intelligent systems involve people, the researcher is confronted with two challenging problems. One problem is to discover

*This research was supported by a grant from Hughes Aircraft Corporation awarded to the first author. We thank Georgia Murray for transcribing the question answering protocols in this research. Requests for reprints should be sent to Arthur C. Graesser, Department of Psychology, Memphis State University, Memphis, TN 38152.

how the person cognitively represents the interacting systems. A second problem is to map out the knowledge structures which evolve while the person learns how to use a new system. Researchers will meet the challenge of both problems to the extent that they have a window to the important content, structures, and procedures in the human mind.

The primary message in this chapter is very simple. We argue that a question answering (Q/A) methodology provides a very rich and informative data base for investigating cooperative interacting systems that involve humans. In order to illustrate this, we will map out the users' acquisition and knowledge of a computer environment. When this Q/A methodology is applied to computer–human interaction, the user performs some benchmark tasks and answers questions about the computer–user interface while performing the tasks. The critical trick in effectively using this methodology lies in *what* questions to ask and *when* to ask them. Although this chapter reports a successful application of the Q/A methodology in the context of computer–human interaction, we believe the Q/A methodology can be applied to many forms of person–system interactions.

Cognitive scientists have recently come to appreciate the value of collecting verbal protocols from individuals while they perform behavioral tasks. The verbal protocols include "think aloud" protocols (Ericcson & Simon, 1980), question-asking protocols (Olson, Duffy, & Mack, 1984), and question-answering protocols (Graesser, 1981; Graesser & Clark, 1985). In previous years, verbal protocols furnished important data for discovering some of the mechanisms involved in problem solving (Newell & Simon, 1972), scientific reasoning (Chi, Glaser, & Rees, 1982; Stevens, Collins, & Goldin, 1982), mathematical reasoning (Ginsburg, Kossan, Schwartz, & Swanson, 1983; Lewis, 1981), writing (Collins & Gentner, 1980; Hayes & Flower, 1980), text comprehension (Graesser, 1981; Graesser & Clark, 1985; Olson et al., 1984), and other cognitive activities. More recently, verbal protocols have been collected for studying conceptual errors and cognitive structures when individuals operate computers and other technological devices (Bainbridge, 1979; Carroll & Mack, 1983; Graesser, Lang, & Elofson, 1987; Kieras, 1982, 1985; La Plat & Hoc, 1981; Mack, Lewis, & Carroll, 1983; Monk, 1984; Wickens & Kramer, 1985).

The content of the verbal protocols to some extent captures the knowledge states and cognitive strategies that underly human performance during person–computer interaction. The extent to which the information in verbal protocols correspond to actual psychological states and processes has been debated in recent years (see Ericcson & Simon, 1980; Graesser & Clark, 1985; Nisbett & Wilson, 1977; Olson et al., 1984). However, we will spare the reader a long discussion on the history of the debate. The correspondence clearly depends on the task, the individual's familiarity with the task, the type of verbal protocols collected, and the timing of the verbal protocols (i.e., during versus after the task is performed). Available evidence suggests that verbal protocols have comparatively high validity when the following conditions are met:

1. The individual is not very familiar with the task
2. There are medium or long pauses between successive actions when the task is performed
3. The translation of knowledge into language is not particularly troublesome
4. The protocols are collected during the execution of the task rather than after the task.

These conditions were indeed satisfied when the individuals learned the new computer environment in the present study. Therefore, we assumed that the Q/A protocols had at least a moderate degree of validity, even though they are not perfectly valid windows to the true knowledge states and processes.

We collected several types of Q/A protocols in the present study. However, we will focus primarily on those Q/A protocols that the users supplied while they performed benchmark tasks on the computer system. The benchmark tasks included (a) logging onto the system, (b) creating a data file with a text editor, and (c) modifying an existing data file with the text editor. We probed the users systematically with questions while they performed these benchmark tasks. The three rules below guided the course of questioning when the experimenter probed the users.

RULE 1: Querying computer prompts and messages
IF ⟨computer prompt or message⟩
THEN ⟨experimenter asks "What does that mean?"⟩

RULE 2: Querying user's actions
IF ⟨user performs action⟩
THEN ⟨experimenter asks "Why did you do that?"⟩

RULE 3: Querying long pauses
IF ⟨user pauses for more than 15 seconds⟩
THEN ⟨experimenter asks "What are you thinking about?"⟩

These rules were selected on the basis of pilot studies and previous psychological research investigating question answering mechanisms (Graesser & Clark, 1985; Graesser & Murachver, 1985).

This chapter is divided into four major sections. The first section briefly introduces some theoretical and practical considerations which furnish a background for investigations of computer–human interaction. The second section describes our methods of collecting data, whereas the third section reports some analyses of the data. Some analyses focus on conceptual errors that users have about the system; for each of these errors, there are recommendations on how to modify the interface to minimize such errors and provide a more cooperative person–computer interaction. The fourth section summarizes the major conclusions of this study and discusses further how the Q/A methodology can be applied to the evaluation and design of interfaces.

BACKGROUND ON THE STUDY OF COMPUTER-HUMAN INTERACTION AND TEXT EDITORS

Our perspective on the study of computer–human interaction has been shaped by three major theoretical and practical foundations. The first foundation is the GOMS model developed by Card, Moran, and Newell (1980, 1983). The second foundation is the research which has recognized the importance of studying the user's conceptual errors during learning and later performance on a task. The third foundation emphasizes that there are tradeoffs in most solutions to design problems which allegedly improve person–system interactions; the costs and benefits of design decisions must be evaluated. This section describes and elaborates on these three foundations.

The GOMS Model

The GOMS model accounts for strategies, actions, and keystrokes that computer users execute when they use a text editor. GOMS stands for *g*oals, *o*perators, *m*ethods, and *s*election rules, the four major components of the model. These components are briefly described below.

Goals. The user has a hierarchy of goals that must be accomplished during the course of editing a text. Some example superordinate goals are editing the manuscript, acquiring a given task, and executing a given task. Some example subordinate goals are inserting text, deleting text, replacing text, and moving text.

Operators. It is convenient to view operators as very subordinate actions. Operators include entering a command, attending to a portion of the screen, and executing a keystroke. Goals are fulfilled by selecting and executing the operators.

Methods. These are sequences or chunks of operators that ordinarily fulfill a goal when executed. These strategies are built up through experience.

Selection rules. The selection rules get activated during the task and

determine what methods are selected to fulfill goals. There ofte
or more methods to achieve a goal, with the best method depending
the states and circumstances that exist. Selection rules are represented as production rules in the GOMS model; these IF-THEN production rules specify that a particular method or operator is executed if a specific state exists. Two production rules are shown below.

IF ⟨more than 2 errors are on a line⟩
  THEN ⟨delete the old line and insert a new line⟩

IF ⟨less than three errors are on a line⟩
  THEN ⟨change erroneous characters on the line⟩

These example production rules illustrate that there are two ways to modify a line and that the method selected depends on how many errors there are on the line.

The GOMS model and its descendents (see Robertson & Black, 1983) go a long way in explaining performance when individuals use text editors. The model accounts for the sequences of methods and operators that are executed in a task. The model accounts for the duration of executing the methods and operators at different grain sizes (i.e., global plans versus specific operators). The model accounts for the errors of experienced users of text editors. The model is also sufficiently broad in scope to serve as a framework for evaluating and designing text editors (Roberts & Moran, 1982). GOMS is therefore a practical tool in addition to a theoretical framework that explains experimental data.

The GOMS model provides a very successful explanation of error-free, expert performance. In contrast, the model falls short of explaining error-ridden performance that exists when a novice starts out learning a new system (Embley & Nagy, 1981). According to Embley and Nagy, 25%–50% of a novice's time is spent committing errors and recovering from errors. A complete account of text editing performance should address the errors of beginning users of a text editor.

The Importance of Studying Errors

Human factors engineers seriously examine the errors that users commit when they interact with a device. A description and explanation of these errors is an important prerequisite for designing an effective interface between a machine and an operator of the machine (Lewis & Norman, 1986; Mack et al., 1983; Norman, 1981; Rouse & Rouse, 1983). Computers are no exception to this generalization.

We have already mentioned that 25%–50% of a novice's early experience with a text editor is consumed by errors. The novice commits errors, tries to recover from the errors, and agonizes over the errors. When the user's errors are difficult to recover from, the user becomes very frustrated, angry, and discouraged. There is a negative feedback loop in which things get worse before they get better. Some users cry and give up after trying to cope with the troublesome errors.

Carroll and Carrithers (1984a,b) have developed a "training wheels" system which was designed to minimize errors of beginning users. The training wheels system is a simplified version of the target system which preserves the major design features of the target system (e.g., menus, spatial layout, messages, prompts, etc.). The training wheels system makes troublesome errors unreachable to the beginning users by cutting out problematic response options. For example, beginning users sometimes pursue exoteric alternatives on a menu (e.g., diagnostics) and end up in "la-la land" trying to understand the alternatives and restoring the main menu. The training wheels system would not allow the user to choose such problematic alternatives. When such alternatives are selected, the computer returns the message "XXX is not available on training system." Carroll and Carrithers conducted a study which demonstrated that the training wheels system is a very effective training method. When the users are transferred to the target system, they complete the tasks faster, they commit fewer errors, and they like the system better than users who were not trained on the training wheels system.

The study of errors should not be restricted to beginning users. Expert users of a text editor also have conceptual errors that occasionally lead to performance problems. An expert's cognitive representation of a system is not an accurate and complete model of the system. Instead, an expert's cognitive representation is incomplete, inconsistent, unstable, oversimple, and superstitious (Lewis & Mack, 1982; Lewis & Norman, 1986). Experts perform satisfactorily on tasks that they are familiar with. However, their conceptual errors are manifested when they try to solve novel problems or learn a new system.

Trade-offs in Design Principles and Decisions

There are a staggering number of factors and constraints to consider when a decision is made about a specific design feature. The designer must consider alternative user populations, alternative design features, and potential changes in technology. The notion of designing an ideal interface is perhaps an impossible goal, because there are both costs and benefits in adopting any particular set of commands, computer prompts, messages, spatial layouts, and help facilities. Instead of trying to find the ideal solution to an interface problem, a more realistic goal is to evaluate the trade-offs in adopting a specific design feature (Black & Sebrechts, 1981; Norman, 1983). The designer should perhaps perform a cost-benefit analysis on alternative design features for a specific system rather than establishing a set of ideal design principles that apply to all computer environments.

In order to illustrate the importance of trade-offs, consider the situation in which the users consistently fail to interpret an error message correctly. The problematic error message is "overflow error." Designer A decides to improve the message by including four lines of text which explains the source of the error and how to correct it. Designer B decides to improve the message by including two lines of text on how to correct the error (without explaining the source of the error). Which designer is correct? There is no straightforward answer to this question. The answer depends on the user population. If the users are system programmers, then designer A is probably correct, because the A-message has more information than the B-message. If the users are novices, then designer B might be correct. The longer A-message would take longer to read and would not improve the likelihood of correcting the specific error. The A-message might also spawn additional misinterpretations and misconceptions that could be avoided in the B-message. The tradeoffs of each message must be evaluated and weighed before the final design decision.

Having introduced the practical and theoretical foundations of our research in computer–human interaction, we are ready to describe the study we conducted using the Q/A methodology.

METHODS OF DATA COLLECTION

This section describes the computer system that we investigated and the methods of collecting data from individuals learning the text editor.

The Computer System and the Text Editor

We investigated a text editor called XEDIT. XEDIT is the normal editor for the NOS operating system on the CYBER 173 computer. In order to create and edit a file, the user must log onto the system, access XEDIT, and use XEDIT. Thus, there is a log-on procedure, a file creation procedure (using XEDIT), and a file modification procedure (using XEDIT). Table 1 shows the log-on procedure for a user with a specific account number (XS077AS) and password (R21KQZQ). Like most operating systems, NOS has several system-level commands (called BATCH commands) which perform major functions. Some of these major functions are listing the files

Table 1. Log-on Procedure.

WHICH SYSTEM? (40) (cr)

CSUF—enter 2 ⟨CR⟩'s after GO
GO (cr)
(cr)

83/02/08. 11.31.57. L167100
CSU FULLERTON CYBER 170/730-2.
NOS 1.4-552A.

FAMILY: (,XSO77AS,R21KOZQ)

TERMINAL: 12,NAMIAF
RECOVER/SYSTEM:
/

Note: The user's responses are in circles. "Cr" refers to the carriage return key.

in a user's account (CATLIST), saving a file (SAVE, fn), replacing an old file with new updated information (REPLACE, fn), and accessing XEDIT (XEDIT, fn).

XEDIT is not the best text editor on the market. However, it was the most frequently used editor among the thousands of students and faculty members who use the NOS system at California State University at Fullerton. XEDIT is a line editor, not a full screen editor. Moreover, the lines in XEDIT are not numbered. Instead, there is an imaginary "pointer" that points to the current line that the user is on. For example, Table 2 shows the content of a file with the pointer positioned at the fifth line. The user must enter specific XEDIT commands to move the pointer and to print out lines in the file. TOP (or simply T) is a command which moves the pointer to the top of the file whereas BOTTOM (or B) is a command that moves the pointer to the end of the file. The command NEXT (or N) moves the pointer to the next line; "NEXT 5" moves the pointer forward 5 lines and "NEXT −3" moves the pointer backward 3 lines. The command

Table 2. Some Text with a Pointer in XEDIT.

	Promptly at 7 the next morning, Jeff reported
	for work. He came to the door that had been
	pointed out to him, and there he waited for
	nearly 2 hours.
POINTER→	12 755 45.83 21.0 1 2 4 45 6
	54 233 23.00 14.8 4 225 66 89
	END OF FILE

PRINT (or P) lists the current line on the terminal; "PRINT 5" lists 5 lines, starting with the line where the pointer is positioned. There are two major commands for adding information into the data file. INPUT adds information at the end of the file whereas INSERT permits the insertion of information in the middle of the text. The DELETE command deletes lines whereas the CHANGE command permits the modification of characters on the current line (where the pointer is positioned). XEDIT has many more commands, but the above commands are satisfactory for completing most editing tasks.

A user must distinguish among three "modes" when performing an editing task. These three modes are listed and described below.

Batch mode. The computer signifies this mode with a slash (/) prompt. Appropriate responses are batch commands, such as CATLIST, SAVE, REPLACE, and XEDIT.

XEDIT command mode. The computer's prompt for this mode is two question marks (??). Appropriate responses are XEDIT commands, such as TOP, BOTTOM, NEXT, PRINT, DELETE, CHANGE, INPUT, and INSERT.

Input mode. The computer's prompt for this mode is a single question mark (?). The user enters the text for the file in the input mode. The input mode is a "submode" within the INPUT and INSERT commands. Users occasionally commit *mode errors* when they edit a text. A mode error occurs when the user loses track of the current mode and enters inappropriate input. Table 3 shows an interaction in which the user committed several mode errors. When the user entered the lines PRINT, LIST, TOP, and so on, the user thought he was in the XEDIT command mode. However, the user was still in the input mode (notice the "?" prompt). When the user entered "SAVE, FILE," the user had inappropriately entered a batch command when he still was in the input mode. The user eventually became very frustrated and discouraged.

Manual

We prepared a short, 9-page manual which explained basic information about the NOS system and XEDIT. The manual covered the log-on procedure (see Table 1), important NOS system commands, computer files, XEDIT, the major hardware components (terminal, printer, computer), and special conventions that a user needs to know in order to communicate with the NOS system (e.g., the cursor, the return key, prompts, modes). The author attempted to write a manual that was easy to comprehend. It was not cluttered with terminology and details that were irrelevant to text editing and major system commands

Table 3. Examples of Mode Errors in XEDIT.

```
/ XEDIT, FILE

XEDIT 3.1.00
EMPTY FILE/ CREATION MODE ASSUMED

?? INPUT

? Promptly at 7 the next morning, Jeff reported
? for work. He came to the door that had been
? pointed out to him, and there he waited for
? nearly 2 hours.
? 12 755 45.83 21.0 1 2 4 45 6
? 54 233 23.00 14.8 4 225 66 89
? PRINT
? LIST
? TOP
? NEXT -5
? BOTTOM
? PRINT
? EDIT
? E,,RL
? SAVE, FILE
? XEDIT, FILE
? This is crazy! Is this computer deaf or something?
```

Note: The user was supposed to hit the RETURN key twice in order to get out of INPUT mode.

Benchmark Tasks

There were three benchmark tasks that the users performed. The first was the log-on procedure, which is shown in Table 1. The second was the creation of a data file. The experimenter handed the subject a sheet of paper and asked the user to create a data file from the material on the sheet. The sheet of data contained both verbal material and columns of numbers (see Tables 2 and 3 for an example of the material). The subject created a file using XEDIT and then saved the file. The third benchmark task involved modifying the data file that was saved. The experimenter handed the subject another sheet of paper that was similar to, but not identical to the first sheet. The experimenter asked the subject to update the old data file with the information on the new sheet of paper.

Subjects

The subjects were 17 college students at California State University, Fullerton, who were enrolled in a course entitled "computer applications in the social sciences." The students participated in the study in order to fulfill a course requirement at the beginning of the semester.

We segregated the subjects into three groups which reflected the amount of previous experience they had with computers. At the beginning of the study, the subjects wrote down the amount and nature of their computer experience. The subjects also specified their sex, age, and level in college (i.e., junior, senior, versus graduate status). Five of the subjects were classified as "novices"; they had never used a computer in the past, except for an occasional computer game. Six of the subjects were classified as "experienced" computer users; they had used a computer for at least 2 years on a fairly continuous basis. We should point out, however, that the computer system that they had used was *not* the NOS system and XEDIT. The remaining six subjects were in an "intermediate" category.

Our classification of the 17 subjects into novice, intermediate, and experienced users proved to be valid because the classification robustly predicted performance on the benchmark tasks. The proportions of subjects who completed all three benchmark tasks (within 2 hours) significantly increased as a function of computer experience, .20, .33, and .83 in the novice, intermediate, and experienced categories, respectively, $F(1, 14) = 5.14$, $p < .05$. The mean time that subjects spent on the task was also computed, using a 2-hour upperbound. The mean times significantly decreased as a function of computer experience, 83, 63, versus 23 minutes in the novice, intermediate, versus experienced categories, $F(1, 14) = 6.88$, $p < .05$. In contrast, the three groups did not significantly differ in sex, in age, and in class level (even when we adopted a liberal .25 level of statistical significance).

Design and Procedure of Study

The entire study had two sessions and several different phases. Session 1 contained four phases, whereas section 2 contained three phases. These phases are listed and described below.

(1) Assessment of computer knowledge. As we mentioned above, the subjects wrote down the amount and nature of computer experience that they had. They also specified their age, sex, and level in college.

(2) Free generation protocols. The subjects provided written protocols that discussed their knowledge of computers. They were given the following three questions to guide them in their free generation protocols: "What is a computer?", "What are computers used for?", and "How does a computer operate?"

(3) Computer term test 1. The subjects were instructed to define and describe the meaning of several terms that were relevant to computers.

These terms included hardware components (*terminal, keyboard, printer*), major software components (*file, program, catalog, subsystem, account, log on*), specific computer symbols and keys (*prompt, cursor, return key, character*), and software functions (*command, edit, delete, insert, list*). The subjects supplied answers in writing.

(4) Reading the computer manual. The experimenter gave the subjects the nine-page manual and asked them to read it very carefully before they showed up at the next session. The subjects had 4 days to study the manual.

(5) Computer term test 2. At the beginning of session 2, the subject again defined and described the meaning of the list of computer terms which had been tested in session 1. The subjects' answers were tape recorded. It should be noted that the subjects read the manual between computer term tests 1 and 2.

(6) Benchmark tasks with question answering protocols. The subjects performed the three benchmark tasks that we described earlier (i.e., log on, file creation, and file modification). The experimenter probed the subjects with questions while the subjects performed the benchmark tasks. The experimenter adopted three rules to guide the course of questioning. These three rules were specified at the beginning of this chapter. The subjects were instructed to give detailed answers to the questions and to supply whatever relevant information they had, even if they were not sure it was correct. The question-answering protocols were tape recorded. In addition to the verbal protocols, the experimenter obtained a printout of the interaction between the subject and computer.

(7) Computer term test 3. The subjects again defined the meaning of the computer terms which had been tested earlier. The subjects' answers were tape recorded. It should be noted that the subjects performed the three benchmark tasks between computer term tests 2 and 3.

Scoring the Verbal Protocols

The free generation protocols, computer term definitions, and question answering protocols involved complex verbal descriptions. We will briefly discuss how we coded and scored these protocols.

For illustration, consider the protocols that the subjects contributed when they were asked to define *program* in the computer term definition task. There were four steps in coding the answers. In the first step of the coding, we segmented each subject's protocol into *statement units*. A statement unit is a basic idea unit that refers to a state, event, action, goal, or style specification (see Graesser & Clark, 1985). There will be many examples of statement units throughout this chapter; the reader should gain a satisfactory impression of statement units through the examples.

In the second step of the coding, we determined which statement units overlapped among the different subjects' protocols, and we combined the overlapping statements. Statement units overlapped if they had the same meaning or a very similar meaning. As a consequence of the second step, we had a list of unique statement units. Associated with each statement was a proportion score measuring the proportion of subjects who generated the statement. A proportion score of .50 means that half of the subjects generated the statement unit. Listed below are a set of statement units and proportion scores for the computer term *program*.

A program is a set of instructions (.60, .83, 1.00)
X wants the computer to do something (.60, .67, .83)
A program analyzes data (.00, .33, .50)

There are three proportions associated with each statement unit. The three proportions correspond to scores for novices, intermediates, and experienced users, respectively.

In the third step of coding verbal protocols, we assigned each statement node to a node category. We adopted four categories: state, event, goal/action, and style specifications. Examples of three of these categories are presented in Table 4, which will be discussed later in this chapter. A state is a component (e.g., the keyboard has function keys), a property of a component (the screen is bright), or a relationship between components (the screen is above the keyboard). States remain unchanged during the time frame when a computer is used. An event is a change of state that occurs within the time frame (e.g., errors occur, the computer processes information). A goal/action unit is either an intention (e.g., X wants to get the computer to do something) or goal that is eventually achieved (e.g., X got the computer to do something). A style specification embellishes the manner in which an event occurs (e.g., X occurred *quickly*). Each statement unit was assigned to one of the four categories.

In the fourth step of analyzing the protocols, we decided whether each statement was true or false. The truth of the statement was evaluated with respect to the question. Thus, a statement was scored as false if it was a true assertion, but an erroneous response to the question.

DATA ANALYSES AND RECOMMENDATIONS FOR DESIGN CHANGES

In this section we report some analyses of the data collected in the free generation task, the computer term definition tasks, and the benchmark tasks in which Q/A protocols were collected. Some tasks were particularly prolific in tapping the conceptual errors that users had about the system. We will devote most of our attention to those tasks that uncover such errors. Some of the findings suggested that there were problems in the computer–user interface. Whenever appropriate, we will suggest ways that the interface can be improved and point out some of the tradeoffs in our recommendations for design changes.

How Useful Are the Free Generation Protocols?

In the free generation task, we asked the subjects to write down what they knew about computers "off the top of their heads." The subjects were asked rather open-ended, global questions, including *what is a computer*?, *what is a computer used for*?, and *how does a computer operate*?. Table 4 presents a sample of the statements which were generated by at least two subjects. Table 4 segregates statements that were generated primarily by novices versus experienced users.

We did *not* find the free generation data to be very useful in identifying conceptual errors and design problems. A very low percentage (2%) of

Table 4. Statements Expressed in the Free Generation Task.

NOVICE USERS
Goals/Actions
The user enters data into the computer
The user retrieves data
Events
The computer analyzes data
The computer solves problems
States
The terminal is like a typewriter
The computer is a machine
EXPERIENCED COMPUTER USERS
Goals/Actions
The user uses a program
The computer uses a disk
Someone writes a program
Events
The computer processes information
The computer performs computations
The computer keeps records
States
The computer is electronic
Code is binary
Information includes numbers

the statement units were false statements. Consequently, the free generation task did not uncover many conceptual errors. We also found most of the statements to be nonspecific and unsophisticated, even those statements generated by experienced computer users (see Table 4). None of the statements in Table 4 were specific, sophisticated, informative statements about computers.

The fact that the free generation statements tended to be very true, but nonspecific and unsophisticated ideas replicates some conclusions reported by Graesser, Hopkinson, Lewis, and Bruflodt (1984). Graesser et al. collected free generation protocols on the topics of economics, cancer, and growing flowers, and found that nearly all of the generated information was true and unsophisticated. The major conclusion to be drawn from these data is very simple. Very little can be learned about a user's knowledge of a system by simply asking the user to talk off the top of his or her head in an unconstrained manner.

How Useful is the Computer Term Definition Test?

We found the computer term definition task to be only moderately useful. A very small percentage (3%) of the statement units were erroneous, so the task was not well suited to uncovering the users' conceptual errors. The users tended to answer "I don't know" rather than fabricating false information. As with the free generation task, the computer term task tends to elicit true and unsophisticated information. However, we did find the answers in the computer term task to be somewhat more specific than the ideas produced in the free generation task.

We had originally hoped to trace the growth of the knowledge structures associated with the computer terms by observing differences among tests 1, 2, and 3. Recall that the subjects read the manual between tests 1 and 2; they performed the benchmark tasks between tests 2 and 3. One change we anticipated was a shift from declarative knowledge to procedure knowledge (Anderson, 1983; Kieras, 1985). Declarative knowledge consists of static facts which describe what is true. Procedural knowledge consists of the goals and intentional actions that are involved with doing something. If there is a shift from declarative knowledge to procedural knowledge as the subject learns how to perform the task, then this shift should be manifested in the node categories across tests 1, 2, and 3. We were disappointed to learn that such a shift was not evident in the computer term test. The proportion of statement units that were goals/actions, events, versus states were .29, .29, and .42, respectively. These proportions stayed constant across tests 1, 2, and 3. We also failed to find any differences in node categories among novice, intermediate, and experienced computer users.

In summary, the computer term test did not show any shift from declarative knowledge to procedural knowledge as a function of computer experience.

We did find the computer term test to be useful in one respect. The computer term test revealed which terms were problematic to subjects before they read the manuals. Both novices and experienced computer users knew the meaning of some of the terms, such as *keyboard, program*, and *command*. However, novices were perfectly ignorant about the meaning of some terms, such as *cursor, prompt, subsystem*, and *catalog*. It is very critical for designers to prepare system messages that are nonproblematic for the beginning users. The computer term test is suited to identifying problematic terms. The computer term test provides informative data for authors of system messages, training manuals, and help facilities (Houghton, 1984; Shneiderman, 1982).

How Useful Are the Benchmark Tasks with Q/A Protocols?

The most useful data came from the benchmark tasks and the Q/A protocols. There were two major sources of data from these tasks. First, we had a printout of the computer-user dialog. Second, we had tape recorded Q/A protocols for the questions the experimenter asked throughout the benchmark tasks. Recall that the experimenter had asked "why did you do that?" whenever the subject performed an action, "what does that mean?" whenever the computer presented a prompt or message, and "what are you thinking about now?" whenever the subject paused for more than 15 seconds.

The printout and Q/A protocols uncovered many of the conceptual errors that the subjects had about the NOS system and XEDIT. Overall, the percentage of statement units in the Q/A task that were erroneous was 18%. This 18% figure was much higher than the percentages for the free generation task and the computer term definition task, 2% and 3%, respectively. Compared to the other tasks that collected verbal protocols, the Q/A task was a goldmine.

Some of the subjects' conceptual errors were easy to identify on the basis of the printout of the computer–user interaction. However, other conceptual errors could only be identified after listening to the Q/A protocols; the printout was not distinctive enough to recognize a conceptual error. We performed some analyses which computed the proportion of errors that were manifested in the printout of the computer–user interaction versus the Q/A protocols (but not the printout alone). We considered all user errors except for minor typing errors.

Table 5 shows the proportion of errors that were exposed by the Q/A

Table 5. Proportion of Errors that Were Manifested in the Question-Answering Protocols, But Not in the Printout of the Computer–User Interaction.

Segment in Benchmark Task	Proportion
XEDIT Commands	
INPUT	.54
INSERT	.25
DELETE	.50
CHANGE	.56
PRINT	.50
TOP/BOTTOM	.60
NEXT	.40
Other Procedures	
Log-on	.56
Naming new file and accessing XEDIT	.52
Re-accessing old file and XEDIT	.33
Saving file	.17

task, but not by the printout. Table 5 reports separate proportion scores for different XEDIT commands, the log on procedure, and different NOS commands. The mean of these proportion scores was .46. Therefore, 54% of the users' errors would be detected from the printout alone, whereas 46% could only be detected only when the researcher had access to the Q/A protocols. In other words, the Q/A protocols were nearly as informative as a printout of the computer–user interaction.

We are convinced that the most productive method of uncovering the users' conceptual errors is to have them perform benchmark tasks and answer questions while they perform these tasks. Between the printout and the Q/A protocols, we were confident that we could reconstruct an accurate explanation of what the users were doing and thinking about while they performed the tasks. Just as important, we could identify ways to improve the interface in order to minimize the errors. We will informally substantiate these claims by analyzing some specific conceptual errors.

A Troublesome Error During the First Step of the Log on Procedure

The subjects committed several errors during the log on procedure (shown in Table 1). The first log-on error was very prevalent and occurred at the first line of the log-on procedure. The subjects failed to hit the return key after they typed in "40". The manual emphasized the importance of hitting the return key at the end of each line the user enters. In fact, the manual included a joke that the computer would "sit there sleeping" until the user hit the return key. Yet most of the subjects, even experienced computer users, made this error.

The lines below show what the printout (of the computer–user interaction) looks like when the subjects fail to hit the return key after typing "40."

CSU Fullerton, Computer Center
WHICH SYSTEM? 40   DISCONNECTED

CSU Fullerton, Computer Center
WHICH SYSTEM? 40   DISCONNECTED

CSU Fullerton, Computer Center
WHICH SYSTEM? 40   DISCONNECTED

•
•
•

The user gets disconnected when the user fails to hit the return key within 20 seconds. Many users go through several cycles when they commit this error, as illustrated above. The proportions of subjects who committed the error during the first cycle were 1.00, .67, and .67, for novice, intermediate, and computer experienced users, respectively. The error proportions for the second cycle were .80, .33, and .00; the error proportions for the third cycle were .40, .33, and .00. One subject went 19 cycles committing the error. This error is very costly in terms of time and frustration. For some subjects, several minutes were expended on this first step of the log-on procedure. This is hardly a satisfactory introduction to computer technology! Experienced computer users also make the error at least once, so there clearly is something wrong with the interface.

The printout alone did not uncover the return key error. The printer does not list a symbol when the return key is entered by the subject. The printout also does not list reaction time latencies for subjects' responses. Consequently, there are no direct clues in the printout as to what the user's error is. We presented to five computer consultants the printout of a subject who had committed the error for four cycles. Only one out of the five computer consultants could correctly diagnose what the user's error was. The return key error is an excellent example of an error which is difficult to detect from a printout of the computer–user interaction.

In contrast, the return key error was easy to identify when we had access to the Q/A protocols in addition to the printout of the interaction. We found out how subjects were interpreting the DISCONNECTED message and the long pause that occurred between their typing "40" and the DISCONNECTED message. The Q/A protocols elicited the subjects' explanations of the problem, their frustrations, and their strategies for correcting the problem. Most of the subjects' explanations and strategies were entirely incorrect. For example, one subject believed that a cord was disconnected from the computer. Troublesome errors are a breeding ground for subjects to fabricate all sorts of misconceptions. The Q/A protocols also revealed the point in time when the subjects correctly diagnosed the problem. We could usually identify the clues which led the subjects to a correct diagnosis.

The return key error is a very severe error. Users expend several minutes struggling with it. Virtually no important information is acquired during the struggle. If anything, the user constructs erroneous explanations of the system. Available data indicated that return key error cannot be prevented by writing a better manual. The manual was not only easy to comprehend, but it also dramatically emphasized the importance of hitting the return key after every line.

The return key error is a product of shabby software design which fails to foster a cooperative interaction. First, the computer did not present a prompt which would prevent the error. Second, the computer did not present an informative error message which explained to the user how to correct the error. Computer prompts and messages should be written so the user can prevent errors and recover from errors (Schneiderman, 1982). With the above considerations in mind, we would recommend the following revised computer prompt (see Table 1 for a comparison of messages).

> WHICH SYSTEM DO YOU WANT?
> Type in the 2-digit number code of the system you want and
> then hit the RETURN key:

Of course, there are tradeoffs whenever a design change is considered. The revised computer prompt is clearly less ambiguous and more informative. However, the screen is cluttered with more information, and some of this information might end up being misinterpreted (leading to additional conceptual errors). After considering the costs and benefits of the revised prompt, the revision would still come out on top. The additional time to read the screen is only 2–3 seconds, according to research in reading time (Card et al., 1983; Graesser & Riha, 1984). Three seconds is negligible compared to several minutes struggling with the return key error. Regarding potential misinterpretations of the expanded computer message, the prompt does not contain any words or ideas that are problematic to the novice user. This claim could be verified in a study that adopts a term definition task. Although there are tradeoffs associated with the recommended revision, a cost-benefit analysis supports the recommended change in the computer prompt.

Other Errors During the Log-on Procedure

There were errors associated with the second step of the log-on procedure, after the computer presented the following prompt.

> CSUF—enter 2 ⟨CR⟩'s after GO

When the log-on proceeds correctly, the computer prints out GO after a few seconds and the user hits the RETURN key twice. Some users misinterpret the above computer message. Some subjects thought that the user should type in GO instead of the computer printing GO. It should be noted that the computer message is entirely ambiguous about who or what should deliver the "GO". The proportions of subjects who misinterpreted the computer prompt was .40, and .17. A few subjects also misinterpreted "⟨CR⟩"s. Two subjects typed in "⟨CR⟩" instead of hitting the RETURN key. Three subjects hit an incorrect key on an auxiliary keyboard that was labeled "CR." These user errors indicate that the above computer message was very cryptic and ambiguous.

The following computer prompt would be an improvement over the existing ambiguous prompt (shown in Table 1).

> CSU Fullerton Computer
> Hit the "RETURN" key twice after the computer prints
> "GO" on the screen.

The verb *hit* is less ambiguous than *enter*; a user can enter input by hitting a single key or by typing several letters. The message specifies that the computer prints GO instead of the user typing GO. In essence, the revised prompt prevents some common ambiguities. Nevertheless, the revised message does suffer one disadvantage. Some terminals do not have RETURN keys. Instead, IBM terminals have ENTER keys and Sperry Univac computers have TRANSMIT keys. Specific and unambiguous computer messages sometimes run the risk of not being portable or transferable to other systems and hardware components.

Mode Errors

As we discussed earlier, the user interacts with three different modes when they perform text editing. First, there is the NOS system's "batch" mode, which has a "/" prompt. Second, there is the "XEDIT command" mode, which has a "??" prompt. Third, there is the "XEDIT input" mode which has a "?" prompt. The subjects often committed mode errors when they interacted with the system. As illustrated in Table 3, the subjects entered XEDIT commands and batch commands when they were in the input mode. Subjects sometimes entered XEDIT commands when they were in the batch mode, and batch commands when they were in the XEDIT command mode. Stated differently, the user and computer were not communicating on the same level of interaction.

We analyzed the mode errors by computing the proportion of subjects who committed such errors. We prepared separate proportion scores for different commands and for different groups of subjects. We found the following distribution of proportion scores among the different XEDIT commands that the subject tried to execute: INPUT (.18), INSERT (.24), DELETE (.12), CHANGE (.00), PRINT (.29), TOP/BOTTOM (.24), and NEXT (.24). The proportion scores for mode errors were .00, .18, and .29 when the subjects named a new file, accessed an old file, and saved a file, respectively. When averaging across these command categories, the mode error proportions were .24, .25, and .07, for novice, intermediate, and experienced users, respectively. Therefore, it appears that mode errors decrease as a function of general computer experience. Users eventually distinguish among the different levels of a command system and to keep track of which (sub)system the user is communicating with.

One factor that contributed to the mode errors was the fact that the prompts were not very distinctive and discriminable. The differences among /, ??, and ? are comparatively subtle. The subjects would be less proned to committing mode errors if the prompts were "Enter system command:", "Enter XEDIT command:", and "Enter text:". The screen would be more cluttered, but the beginning users would make fewer mode errors. Perhaps the best solution is to present these wordy prompts to beginning users, and to allow experienced users to enter a special command which suppresses the wordy prompts.

Losing Track of the Pointer

As we discussed earlier, XEDIT is a line editor without numbered lines. There is an imaginary pointer which refers to the current line in the text. The user must keep track of where the pointer is and must frequently enter XEDIT commands which move the pointer. For example, TOP moves the pointer to the top of the file and NEXT 5 moves the pointer down 5 lines. Although some of the XEDIT commands print out the current line, very often the location of the pointer is in the mind of the user rather than being on the display screen.

We found that users frequently lost track of where the pointer was.

Table 6. Probability of Losing Track of the Current Line.

	Group of Subjects		
XEDIT Command	Novice	Intermediate	Experienced
INSERT	.40	.50	.33
DELETE	.40	.67	.50
CHANGE	.20	.00	.17
PRINT	.40	.17	.33
NEXT	.60	.50	.50
Overall	.40	.37	.37

Table 6 presents the proportion of users who lost track of the current line when they entered various XEDIT commands. Table 6 also segregates proportion scores for novice, intermediate, and experienced computer users. When averaging over subjects, the likelihood of losing track of the pointer was .41, .52, .12, .30, and .53, for the INSERT, DELETE, CHANGE, PRINT, and NEXT commands, respectively. When averaging over commands, the proportions were .40, .37, and .37, for novice, intermediate, and experienced computer users, respectively.

The problem of losing track of the pointer is comparatively serious. The users often ended up deleting the incorrect lines. They often inserted information after the current line rather than before the current line. Sometimes it took several minutes to recover from these errors, because the subjects never quite figured out how to print out several lines where they had been working. The proportion scores in Table 6 are rather high, so the problem of losing track of the pointer is quite pervasive. Moreover, the error likelihoods do not decrease as a function of computer experience. Everyone seems destined to make these errors.

Fortunately, the unnumbered line editors are losing popularity in the technology of text editing. Indeed, most commercial text editors are screen editors rather than line editors. Screen editors (e.g., WORDSTAR) permit the user to view several lines of text simultaneously, which is easier on the user (Neal & Darnell, 1984). In screen editors, there is a cursor at one point on the screen. Perhaps the only tradeoff in adopting the screen editing feature is that extra time is needed to search for the cursor in a congested screen. However, this scanning time is negligible compared to several minutes of recovery time when the frequent errors are made in unnumbered line editors.

We should note that the Q/A protocols were usually needed to identify instances when the user lost track of the pointer. It was usually impossible to infer this error from a printout of the computer-user interaction. This errors constitutes an excellent example of a frequent, serious error which would not be detected by a printout (or even a videotape) of the computer–user interaction.

To What Extent do the Users Consult the Manual?

The users were permitted to reread the manual while they performed the benchmark task. Whenever they consulted the manual, the experimenter asked the user "why are you reading the manual?" Table 7 presents data on the likelihood that the users consulted the manual when they entered different XEDIT commands, batch commands, and log-on information. These proportion scores varied from .20 for the NEXT command to .71 for the CHANGE command. The proportion scores substantially decreased as a function of computer experience, .52, .39, versus .26, for novice, intermediate, and experienced users.

It would perhaps be more convenient for users to use an online help facility than to consult a manual. It takes a measurable amount of time to locate the correct manual, to search through its index and pages, and to select the information that is relevant to the task at hand. With an online help facility, the user would not need to find the correct manual and search through the pages. However, available research indicates that users are more inclined to consult manuals than to take advantage of online help facilities (Rosson, 1984). Users are often reluctant to use help facilities when the help messages replace the text that they are working on. Users want to keep the information in their working space intact. Perhaps the best help facilities are those that present information in a separate window that is segregated from the user's working space.

Table 7. The Likelihood that the User Consults the Manual.

	Group of Subjects		
	Novice	Intermediate	Experienced
XEDIT Commands			
INPUT	.60	.33	.17
INSERT	.40	.83	.67
DELETE	.60	.33	.00
CHANGE	.80	.83	.50
PRINT	.40	.50	.33
TOP/BOTTOM	.20	.50	.17
NEXT	.60	.00	.00
Other Procedures			
Log-on	.80	.17	.17
Naming new file and accessing XEDIT	.20	.33	.17
Re-accessing old file and XEDIT	.40	.33	.67
Saving file	.80	.17	.00

Did the Subjects Understand the Goals of the Benchmark Tasks?

A cooperative interaction between a computer system and a user requires that the user construct an adequate goal structure. We performed some analyses which assessed the extent to which the subjects understood the goals associated with the benchmark tasks. In the spirit of the GOMS model, we assumed that the execution of the benchmark tasks was guided by a hierarchical goal structure. A user could not satisfactorily complete the tasks if such a goal structure were not constructed.

Before we collected the data in this study, we constructed an ideal goal structure that was necessary for the user to complete the benchmark tasks. We segregated the goals into superordinate (main) goals, midordinate goals, and subordinate goals. The ideal goals are listed and categorized below.

SUPERORDINATE GOALS—(log-on computer system, create a file, modify a file, use XEDIT)

MIDORDINATE GOALS—(give file a name, input information into the file, modify information in file, correct errors in file, save the file)

SUBORDINATE GOALS—(use INPUT command, use INSERT command, delete a line, change a word on a line)

We found that the subjects often articulated these goals when they supplied answers in the Q/A task. When averaging over subjects, the likelihood of articulating a given superordinate goal was .96; the corresponding likelihood scores for midordinate goals and subordinate goals were .80 and .68, respectively. These scores support three conclusions. First, the subjects understood the goals of the benchmark tasks because the proportions were very high. Second, the subjects' major challenge was constructing a plan to achieve the goals. This conclusion is supported by the trend that subordinate goals (which are at the plan level) had lower proportion scores than did the superordinate goals. Third, the Q/A protocols tapped many of the goals that are part of the goal structures of users.

Planning Networks

We were able to prepare planning networks from the Q/A protocols and the actions in the benchmark tasks. A *planning network* is a structured set of goals and actions, along with states and events that initiate the goals. We prepared a planning network for each of the XEDIT commands. The *content* of each planning network was extracted from the Q/A protocols associated with the XEDIT commands. This content was *structured* according to a theory of knowledge representation developed by Graesser (Graesser, 1981; Graesser & Clark, 1985).

According to the theory, knowledge is represented in the form of conceptual graph structures. A conceptual graph structure is a set of categorized statement nodes that are interrelated by a set of categorized, directed arcs. The node categories refer to states, events, goals, actions, and style specifications, as discussed earlier in the chapter. The arcs are higher-order relations which specify the conceptual relationships between statement nodes. Graesser's representational system includes nine arc categories altogether: REASON, INITIATE, OUTCOME, MANNER, CONSEQUENCE, IMPLIES, PROPERTY, SET-MEMBERSHIP, and REFERENTIAL-POINTER. Five of these categories are briefly described below.

REASON (R). This arc category links goal nodes. One goal is a motive or reason for another goal node.

INITIATE (I). Goals are initiated (i.e., created, activated) by states, events, and actions.

OUTCOME (O). A goal may or may not be achieved. This arc links a goal with a state or event which specifies whether the goal is achieved.

MANNER (M). An event or action is embellished with another node specifying how the event or action occurred. This arc may also link goal nodes.

CONSEQUENCE (C). One event, state, or action causally leads to another. The nodes in causal chains and networks are linked by this arc category.

Figure 1 shows an example planning network for the PRINT command in XEDIT. All of the nodes in Figure 1 were extracted empirically from either the actions in the benchmark task or the answers in the Q/A protocols (elicited when PRINT commands were queried with why-questions). The structure contains several goal/action nodes which are interrelated by Reason (R) arcs and Manner (M). The Reason arcs are directed and the direction is nonarbitrary. For example, it is appropriate to say that the user gets the computer to print a line *in order to* verify that the line was modified; it does not make sense to say that the user verified that a line was modified *in order to* get the computer to print the line. The goal nodes are initiated (activated) by events and states via Initiate (I) arcs. Each node in Figure 1 contains three proportion scores, which correspond to the proportion of novice, intermediate, versus experienced users who generated the statement node in the Q/A protocols and benchmark tasks.

The purpose of presenting Figure 1 is to illustrate that the Q/A methodology provides a very rich database which can be structured according to available theories of knowledge representation. Although it is beyond the scope of this chapter to discuss the representational theories in detail, we do want to acknowledge that such analyses can be performed. Thus,

Figure 1. Planning Net for the PRINT Command in XEDIT. (U = user; C = Computer). The arc categories include Reason (R), Manner (M), and Initiate (I) arcs.

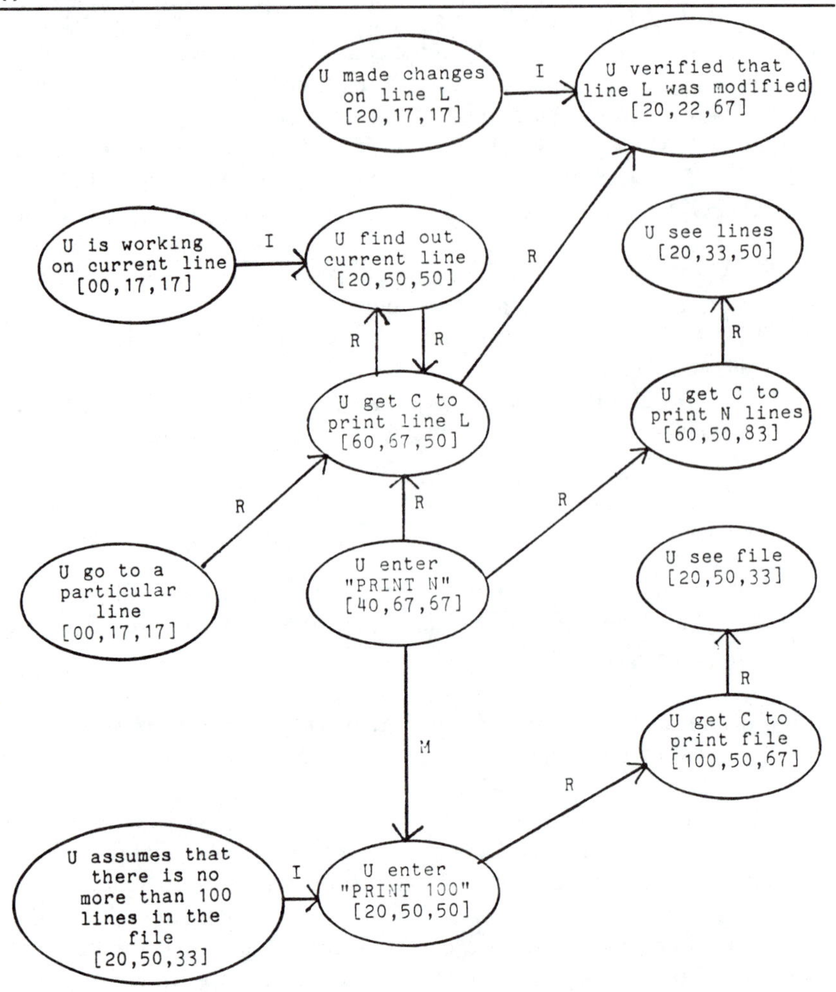

the Q/A methodology uncovers detailed knowledge structures in addition to the users' conceptual errors.

GENERAL DISCUSSION AND CONCLUSIONS

We were very encouraged with the Q/A methodology as a tool for exploring the users' knowledge as they learn how to interact with computers. Investigators can learn a great deal about the interface by asking users to perform representative benchmark tasks and to answer questions while they perform the tasks. The Q/A protocols uncovered several kinds of conceptual errors that the users had while they performed the tasks. We could identify when users misinterpreted computer prompts and messages. We could identify the users' mode errors and instances when they lost track of the pointer in the XEDIT line editor. An analysis of these errors is obviously critical for designing cooperative computer–human interfaces, for writing effective training manuals, and for designing online help facilities. The Q/A methodology exposed the detailed knowledge structures which explain the users performance in these tasks. These knowledge structures include the goal hierarchies and planning networks which direct organized behavior. We could also identify points in benchmark tasks when the users needed help and consulted the manual. The Q/A protocols provide a very rich and informative database for software engineers.

Selecting the appropriate questions is obviously very critical in the Q/A methodology. The three rules in the introduction specify what questions the experimenter asked and when the experimenter asked them. We did not arbitrarily or intuitively select these questions. Our selection was based on previous psychological research on question answering and verbal protocol methodology (see Graesser & Clark, 1985). For example, when a user's actions are probed with why-questions (i.e., *why did you do that?*), the answers sample superordinate goals in the goal/action hierarchy. The superordinate goals are very critical in directing and in explaining the user's behavior. In contrast, when a user's actions are probed with how-questions (*how did you do that?*), the answers sample subordinate goals in the goal/action hierarchy. The subordinate goals are comparatively less important. When a user types in a word, it is more important to find out why the user typed it in than finding out how the user typed it.

In our judgment, the Q/A protocols have certain advantages over think aloud protocols. When subjects are asked to think aloud while they perform tasks, they are sometimes unsure about the relevant level or dimension for pitching their verbalizations. In other words, the instruction to "think aloud" is often too open-ended. The "why" questions and "what does that mean?" questions are comparatively more focused. When subjects answer these questions, their responses are quicker and they seem more comfortable with the task.

Our three rules for asking users questions were incomplete in one respect. The questions did not uncover alternative strategies that users could apply in order to achieve task goals and subgoals. Our Q/A rules tapped the goal structures corresponding to the strategies the subjects ended up selecting, but not alternative strategies that could have been followed. Therefore, a fourth rule for asking questions would perhaps be beneficial.

RULE 4: Querying Goals.
IF ⟨user claims he has a goal G⟩
THEN ⟨experimenter asks "what are some alternative methods of achieving goal G⟩

Rule 4 would be important if the investigator wanted to trace the profile of user strategies in achieving task goals.

The benchmark tasks with Q/A protocols provided more informative data than did the free generation task and the computer term definition task. The latter two tasks exposed very few conceptual errors and most of the statement units were unsophisticated and nonspecific. Very little can be learned by asking a user to talk about a computer system off the top of the head. The computer term definition task did reveal which terms and concepts the users did not understand. However, the term definition task did not penetrate the procedural knowledge that is associated with the computer terms and concepts. In contrast, the benchmark tasks (with Q/A protocols) exposed both the users' conceptual errors and the users' procedural knowledge.

When users performed the benchmark tasks with Q/A protocols, there were two major sources of data. First, there was the printout of the computer–user interaction. Second, there were the Q/A protocols. We found that approximately half of the users' conceptual errors could be identified by the printout alone. The other half of the errors could only be identified with the aid of the Q/A protocols. Investigators are severely limited when they restrict their observations to printouts or even videotapes of the interactions. They miss out on identifying (a) conceptual errors that have complicated behavioral manifestations, (b) goal structures that guide the behavior, and (c) planning networks that characterize the users' plans and strategies. The Q/A protocols uncover this information.

When we conduct our future studies in computer–human interaction, we plan on revising our data collection procedures somewhat. We plan on including four phases. Phase 1 is an assessment of the user's prior knowledge about computers, with the same questions that we used in the present study. Phase 2 is a term definition test on the words and concepts that would potentially be problematic for the user. In phase 3 we would videotape the user performing some representative benchmark tasks, but without asking the user questions. In phase 4 we would videotape the user performing benchmark tasks and we would also collect Q/A protocols. Phases 3 and 4 could alternate, such that Q/A protocols are collected during some trials or segments and not during others. The purpose of having trials without Q/A protocols is to observe the user performing the tasks under normal conditions without the interruption of the questions. We could then assess the extent to which the data in phase 4 (with Q/A protocols) can explain performance measures in phase 3 (without Q/A protocols). If the phase 3 performance can be explained to some extent by the Q/A protocols in phase 4, then the Q/A methodology has some validity in tapping the functional conceptualizations that underly behavior.

One of the practical virtues of the Q/A methodology is that a single, well designed study can supply a rich and very informative set of data. When the Q/A methodology is applied to 1–2 dozen users, many (if not most) of the interface problems are identified. The data base is rich and informative enough to suggest specific recommendations for interface modifications. Moreover, it takes only 1 or 2 months to collect and analyze the data. The Q/A methodology is very cost effective in the sense that it uncovers a wealth of informative data in a very short amount of time. The methodology is particularly suited to a technology that moves quickly, with specific products that have a very short half life.

Given that the Q/A methodology is new, we are uncertain whether it will survive the test of time. However, researchers have not been satisfied with the available methods of evaluating software, such as checklists and other scaling procedures. The standard behavioral methodologies are also not well suited for discovering solutions to design problems; they can measure imperfections in existing software, but the data are not rich enough for discovering design solutions. With this context in mind, we would like to see the Q/A methodology and other verbal protocol methodologies have their decade in the methodological arena.

REFERENCES

Anderson, J. R. (1983). *The architecture of cognition*. Cambridge, MA: Harvard University Press.

Bainbridge, L. (1979). Verbal reports as evidence of the process operator's knowledge. *International Journal of Man-Machine Studies, 11,* 411.

Black, J. B., & Sebrechts, M. M. (1981). Facilitating human-computer communication. *Applied Psycholinguistics, 2,* 149–177.

Card, S. K., Moran, T. P., & Newell, A. (1980). Computer text-editing: An information processing analysis of a routine cognitive skill. *Cognitive Psychology, 12,* 32–74.

Card, S. K., Moran, T. P., & Newell, A. (1983). *The psychology of human-computer interaction*. Hillsdale, NJ: Erlbaum.

Carroll, J. M., & Carrithers, C. (1984a). Blocking learning error states in a training wheels system. *Human Factors, 26,* 377–391.

Carroll, J. M., & Carrithers, C. (1984b). Training wheels in a user interface. *Communications of the ACM, 27,* 800–806.

Carroll, J. M., & Mack, R. (1983). Actively learning to use a word processor. In W. Cooper (Ed.), *Cognitive aspects of skilled typewriting*. New York: Springer-Verlag.

Chi, M. T. H., Glaser, R., & Rees, E. (1982). Expertise in problem solving. In R. J. Sternberg (Ed.), *Advances in the psychology of human intelligence.* Hillsdale, NJ: Erlbaum.

Collins, A. M., & Gentner, D. (1980). A framework for a cognitive theory of writing. In L. W. Gregg & E. R. Steinberg (Eds.), *Cognitive processes in writing.* Hillsdale, NJ: Erlbaum.

Embley, D. W., & Nagy, G. (1981). Behavioral aspects of text editors. *ACM Computing Surveys, 13.*

Ericsson, K. A., & Simon, H. A. (1980). Verbal reports as data. *Psychological Review, 87,* 215–251.

Graesser, A. C. (1981). *Prose comprehension beyond the word.* New York: Springer-Verlag.

Graesser, A. C., & Clark, L. F. (1985). *Structures and procedures of implicit knowledge.* Norwood, NJ: Ablex Publishing Corp.

Graesser, A. C., Hopkinson, P. L., Lewis, E. W., & Bruflodt, H. A. (1984). The impact of different information sources on idea generation: Writing off the top of our heads. *Written Communication, 1,* 341–364.

Graesser, A. C., Lang, K., & Elofson, S. C. (1987). Some tools for redesigning system-operator interfaces. In D. E. Berger, K. Pezdek, & W. Banks (Eds.), *Applications of cognitive psychology: Computing and education.* Hillsdale, NJ: Erlbaum.

Graesser, A. C., & Murachver, T. (1985). Symbolic procedures of question answering. In A. C. Graesser & J. B. Black (Eds.), *The psychology of questions.* Hillsdale, NJ: Erlbaum.

Graesser, A. C., & Riha, J. R. (1984). An application of multiple regression techniques to sentence reading times. In D. Kieras & M. Just (Eds.), *New methods in comprehension research.* Hillsdale, NJ: Erlbaum.

Ginsburg, H. P., Kossan, N. E., Schwartz, R., & Swanson, D. (1983). Protocol methods in research on mathematical thinking. In H. P. Ginsburg (Ed.), *The development of mathematical thinking.* New York: Academic Press.

Hayes, J. R., & Flower, L. S. (1980). Identifying the organization of writing processes. In L. W. Gregg & E. R. Steinberg (Eds.), *Cognitive processes in writing.* Hillsdale, NJ: Erlbaum.

Houghton, R. C. (1984). On-line help systems: A conspectus. *Communications of the ACM, 27,* 126–133.

Kieras, D. E. (1982). *What people know about an electronic device: A descriptive study* (Tech. Rep. No. 12). Tucson, AZ: Department of Psychology, University of Arizona.

Kieras, D. E. (1985). *The role of prior knowledge in operating equipment from written instruction: Final report* (Tech. Rep. No. 19). Ann Arbor, MI: Psychology Department, University of Michigan.

LePlat, J., & Hoc, J. (1981). Subsequent verbalization in the study of cognitive processes. *Ergonomics, 24,* 743–756.

Lewis, C. (1981). Skill in algebra. In J. R. Anderson (Ed.), *Cognitive skills and their acquisition.* Hillsdale, NJ: Erlbaum.

Lewis, C., & Mack, R. (1982). *The role of abduction in learning to use a computer system* (Tech. Rep. No. 41620). Yorktown Heights, NY: IBM Watson Research Center.

Lewis, C., & Norman, D. A. (1986). Designing for error. In D. A. Norman & S. W. Draper (Eds.), *User centered system design.* Hillsdale, NJ: Erlbaum.

Mack, R. L., Lewis, C. H., & Carroll, J. M. (1983). Learning to use word processors: Problems and prospects. *ACM Transaction on Office Information Processing, 1,* 254–271.

Monk, A. (1984). How and when to collect behavioral data. In A. Monk (Ed.), *Fundamentals of human-computer interaction.* London: Academic Press.

Neal, A. S., & Darnell, M. J. (1984). Text-editing performance with partial-line, partial-page, and full-page displays. *Human Factors, 26,* 431–441.

Newell, A., & Simon, H. A. (1972). *Human problem solving.* Englewood Cliffs, NJ: Prentice-Hall.

Nisbett, R. E., & Wilson, T. D. (1977). Telling more than we know: Verbal reports on mental processes. *Psychological Review, 84,* 231–279.

Norman, D. A. (1981). Categorization of action slips. *Psychological Review, 88,* 1–15.

Norman, D. A. (1983). Design rules based on analysis of human error. *Communications of the ACM, 26,* 254–258.

Olson, G. M., Duffy, S. A., & Mack, R. L. (1984). Thinking-out-loud as a method for studying real-time comprehension processes. In D. Kieras & M. Just (Eds.), *New methods in the study of immediate processes in comprehension.* Hillsdale, NJ: Erlbaum.

Roberts, T. L., & Moran, T. P. (1982). Evaluation of text editors. *Proceedings of the Conference on Human Factors in Computer Systems,* Gaithersburg, MD.

Robertson, S. P., & Black, J. B. (1983). Planning units in text-editing behavior. *Proceedings of CHI'83 Conference on Human Factors in Computing Systems* (pp. 217–222). Boston, MA: ACM.

Rosson, M. B. (1984). Effects of experience on learning, using, and evaluating a text editor. *Human Factors, 26,* 463–476.

Rouse, W. B., & Rouse, S. H. (1983). Analysis and classification of human error. *IEEE Transactions on System-Man Cybernetics, 13,* 539–549.

Shneiderman, B. (1982). System message design: Guidelines and experimental results. In A. Badre & B. Shneiderman (Eds.), *Directions in human–computer interaction.* Norwood, NJ: Ablex.

Stevens, A., Collins, A., & Goldin, S. E. (1982). Misconceptions in students' understanding. In D. Sleeman & J. S. Brown (Eds.), *Intelligent tutoring systems.* New York: Academic Press.

Wickens, C. D., & Kramer, A. (1985). Engineering psychology. *Annual Review of Psychology, 36,* 307–349.

10

Knowledge Transformations During the Acquisition of Computer Expertise*

Dana S. Kay

Department of Psychology
Carnegie-Mellon University
Pittsburgh, PA

John B. Black

Teachers College
Columbia University
New York, NY

As people become more expert with a procedural skill like using a computer, the way they think about the skill changes qualitatively. In fact, there are several qualitative changes with increasing expertise. In the research reported in this chapter, we examined how people of varying expertise conceptualized computerized text-editing systems and found that their knowledge representations for information about the systems progress through four distinct phases as they become increasingly expert in using the system.

Recently, there have been studies that examined the differences in the knowledge representations of expert and novice programmers (e.g., Ehrlich & Soloway, 1984; McKeithen, Reitman, Rueter, & Hirtle, 1981; Adelson, 1981), and other studies that have examined the knowledge and performance of expert users of command languages (e.g., Card, Moran, & Newell, 1983), but there has been little previous investigation of how people's knowledge representations change as they learn to use command

*This research is supported by grants from IBM, but represents the views of the authors and not necessarily the views of IBM.

Portions of this chapter have been presented at the Seventh Annual Conference of the Cognitive Science Society, 1985, and the Human Factors Society 28th Annual Meeting, 1984.

language systems. Our research has focused on how people learn to use text editors and, in particular, how their knowledge representations change as they become increasingly expert at using text editors. The assumption behind our approach is that, as learning progresses, the content and organization of the knowledge changes to accommodate the acquired information. In our discussion here, we will present a study that traces the changes in knowledge representation, and then we will use the results of this study in conjunction with the results of previous studies as evidence for our model of the acquisition of text-editing expertise.

Before investigating *how* computer knowledge is acquired, we must understand *what* knowledge is necessary to produce skilled performance. That is, we need to characterize the knowledge possessed by expert computer users and thus, establish a goal or end-point for the acquisition process. Past research has accomplished this task for both text-editing experts and programming experts.

Card, Moran, and Newell (1983) provide us with the most comprehensive accounts of expert knowledge representations of text-editing systems. From a series of studies using different text editors and different editing tasks, they were able to extract a model of the knowledge representations for expert users. This model has been named the GOMS model, based upon the four components that are used to represent the cognitive structures of expert users. The four components that they propose are *g*oals, *o*perators, *m*ethods, and *s*election rules. *Goals* define the set of editing tasks to be accomplished. *Operators* are the elementary actions that are performed. *Methods* are procedures for accomplishing goals and finally, *selection rules* are conditional statements that are used to choose the most appropriate method.

Card, Moran, and Newell used these components to decompose editing tasks from the highest level of analysis (e.g., GOAL-EDIT-MANUSCRIPT) to the lowest level of analysis (e.g., TYPE-d). Using a small number of the GOMS components, they were able to describe the sequence of actions used in text-editing tasks and predict the time necessary to accomplish these tasks. Thus, the GOMS model provides us with a characterization of error-free, expert text-editing performance.

Recent work by Soloway and his colleagues also provides us with a characterization of expert knowledge representation; however, their analysis is for expert programming knowledge and is used to diagnose the bugs present in novice Pascal programs. Soloway et al. propose that experts use a complex set of goals and plans to guide the creation and comprehension of computer programs. They argue that goals and plans are the most plausible means of representing the knowledge underlying programming and propose the goal-and-plan (GAP) tree as a mechanism for representing the goals and plans relevant to solving a specific programming problem

(Spohrer, Soloway, & Pope, 1985). The goals in the GAP tree are extracted from the specification of the problem, and the plans are inferred from the various methods that programmers use to accomplish the goals. The GAP tree has been successfully used to analyze buggy programs and provide explanations for the existence of the bugs. Thus, Soloway and his colleagues have used their conceptualization of expert knowledge representations to localize the misconceptions that novices have when learning to program.

Card, Moran, and Newell, and Soloway and his colleagues, provide us with goal and plan representations of the knowledge possessed by expert users. Having characterized the expert knowledge representations, the next issue to address is how is the expert knowledge acquired? That is, what is the underlying learning process that results in these expert knowledge representations?

In 1984, Anderson, Farrell, and Sauers presented their account of how computer knowledge is acquired. Using the mechanisms of Anderson's ACT* theory of cognitive architecture (Anderson, 1983), they were able to simulate the process that underlies learning to program in LISP. In this simulation, Anderson et al. argue that, initially, computer instructions are encoded declaratively (i.e., as a set of facts for computer programming). However, to write a program, users need procedures that are represented as productions. Thus, novices must use general productions to interpret the information that has been encoded. They propose *structural analogy* (i.e., using the structure of previous problems to solve current problems) as one of the mechanisms for mapping known procedures onto new information.

With experience, knowledge is compiled into task-specific procedures. This compilation process is composed of two subprocesses—composition and proceduralization. *Composition* is the process by which two productions are collapsed into a single rule, and *proceduralization* creates productions that incorporate domain-specific information. Together, these processes allow for the acquisition of domain-specific procedures. To direct the composition and proceduralization of the task-specific productions, Anderson et al. propose that overall performance is guided by a hierarchically organized goal structure that decomposes the programming problems into individual problem-solving episodes.

Although Anderson et al. provide a comprehensive account of the acquisition of programming knowledge, their analysis is not at the GOMS or GAP tree level of analysis. In particular, other than having simple productions learned before complex ones, they do not address what kinds of productions are learned first and what kinds are learned later. This is the goal of our study. By examining users at varying levels of expertise, we are able to probe the development of goal and plan representations of text-editing information and establish the order in which different kinds of computer knowledge are acquired.

Two recent studies are closely related to the one we report here, and we will combine their results with ours when making our conclusions. In their investigation of people learning to use the IBM Displaywriter and the UCSD p-system text editor, Sebrechts, Black, Galambos, Wagner, Deck, and Wikler (1983) traced the transition from naive to novice user. By examining the changes in people's perceptions of command similarity as a result of training, they found that, before training, the similarity of the commands was based upon prior knowledge about the command terminology. For example, the commands CANCEL and DELETE were considered to be similar, because both of these words refer to aborting something. After training, users no longer judged command similarity using prior knowledge associations, but used text editing goals to organize the commands. For example, DELETE was now perceived as similar to BACKSPACE (and not CANCEL) because, in the Displaywriter, either of these commands can be used to accomplish the goal of erasing a piece of text.

Robertson and Black (1983) provided supplementary results to those of Sebrechts et al. in that they investigated the transition from novice to expert in addition to the transition from naive to novice. Using a simple experimental text editor, they observed the timing changes between keystrokes as people learned to use the system. They used interkeystroke time to examine the formation and use of plans in text-editing tasks. A *plan* is the sequence of actions that a user performs with a system to accomplish the goal of completing a task. Robertson and Black found that, with increased experience, the time spent pausing between simple plans decreased. This suggested that users were beginning to combine the simple plans that they had learned to form compound plans. In addition, the decision time to initiate a compound plan decreased as the editing session progressed, suggesting the acquisition of selection rules that facilitate accessing the most appropriate plan. Further evidence for selection rule acquisition was provided by observed changes in what plans were used most frequently to accomplish goals as the user became more experienced.

Although these studies provide insight into the development of text-editing performance, the behavior that they study is short-term, experimentally controlled learning. The current study investigates the complete transition from naive to novice to expert among people who naturally acquire expertise as part of their work. We will use a methodology similar to that of Sebrechts et al. (1983) to understand the changes in the organization and content of knowledge representations as users learn to use a text-editor. After presenting the results of our study, we will combine them with the results of the other two studies to provide evidence for a four phase model of the acquisition of text editing expertise. In this model, the user builds a complex knowledge representation by acquiring different types of knowledge during different learning phases.

METHOD

Subjects

Three groups of subjects participated in this study. The first group, the expert group, were 10 Yale graduate and undergraduate students in computer science who used the computer every day throughout the day. These subjects participated in this study as volunteers. The second group, the novice group, were also 10 volunteer Yale graduate and undergraduate students. These subjects, however, were psychology majors who use the computer for word processing and data analysis and, therefore, only use the computer once or twice a week. The last group of subjects, the naive group, were 10 Yale graduate and undergraduate students who had never used a computer. These subjects were paid three dollars for their participation in the study.

Both of the computer groups used EMACs-style editors developed locally at Yale. These editors were full-screen, WYSIWYG editors with commands that use control and escape keys in conjunction with character keys (e.g., holding down the control key and the *s* key for deleting a character). A typical editing session using these systems entails invoking the editor for a given file, performing a series of editing tasks (using the control or escape key in sequences with other keys), and then saving the edited version of the file.

Materials

Fifteen text-editing commands were chosen. These commands were picked because of their general use in text-editing. In addition, the generic form of the command was used. That is, if there were two types of a command, one for characters and one for lines of text, the command was used as a single item. The commands used were CANCEL, HELP, EXIT, ARGUMENT, MARK, SEARCH, REPLACE, DELETE, INSERT, PICK, PUT, BALANCE, JUSTIFY, HOME, and CENTER. Table 1 presents a list of the commands used and a brief definition of each command.

Each subject was given a stack of 105 4 x 6 inch index cards. A pair of nonidentical editing commands were written on each index card. In the

Table 1. Fifteen Commands and Brief Definitions.

Command	Definition
CANCEL	Cancel current edit operation
HELP	Display editor help file for specified command
EXIT	Exit the editor
ARGUMENT	Enter argument
MARK	Move cursor to a specific location
SEARCH	Search through text
REPLACE	Replace with query
DELETE	Delete a character or word
INSERT	Insert a character or word
PICK	Copy text
PUT	Insert contents of most recently filled buffer
BALANCE	Parenthesis balance
JUSTIFY	Justify text margins
HOME	Move cursor to upper left hand corner
CENTER	Center text

entire set, each command was paired once with each of the other commands. Each pair was centered on the index card with a short dash separating the two items. The items were printed in black ink and the letters were one inch high. There was also a number assigned to each item pair which was found in the upper-right-hand corner of the index card and used by the subject to refer to each item pair.

Procedure

Each subject was seated at a desk and given a blank sheet of paper and the stack of 105 index cards, the order of which had been randomized for each subject. The subjects rated the similarity of the two items on a scale of one to seven, one being highly dissimilar, seven being highly similar. Intermediate values reflected intermediate similarity. Although subjects were not told a specific characteristic on which to base their judgments, they were told that they were to make use of the entire scale (i.e., that they were not to use only the numbers at the extreme ends of the scale or only those numbers in the center of the scale).

RESULTS

As is common practice for many multivariate studies, a mean similarity rating for each command pair was calculated for each group of subjects and placed into the lower half of a matrix without the diagonal. Order of presentation of the items was disregarded. Three multivariate analyses were performed on the data for each group: multidimensional scaling, hierarchical clustering, and overlapping clustering. The multidimensional scaling and the hierarchical clustering were performed for exploratory purposes, and the overlapping clustering was used for confirmatory purposes. From the exploratory analyses, we hoped to be able to probe how the subjects conceptualized the editing commands and to obtain evidence for changes in this conceptualization with the development of expertise. The confirmatory analyses allowed us to assess whether or not the changes that we observed in the exploratory analyses resulted from general knowledge changes or domain-specific changes. Each of the analyses provided complementary information for the interpretation of the data. A general discussion of the results across analyses will follow the report of the individual analyses.

Multidimensional Scaling

We chose to use multidimensional scaling to examine the overall structure of the users' representations of the 15 editing commands. The results of this analysis provided us with information about the primary categories of organization for the commands and the relative statistical distances in this representation could be used to judge the psychological distances between the commands in the mental representations of the commands. Nonmetric, Euclidean multidimensional scaling (Kruskal, 1964; Shepard, 1962), using the SAS ALSCAL (Takane, Young, & deLeeuw, 1977) program, was used with Kruskal's stress formula 1 and the primary approach to ties. For each group of subjects, the input to the program was the lower half of the similarity matrix without the diagonal.

Dimensionality. For each group, three criteria were used to decide the best number of dimensions on which to plot the points. The first criterion was the stress value or the badness-of-fit of the data to a computer generated model. This value measures the discrepancy between the original similarity data and predicted interpoint distances that were used to statistically generate the best representation of the data. Although stress decreases with an increase in dimensionality, one can establish the number of dimensions to use by plotting the stress against the number of dimensions and looking for an *elbow* in the curve, which denotes the point at which there is no longer a substantial decrease in stress. This criterion was used for two, three, and four dimensions for the solution of each subject group. In all three cases, a small elbow appeared at three dimensions, suggesting that three dimensions provided the best fit of the data.

The second and third criteria are more subjective, but as important in deciding the dimensionality of the solution. The second criterion is the interpretability of the dimensions in terms of the actual items being plotted. This criterion is very important in psychological research, as one of the main goals in using multidimensional scaling is to arrive at latent characteristics of the data. For each group, a three-dimensional solution seemed to be the most informative, as it partitioned out information that did not appear in the two-dimensional solution.

The third criterion is the visualizability of the solution. As a rule, beyond three dimensions the solution becomes difficult to visualize, and one- and two-dimensional solutions are easiest to visualize. However, in each group, the previous two criteria resulted in three dimensions as the best level of dimensionality. Therefore, a three-dimensional solution was chosen to be presented here.

Interpretation of dimensions. Since there was some overlap in the results for the expert and novice groups (i.e., the computer user groups), these results will be discussed first. For the expert group, the stress value of the solution was 0.126. The plot of the commands for this group can be seen in Figure 1. In the solution figure, the three dimensions are labelled at each end of the axes and in the table, the dimensions labels are presented as headers for each column of numbers.

For each command represented in the solution figure, there are two points of reference that must be considered—an *X* and a dashed line that connects the *X* to a point that is labeled with the command name. The *X* represents the values of the command on the horizontal and diagonal dimensions. The values for these dimensions can be estimated by imagining a straight line that connects the *X* to each of the axes. For example, in Figure 1, the *X* for the Cancel command shows that this command has a strong value on the *achievement* (or upper-right) side of the diagonal axis and a moderately strong value on the *system* (or left) side of the horizontal axis. From each *X*, there is a dashed line connecting it to a point located above or below the *X*. The length of this line represents the magnitude of the value of the command for the vertical axis and the direction of the line (up or down) represents whether the value refers to the top or bottom half of the axis. To continue our previous example from Figure 1, the long dashed line that connects the *X* associated with the Cancel command to a point below the *X* means that this command has a strong value along the *nonformatting* (or bottom) side of the dimension.

For the expert multidimensional representation, the first dimension (the horizontal axis in Figure 1) has Put, Insert, and Replace on the right portion and Exit, Cancel, Help, and Home on the left. This dimension suggests an *editor/system* dimension. Commands that are located on the right are used within the actual editing session, and commands on the left are more abstract, system-level commands. For example, the Put and Insert commands are used in the editing session to add new text to the file. However,

Exit and Help are commands directed to the system, because Exit ends the editing session and saves the edited text, and Help exits the session into special help files that provide information about the various editing commands.

The second dimension (the vertical axis) has Center, Justify, and Balance on the top and Cancel, Argument, Pick, and Insert on the bottom. This characterization suggests a *formatting/nonformatting* interpretation for this dimension: Commands on the top part of this dimension are used in formatting text, while commands on the bottom are not used in formatting. This dimension represents a breakdown of the commands within the formatting task that allows users easy access to the specific formatting commands such as the Center command. For example, Justify is used to justify the margins of the text (a typical formatting task), whereas, Delete erases a piece of text (a nonformatting task).

The diagonal (or third) dimension in the figure has Home, Mark, Search, and Argument in the lower left and Delete, Justify, Balance, and Insert in the upper right. This configuration of commands suggests a *instrumental/achievement* dimension. These two labels are taken from the goal classification scheme presented by Schank and Abelson (1977). *Achievement* refers to goals that, when realized, result in the attainment of something. *Instrumental* refers to goals whose achievement is a precondition for the satisfaction of another goal. If the command is located in the upper right part of this dimension, then invoking the command results in the immediate satisfaction of an editing goal. For example, the Replace command satisfies a substitution goal by immediately substituting a new piece of text for an old one. If the command is located in the lower left, then it is instrumental to the satisfaction of another goal. For example, the Mark command moves the cursor to a specified location where there are other editing goals that are to be satisfied.

In the novice solution, two of the three dimensions are the same as those dimensions found in the expert solution. For the novice group, the stress value of the three dimensional solution was 0.120. A graphic representation of the solution is seen in Figure 2.

The first dimension (the horizontal axis) has Justify, Center, Put, and Insert along the right side and Cancel, Exit, and Help on the left side. This configuration of commands indicates an *editor/system* dimension that is similar to the editor/system dimension of the expert solution. If a command is located to the right of this dimension, then the command is used in the actual editing of the text, while, if the command is located on the left, then it is used for global system actions independent of the text. For example, Insert adds text during an editing session, and Exit takes the user out of the editor while simultaneously saving a copy of the edited text.

The second dimension (the vertical axis) had Center, Justify, and Bal-

Figure 1. Multidimensional scaling for expert users.

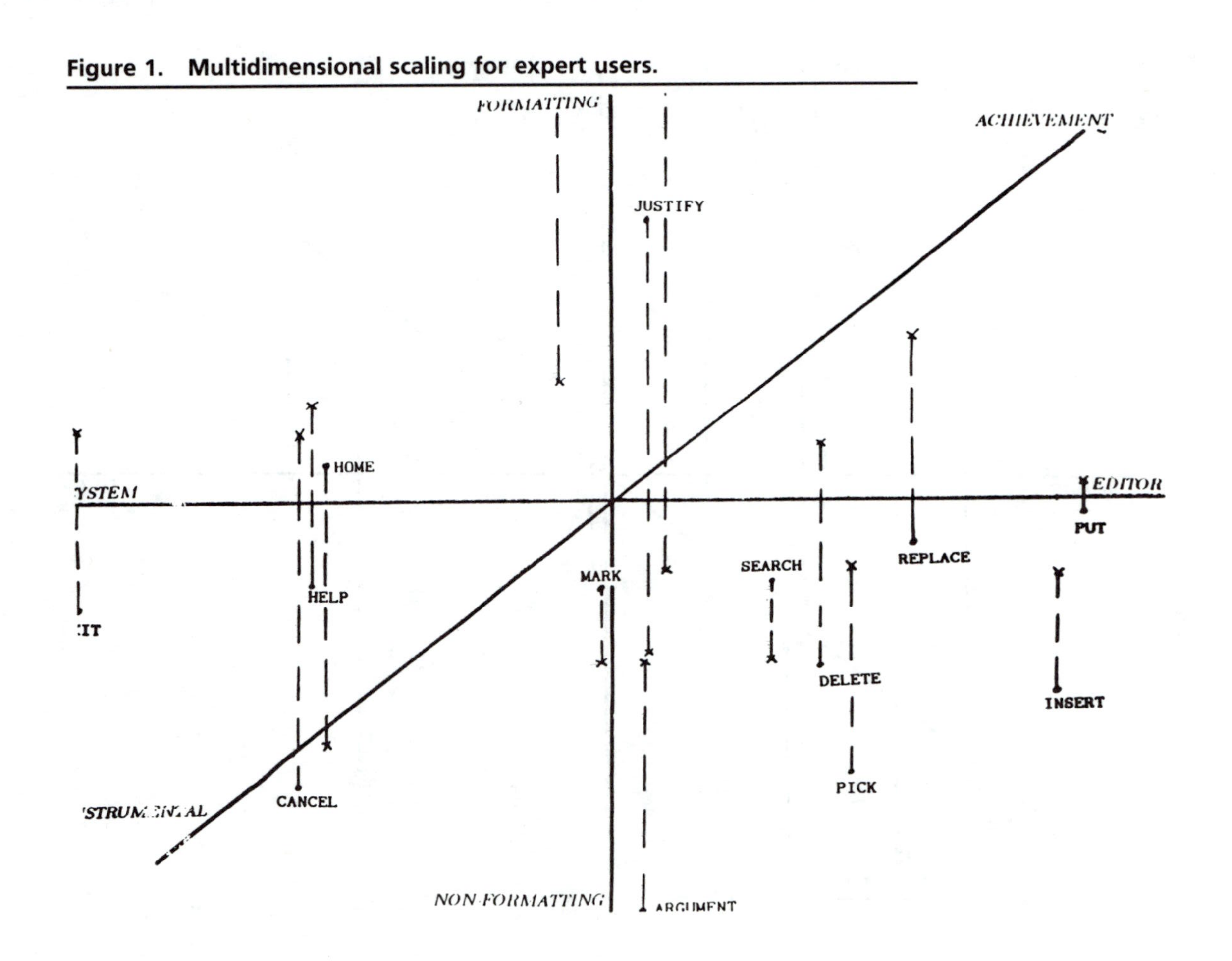

Figure 2. Multidimensional scaling for novice users.

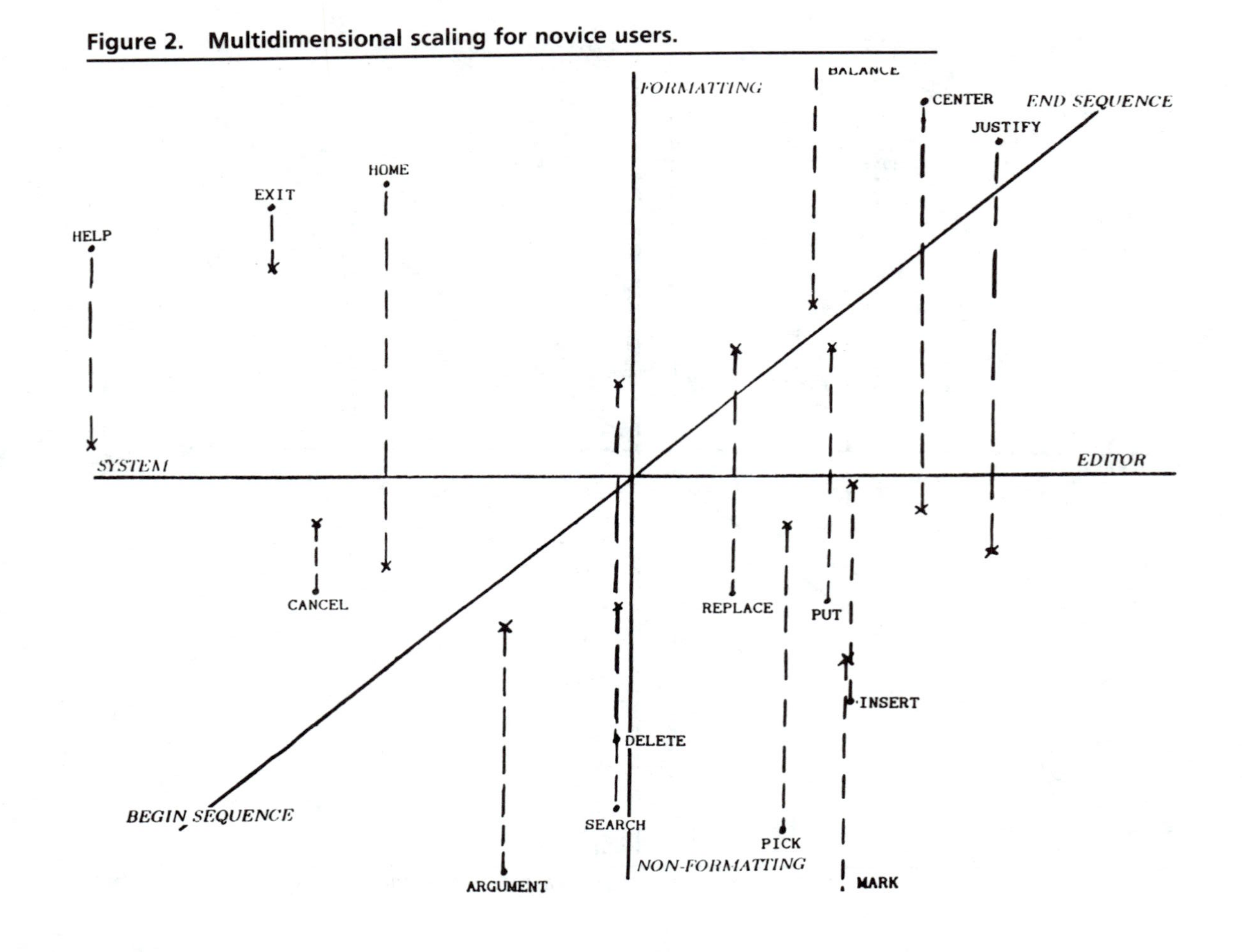

ance on the top of the dimension and Pick, Delete, and Mark along the bottom. The location of these commands suggests a *formatting/nonformatting* dimension. This dimension is defined in the same manner as the formatting dimension for the expert solution results. That is, the commands are divided into those commands that format the text, such as the Center command that centers a piece of text, and those commands that have nothing to do with formatting, such as the Pick command that copies text into the buffer of the editor.

The third dimension in the novice solution is different from the third dimension in the expert solution. This dimension, (the diagonal axis) has Exit, Balance, Put, and Replace in the upper-right quadrant and Argument, Search, Mark, and Home in the lower left. The placement of these commands can be explained by positing a *sequence* dimension. Commands in the lower left are characterized as being used in the beginning of a command sequence or editing session, and commands in the upper right are the ending commands in a sequence. For example, users Search for things and Mark things in the beginning of editing sequences, and Put and Replace information at the end of a sequence.

The multidimensional solution for the naive group did not lend itself to a straightforward interpretation as in the results for the novice and expert groups. Because most of the commands clustered around the origin of the solution, there were no clear dimensions that emerged. If we consider that the prominent organization of these command names comes from their computer references, this result is not surprising. Thus, it appears that, for the naive group, there is no overall space to organize the given commands. However, we will see in the hierarchical results that the command names are organized at a local level.

From the multidimensional scaling, we found that (a) there is a clear distinction between how the computer users conceptualize the commands and how naive subjects perceive the words that refer to the commands, and (b) the experts and novices have similar command conceptualizations. For the computer users, the dimensions suggest the development of a goal space for the commands. That is, each of the dimensions appears to represent a core set of general goal categories that can be used to describe the editing commands.

The dimensions that were present in the solutions for the two computer user groups suggest that there are general editing goals common to all users, regardless of level of expertise. The use of these goals to classify the commands helps the user in accessing the correct commands during an editing session. For example, if a user is formatting text, then the formatting dimension provides access to goals like *justifying text* that are members of the general goal category of formatting.

Although two of the three dimensions were the same for the novice and expert solutions, the difference found for the third dimension suggests that there is a change in the goal space with the development of expertise. The novices have a *sequence* dimension that indicates what commands fulfill the goals of being in the first parts of command sequences and what commands fulfill the goals of being in the last parts of command sequences. These general goal categories prevent users from errors due to incorrect sequencing. For example, to use the Put command, there must be something in the buffer that can be inserted into the text, and, in order to have something in the buffer, a command such as Pick that copies text into the buffer, must be used. If the novice did not have the sequence information that Pick is used in the beginning of a sequence and Put is used in the end, then he or she might try to use the Put command first and as a result, get an error message or an unwanted result.

The experts appear to have an achievement goal category and an instrumental goal category that classifies the commands according to whether or not the result of invoking the commands achieves an editing goal or is merely instrumental in the pursuit of another editing goal. In some sense, these goal categories represent a goal/subgoal distinction in which the instrumental commands accomplish a subgoal of a task and achievement commands accomplish a specific goal. This breakdown provides the experts with a general problem solving structure that can be used in planning the sequence of goals and subgoals necessary to accomplish an editing task.

Hierarchical Clustering

The results of the multidimensional scaling analysis provided us with a global framework or goal space for the commands. However, we are also interested in a local level of analysis that would address the relationships between the individual commands. To accomplish this goal, we performed a hierarchical clustering on the data for each of the groups. The clustering was performed using a BMDP version of the diameter or maximum method (Johnson, 1967). This one is a more conservative method of hierarchical clustering and often leads to more interpretable results than other clustering methods.

The clusters for the expert group are presented in Figure 3. Read this figure from the bottom to the top. By following up from each command to the node that it is connected to, you can see which commands clustered together. The lower down in the figure two commands are connected, the more related these commands are according to the subjects similarity ratings. Thus, the lower clusters represent highly related groups of commands,

Figure 3. Hierarchical clustering for expert users.

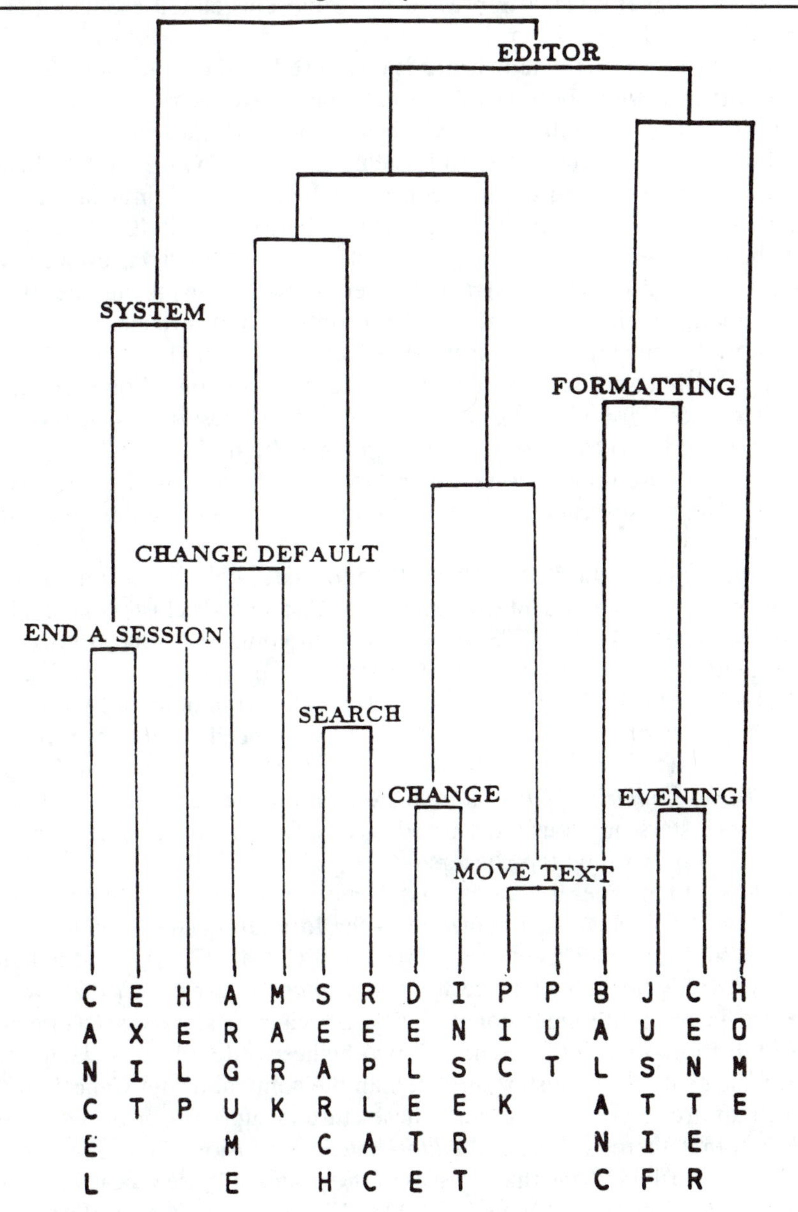

while the higher clusters represent more loosely related groups of commands.

The most highly related cluster is composed of the Pick command that copies text into the buffer and Put command that inserts the contents of the buffer. Together, these commands are used to move a piece of text and, therefore, we have labeled this cluster as the MOVE TEXT cluster. The next most related clusters consist of Justify and Center and Delete and Insert. The Justify and Center cluster (labeled EVENING in the figure) can be interpreted as an *evening out* cluster, because Justify is used to even out the margins of the text and Center is used to even out the blanks surrounding a title or heading. The Delete and Insert cluster (labeled CHANGE), and the next most related cluster (labeled SEARCH), that groups Search and Replace together represent common editing sequences similar to that found in the Move Text cluster. These two sequences accomplish some type of changing a piece of text. In the CHANGE sequence, old text is erased and new text is inserted. In the SEARCH sequence, a piece of text is searched for and then replaced with a more desirable piece of text.

If we move up the figure, we find the formation of two clusters that are similar to the Justify/Center cluster. The Cancel/Exit cluster, labeled on the figure as "END A SESSION," are two commands that are used to abort either a command sessions (Cancel) or an editing session (Exit). The "CHANGE DEFAULT" cluster consists of the Argument and Mark commands, both of which are used to specify a specific value for another command. These two clusters and the Justify/Center cluster can be categorized as *similarity of function* clusters and are contrasted with the *sequence* clusters previously discussed. Overall, there were three sequence clusters and two function clusters.

Three of the general, loosely related clusters are particularly interesting. In one of these clusters, Balance joins the formatting cluster consisting of Center and Justify. This cluster is labeled "FORMATTING" in the figure and provides further evidence for the *formatting* dimension that we observed in the multidimensional analysis, because Balance is a command used to format parentheses. In the two highest order clusters, Help joins the Cancel and Exit cluster, and then all the commands but Cancel, Help, and Exit are clustered together. These clusters support our previous conclusion that there is a general *editor/system* dimension. That is, there are two large clusters—one that consists of the commands used locally in actual word-processing tasks (labeled "EDITOR" in the figure), and another that consists of global commands that are not directly related to the text being edited (labeled "SYSTEM" in the figure).

The hierarchical results for the novice group can be seen in Figure 4. The most highly related cluster in this figure groups the Insert command,

Figure 4. Hierarchical clustering for novice users.

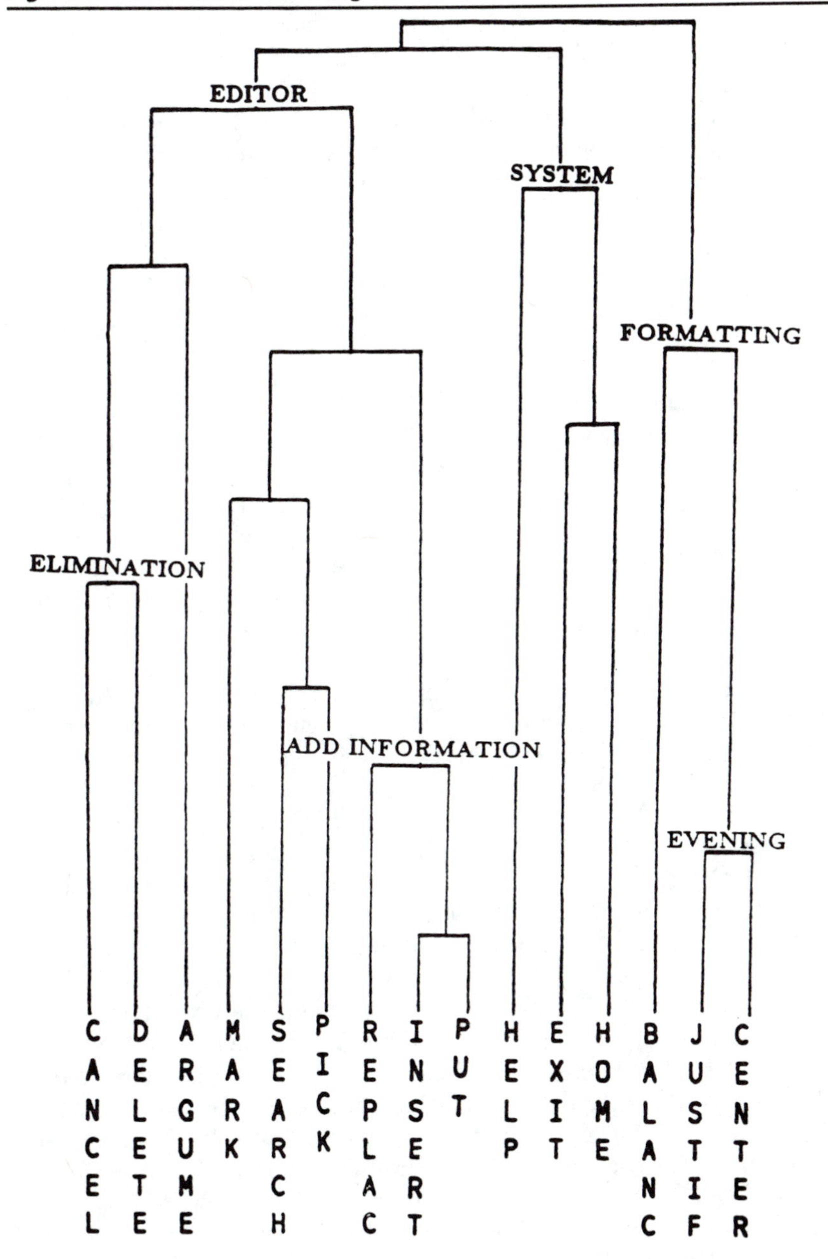

which allows the user to type in additional text, and the Put command, which is used to input text from the buffer. Both of these commands are used to add information to the text. The next most related cluster was also found in the expert hierarchical clustering. This cluster is located to the far right of the figure and consists of the Justify and Center commands. As previously discussed, both of these commands are used for "evening" a piece of text, and, therefore, the cluster is labelled EVENING. At the next node, which is labeled "ADD INFORMATION," the Replace command, which is used to substitute new text for old text, joins the Insert/Put cluster. We have labeled this cluster Add Information because all three of these commands accomplish the goal of adding information.

The cluster that is labeled "ELIMINATION" groups together the Cancel command that aborts a command and the Delete command that erases text. This cluster is particularly interesting when compared to the expert cluster in which Cancel was grouped with Exit. It appears that, for the novices, Cancel is classified as an *editor* command, while, for the experts, Cancel is a *system* command. We can also compare the loosely related clusters of the novice figure to the loosely related clusters of the expert figure. As in the expert results, the most loosely related clusters correspond to the dimensions found in the multidimensional scaling. These clusters are labelled in the figure with the "FORMATTING" cluster being more related that the "EDITOR" and "SYSTEM" clusters.

The clustering for the naive group can be seen in Figure 5. Although the multidimensional scaling for the naive group did not lend itself to a clear interpretation, the hierarchical clustering for this group yielded various interpretable clusters that allow us to tap the preconceptions that the users bring to the editing domain. The most highly related cluster in this figure groups the Cancel and Delete commands together. This cluster is labelled ABORT, because, if we consider the natural language definitions of these words, they are similar in that both refer to aborting an action. Recall that, in the novice clustering results, there was also a Cancel and Delete cluster. This suggests that some of the preconceptions that the users bring to the text-editing domain can be used in initial text-editing tasks.

As we move up the figure, we find that Balance and Center and Insert and Put are the next most related clusters. Again, we can interpret these clusters according to their similarity in natural language connotation. The Balance/Center cluster groups together words that refer to actions that result in making something even. Therefore, we have labeled this cluster "MAKE EVEN." The Insert/Put cluster suggest a similarity based upon the fact that both Insert and Put result in placing something, and this cluster is labeled PLACEMENT in Figure 5. The Placement cluster was also found in the novice results, giving stronger support to our proposal that some of the naive preconceptions are used by novice users.

Figure 5. Hierarchical clustering for naive users.

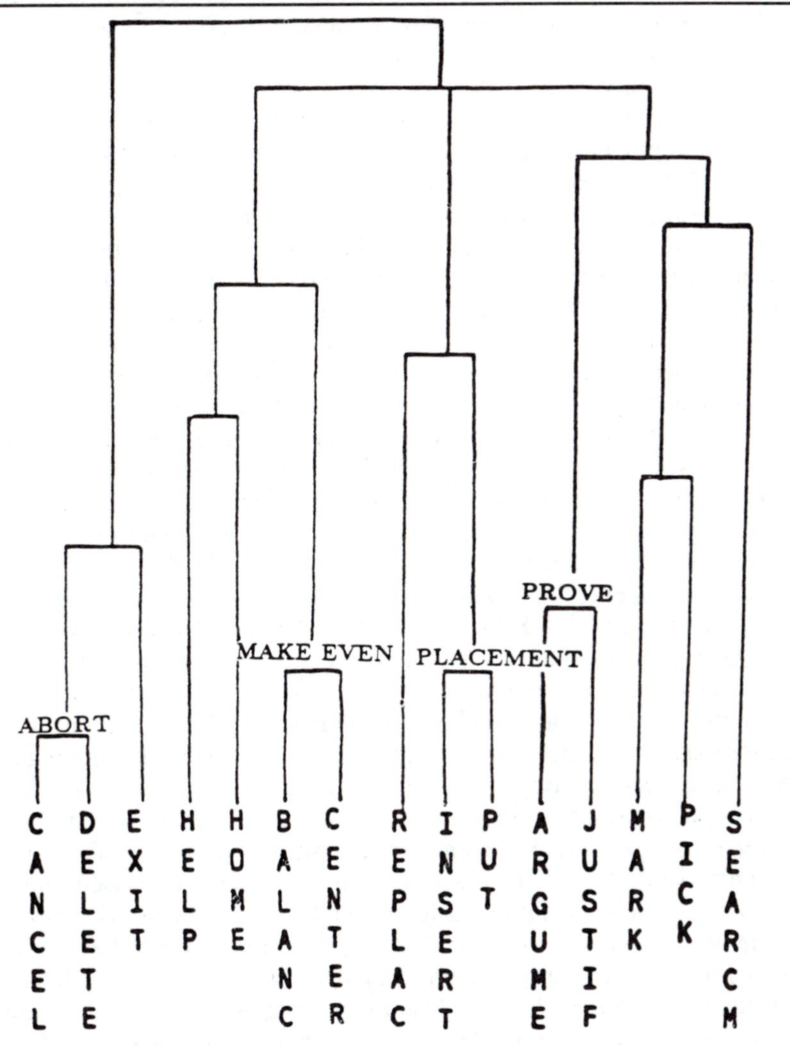

We found the next most highly related cluster to be of particular interest. This cluster is labeled "PROVE" in the figure and consists of Argument and Justify. These two words are related, because one justifies oneself in an argument. That is, justification is used to prove an opinion in an argument. This suggests that the naive subjects were not only considering the natural language definitions of the words, but also the relationships between the words in general knowledge about the world. Contrary to our ability to interpret the most loosely related clusters in the expert and novice results, we were not able to provide specific interpretations for the most loosely related clusters in the naive results. However, this lack of interpretation supports the corresponding lack of interpretation that was found in the multidimensional scaling.

The hierarchical clustering results provide us with snapshots of how the users' conceptualizations of the commands changes with experience. The clusters that resulted from the ratings by the naive group appeared to be based upon prior knowledge associations between the words. These associations reflected, not only similarity in natural language definitions (e.g., Balance and Center, and Cancel and Delete), but also similarity along more general knowledge action associations (e.g., Justify and Argument).

In the results of the hierarchical clustering for the novice users, commands appeared to be clustered based upon similarity of function in the editor. That is, two or more commands were similar if they were alternative ways to accomplish a similar goal. For example, Insert, Put, and Replace clustered together, because each of these commands can be used to accomplish the goal of adding information to a text, and Cancel and Delete clustered together, because they are each used to accomplish the goal of getting rid of a command or piece of text.

Further evidence for these naive and novice group differences is provided by the results of Sebrechts et al. (1983) with their subjects before and after training. That is, users initially cluster commands together using prior knowledge associations that they have for the words that refer to editing commands. However, with experience, the organization of commands begins to reflect domain-specific knowledge about the editing goals that are associated with the specific commands.

For the expert group, the clusters are somewhat different. Expert users appeared to use a more sequence- or plan-oriented organization. That is, the perceived similarity between the command pairs was based upon the use of the commands in a plan to accomplish a given goal. For example, in the expert clustering Insert and Delete appeared together, because these two commands are part of a plan to change a word or letter. In the novice results, these commands were clustered separately, because, when considered individually, they are related to different editing goals—namely, adding material and removing material. Corroborating evidence for this shift from commands to plans is provided by the Robertson and Black study (1983). In that study, the development of text-editing plans was observed using changes in the pauses between commands.

In addition to the differences between the hierarchical clustering results for the three groups, there are also similarities that suggest that the transitions between the levels of expertise are not all-or-none changes. In the naive and novice groups, there were two clusters that were the same: the Insert/Put cluster and the Cancel/Delete cluster. In both of these cases, we find that the natural language connotations of the words can be used to help in understanding the text-editing actions that these words invoke. Thus, it appears that, in order to become novices, naive users do not have to totally reorganize the knowledge that they bring to the text-editing domain.

In the novice and expert groups, there is only one cluster that is the same for these two groups. This cluster is labelled "formatting" in the figures for both the groups and consists of Balance, Justify, and Center, the three main formatting commands. This result suggests that formatting text is a simple task that is acquired early in the learning process and remains throughout the development of expertise.

Summary of Exploratory Analyses

From the multidimensional scaling and hierarchical clustering results, we can extract general pictures of how each group conceptualized the commands. For the expert group, we found that the dimensions resulting from the multidimensional scaling represent a general goal classification scheme for the editing commands. This classification provides categories for describing general properties of the commands. The hierarchical clustering results for the expert group tap a more specific conceptualization of the commands. That is, the hierarchical clusters represent the local, individual relationships between the commands as they relate to specific goals. In general, there were two types of relationships. One type represents similarity in use in a command sequence (e.g., Pick and Put), and the other type represents similarity of action (e.g., Justify and Center).

The multidimensional scaling for the novice group is similar to the expert results in that two of the three dimensions (i.e., formatting and editor/system) were found in both the expert and novice solutions. The third dimension, on which these two groups differed, can be explained by considering the hierarchical results for the expert group. As previously mentioned, many of the expert clusters were interpreted as common command sequences. In the novice multidimensional solution, the third dimension was begin/end sequence. We believe that this dimension dropped out in the expert solution because the sequence information that it provided was replaced by a more specific conceptualization of each individual command sequence.

If we examine the clusters that resulted in the novice hierarchical clustering, we find that the majority of the clusters were grouped according to similarity in action. There were no clusters that represented common command sequences. The presence of the begin/end sequence dimension in the novice multidimensional solution can be used to explain this lack of sequence clusters. It appears that novices possess general sequence information that is not specific enough to influence the relationships between individual commands. Their conceptualization of the relationships between the individual commands reflects only information pertaining to the use of the commands as alternatives to accomplishing simple text-editing goals.

In the multidimensional scaling and hierarchical clustering results for both the expert and novice groups, we proposed a global goal space and local between-command relationships. That is, the multidimensional scaling gave us a view of how editing goals are organized, and the hierarchical clustering gave us a view of how the actions (commands) are organized by the goals. In Figure 6, we present an example of how these results can be combined into a command conceptualization scheme for experts.

The top half of this figure presents the high level goal space that was extracted from the multidimensional scaling. We have represented the formatting/nonformatting distinction that was present in both the expert and novice solutions. The bottom half of Figure 6 was extracted from the hierarchical clustering results. In this part of the figure, we find more specific goals (e.g., centering-text and move-text) embedded within the formatting/nonformatting distinction, and then, linked to these specific goals, are the individual editing commands that are used to accomplish the goals. The specific clustering goals are embedded within the formatting and nonformatting goals, because we believe that, at the expert level, the direct general goal/action link is now divided into a general goal/specific goals link and a specific goal/action(s) link.

The links that connect the general goals to the specific goals are ISA links because these goals are instances of the general goal category. For example, "centering-text" is an instance of a formatting goal. The links that connect the specific goals to the actions can be either AND links or OR links. The OR links (as in the centering-text goal) mean that the actions

Figure 6. Example of goal space organization.

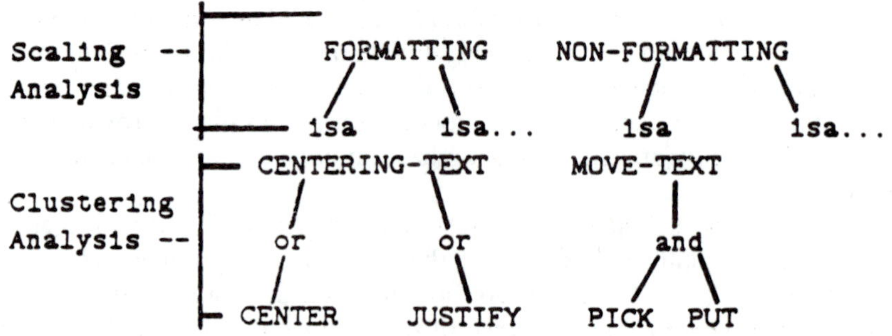

are related to the goal in that they are alternatives for accomplishing the goal. The AND links (as in the move-text goal) mean that the actions form a sequence that accomplishes the specific goal.

The scaling and clustering results for the naive group were not as interpretable as the expert and novice results. However, the clustering results for this group did provide us with insight into the preconceptions that users bring to text-editing. That is, we found that naive people have prior knowledge associations for the words that will later refer to editing commands.

Overlapping Clustering

Thus far, the analyses presented have been exploratory. To confirm our hypothesis that there were genuine differences in the ways that the groups conceptualized the commands and that, with experience, the conceptualizations became more similar to the actual design of the system. We analyzed the data for each group using the MAPCLUS program for confirmatory, overlapping clustering (Shepard & Arabie, 1979; Arabie & Carroll, 1980).

The purpose of this analysis was to examine the relationship between each of the three groups of subjects and an externally accepted expert model. This model was composed by asking one of the designers of the text-editing system to describe various properties that he felt could be used to characterize the experimental set of commands. Using these commands, he was to note which of the commands had each of the properties. Of the eight properties defined by the expert, five were chosen because they each contained more than one command possessing the property. The five properties and their associated commands can be seen in Table 2.

The first property, destructive editing, categorizes the commands that destructively alter the text. The second property, language oriented, refers to command naming and pertains to the commands whose names have a natural language orientation. The third command property, buffer ori-

Table 2. Expert Model.

Property	Commands
1. Destructive Editing	Replace, Delete, Insert, Put, Balance, Justify, Center
2. Language Oriented	Balance, Justify, Center
3. Buffer Oriented	Delete, Pick, Put
4. Session Control	Cancel, Help, Exit, Search
5. Search	Search, Replace

ented, refers to those commands that either place information into or take information out of the buffer of the text editor. The use of this property and the destructive editing property suggests that the designer possesses specific knowledge about the commands that includes information about the side effects of invoking the commands. The fourth property, session control, and the fifth property, search, both refer to text-editing goals. The session control property refers to those commands that are used to accomplish general editing goals such as ending an editing session. The search property consists of the two primary commands that are used to accomplish the goal of searching for something.

The designer model was fit to the mean similarity ratings of each group. Our hypothesis was that the match between the designer model and each subject group should increase as computer experience increased. The percentage of the variance accounted for in the expert group was 33.1%. This percentage was significant at the $p < .001$ level ($F(5,99) = 9.815$). For the novice group, the percentage of variance accounted for by the designer model was 15.8%, which was significant at the $p < .005$ level ($F(5,99) = 3.729$). For the naive group, the percentage of the variance accounted for was 7.6%, which was not significant ($F(5,99) = 1.637$). Because of the cross-sectional design of this study, we are not able to directly compare the percentage accounted for in each of these groups. However, our results do suggest that there was more variance between the designer model and the novice group than between the designer model and the expert group.

As we predicted, the designer model fit best with the expert groups and did not fit at all with the naive group. In addition, the designer model was able to account for a significant amount of the variance in the novice group, which suggests that the novices have begun to establish representations of the text-editing information that are similar to the way that the system was designed. However, the most interesting aspect of these results is that, although a significant amount of the variance was accounted for in the expert group, this was only one third of the possible variance for this group. It appears that, although the users in the expert group are experts at using this system, the mental representation that they have for the system does not completely match the true design of the system. For example, in the designer model, there is a buffer-oriented cluster in which Delete, Pick, and Put are viewed as similar, while, in the hierarchical clustering analysis for the expert group, Delete is clustered with Insert, because the two commands are frequently used together in a sequence. From the way the system is implemented, the buffer concept is invisible to the user, so it never becomes important in their conceptualization of the system. Perhaps different implementations would have made such underlying concepts more apparent and then made the system less confusing to the user.

THE PHASES OF LEARNING A SYSTEM

The results of this study, in conjunction with the results of the Sebrechts et al. (1983) and the Robertson and Black (1983) studies led us to conclude that the knowledge representations of users evolve through four phases as they become expert with a system. In the beginning, user knowledge is organized by preconceptions that the user has for the system terminology, but with a little experience this knowledge is reorganized to reflect what commands are relevant to accomplishing goals with the system. However, with more experience the representation of the knowledge changes again to form complete plans (i.e., sequences of commands) for accomplishing goals. Furthermore, with increasing expertise the plans become more complex and the user learns when each of these plans is most appropriately selected to accomplish a goal. In the following, we explain each of these phases in detail, using the results of the three studies as evidence for our claims.

Phase One: Preconceptions

In Phase One, we have the completely naive users who have no experience using a text editor. At this stage of learning, users have preconceptions about the system based on applying their prior knowledge of the world to how the system is described to them. Prior experience with the terminology to be used in text-editing results in users coming to the text-editing domain with a knowledge representation that may or may not correspond to the knowledge representation that will develop as text-editing experience increases. Figure 7a presents an example of the type of knowledge structures that exist before any learning has taken place. In this structure, there are two actions that are related by prior knowledge.

Figure 7b gives an example of the Phase One knowledge structures. In this knowledge structure the Center and Balance commands are related by their natural language connotation of *making something even.*

Figure 7a. Phase One—Preconceptions.

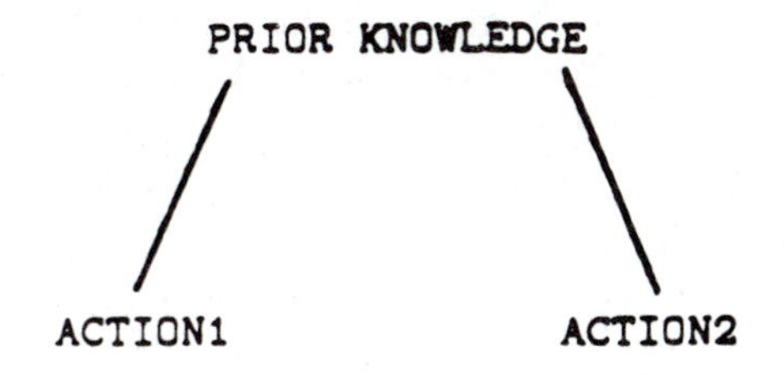

Figure 7b. Phase One—Example Structure.

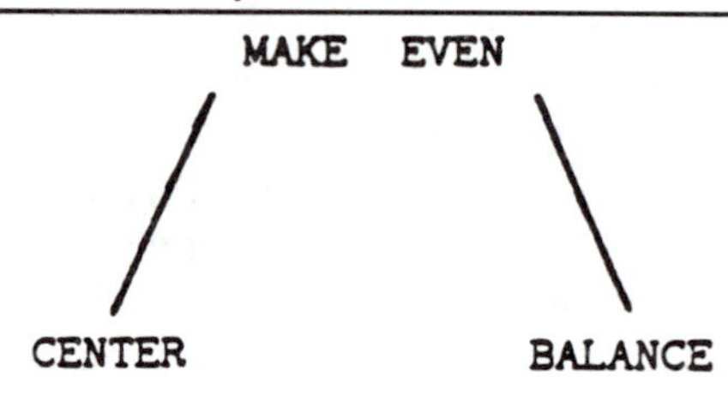

Evidence of this phase is found in both the current study and the Sebrechts et al. study. From the hierarchical clustering results of the studies, we find that the clusters formed are based upon prior knowledge about the command terminology. For example, in the clustering results for the naive subjects, Center and Balance clustered together, because these two words refer to evening something out. Similar results were obtained in the Sebrechts et al. study for the hierarchical results of the pretraining subjects. For example, Cancel and Delete for the Displaywriter were seen as similar (both suggest aborting something) and so were Find and Get for the UCSD p-system (both involve obtaining something).

Although there may be some correspondence betwen the prior knowledge representations of the commands and the text editing representations, in most situations, the two representations are quite different. Therefore, users in Phase One of the learning process must overcome a bias toward interpreting the commands in terms of their prior knowledge associations. To accomplish this task and achieve some level of expertise, the previously existing knowledge representations must be reorganized to accommodate the acquisition of text-editing knowledge.

Phase Two: Initial Learning

Initial text-editing knowledge can be acquired from a manual, a class, self-teaching, or any combination of these. No matter what the learning method, the main goal that the user has is to overcome the prior knowledge bias that exists for the text-editing commands. We believe that the accomplishment of this goal entails (a) learning the goals relevant to text editing and (b) learning the commands that can be used to accomplish these goals. The first part of the learning process takes place as soon as the user begins to edit and is exposed to various editing tasks. Knowledge of general goals allows for the generation of high-level goal structures that can be used to organize the editing commands. For example, if a user types *th* instead of *the*, then the goal or task that is instantiated is to add the extra *e*. However, before this goal can be accomplished, the user must learn the actual editing

Figure 8a. Phase Two—Initial Learning.

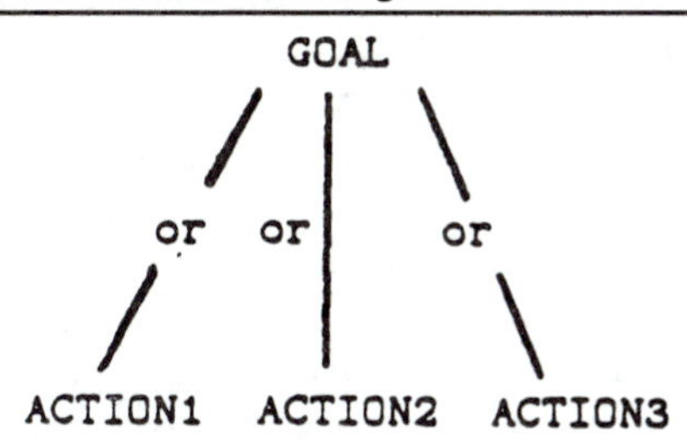

command(s) that are used. That is, the user must learn the functions of the commands and selectively choose the most appropriate command(s). A user might learn that the command Insert serves to add text and decide to use Insert to add the desired letter. At this time, the user learns that one of the possible goals to be found in text editing is to add information, and that one command that accomplishes this goal is Insert. As a result of this learning, the user develops knowledge structures that link specific goals and commands. Figure 8a presents the knowledge structures that exist at this phase of learning. Here the goal is linked to the actions by OR links, because one can use ACTION1 or ACTION2 or ACTION3 to accomplish the goal.

Figure 8b presents the representations for our previous example. The GOAL is ADD-INFORMATION, and ACTION1 refers to Insert. As more commands that add information, such as Replace and Put, are learned, they are added to the representation with OR links, because the user knows that, if you want to ADD-INFORMATION, you can use Insert or Replace or Put.

It is important to note that, although these users have acquired some text-editing knowledge, the organization of this knowledge is based only upon the *result* of the commands and not the procedure that leads to this result. Because of this narrow focus, the users do not possess the knowledge necessary to develop more complex structures such as plans. In particular, while the users know that Put can be used to ADD INFORMATION, they

Figure 8b. Phase Two—Example Structure.

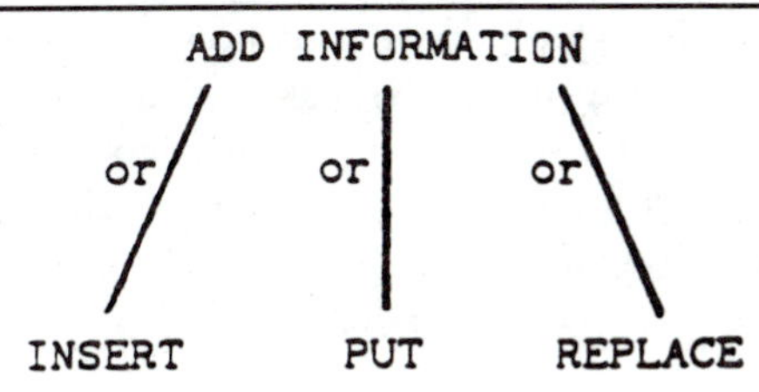

do not readily know that Put is merely the main action in a multi-action plan to accomplish this goal.

Because novice users have not refined their definitions of commands, several commands are often linked to a single goal by nondiscriminating OR links. At this level of understanding, the user will employ any one of the actions linked to the goal to accomplish the goal. For example, in the knowledge structure presented in Figure 8b, Put, Insert, and Replace are conceptualized as similar, because they all accomplish the goal of adding information to the text. With experience, the user will develop more complex representations of these commands that include knowledge of the full plans that accomplish the desired results, and will learn that, although these commands all add information to the text, the added information comes from different sources. Evidence for this change in knowledge organization will be presented in Phase Three. For example, Put uses information from a buffer and needs Pick or Delete to put the information into the buffer before the information can be Put somewhere. However, Insert needs only the information that is typed by the user after the Insert command is issued.

In addition to providing evidence for Phase One, the results of the hierarchical clustering in the present study and the Sebrechts et al. study also illustrate the transition from Phase One to Phase Two. With experience, users no longer cluster commands by prior knowledge associations. The clusters change to be based more on the function of the system. For example, Justify which clustered with Argument for the naive group, was clustered with Center for the novice group. In this system, Center and Justify are two formatting commands that perform similar functions.

In the results of the multidimensional scaling for both the current study and the Sebrechts et al., we see the global organization of the commands for Phase Two users. The results of these analyses were the same in both studies. The commands were organized along (a) a dimension that differentiated system and editor commands, (b) a dimension that distinguished formatting from nonformatting commands, and (c) a dimension that differentiated commands to begin a sequence from commands to end a sequence. As mentioned earlier, these dimensions suggest the development of a goal space for organizing the commands, because they provide a classification of the goals that can be accomplished with the system. The multidimensional analysis gives a view of how the goals are organized, while the hierarchical clustering analysis gives us a view of how the actions (commands) are organized by the goals. Knowledge of this goal space helps the user in accessing the correct goals during an editing session. For example, if one is formatting a document, it is more appropriate to have the goal of centering the text (and accessing the CENTER command through the goal) than it is to have the goal of changing a misspelled word (a nonformatting goal). The third dimension (begin/end of sequence) is particularly interesting, because it suggests that, although users do not organize the commands by specific plans sequences at this phase of learning, they do understand that commands are used in sequences at the abstract goal level.

As previously mentioned, Card, Moran, and Newell (1983) proposed the GOMS model to account for the text-editing behavior of experts performing routine tasks. Using our account of initial learning, we propose that it is the Goals and Operators components of the GOMS model that are acquired first. That is, novices are able to understand the general goals involved in text editing and the individual commands (operators) that are related to these goals. It is reasonable that the goal/action link would be the first link to be formed. During the initial learning of a system, the user is introduced to numerous command names and definitions with little reference made as to when each command should be used. If, instead of forming links between the commands and the goals that they accomplish, the first links to be formed were between commands, then the user would never know when to use each command and therefore, would have to resort to a trial-and-error method of achieving a goal. Thus, the linking of the command to a goal provides the user with some aid in using the correct command in the correct situation.

Novices seem to conceptualize the commands merely by what goals they are relevant for accomplishing. They have not yet acquired the actual procedures or plans that are associated with text editing, and, thus, each text-editing task becomes a problem-solving task in which they must actively search through their representations of the commands and determining the sequence of commands necessary to accomplish the task.

Phase Three: Plan Development

Once the users have acquired the basic editing commands and goals, they learn that there are combinations of commands that are often used together to accomplish goals. In Phase Three, users develop the ability to form plans by combining the actions that were represented separately in Phase Two. These plans correspond to the Methods of the GOMS Model. There are various ways that the transition from Phase Two to Phase Three takes place. For example, with the system that we studied, users realize the inefficiency in repeating a command numerous times. To overcome this inefficiency, they learn to use the Argument command that automatically repeats another command for the number of times specified as the argument. Once they have this knowledge, users begin to notice that there are other commands that can be used together in a sequence. This realization leads to a reorganization of the knowledge representation to accommodate the command sequences or plans that are used to accomplish goals.

This reorganization process entails the modification of the goal/action links that were formed in Phase Two. In this third phase, two types of links are formed. At one level, links are formed between the commands or actions that are used in a plan. At a higher level, links are formed between the goals and the plans that are used to achieve the goal. It is also possible to represent this knowledge as productions and explain the transition from Phase Two to Phase Three using the composition process proposed by Anderson (1983). However, since we are primarily interested in specifying the knowledge that is acquired at each phase of learning, we chose to use a network representation in which we trace the development of the links in the network because we think the networks are more perspicuous than productions.

Figure 9a presents the type of knowledge structures that exist at Phase Three of the learning process.

In these structures, just as in the Phase Two structures, the goals will guide the use of the commands. However, this guidance is provided by plans consisting of actions, rather than by each action individually. For example, the GOAL is accomplished by accessing PLAN1 that consists of performing ACT1 and then ACT2. In Phase Two, we would merely have known that ACT1 and ACT2 are related to the GOAL, but now we know how to use them to accomplish the GOAL.

Using the results of the hierarchical clustering analysis from our study and the Sebrechts et al. study, we find evidence for the development of Phase Three knowledge structures. As previously mentioned, the novice users organized commands as groups of actions related to goals. On the other hand, expert users appeared to use a more sequence-oriented or-

Figure 9a. Phase Three—Beginning Expertise.

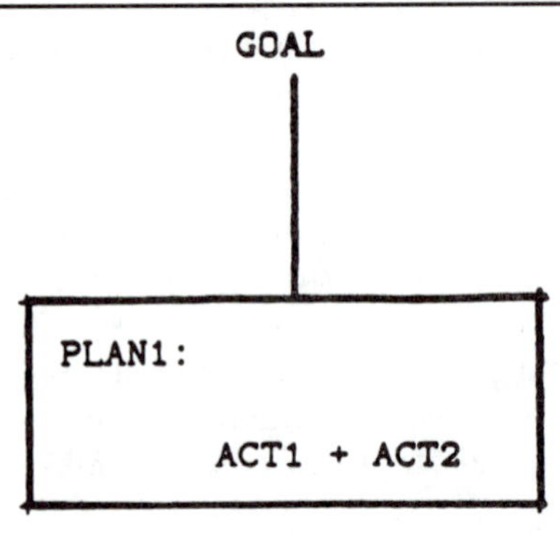

Figure 9b. Phase Three—Example Structure.

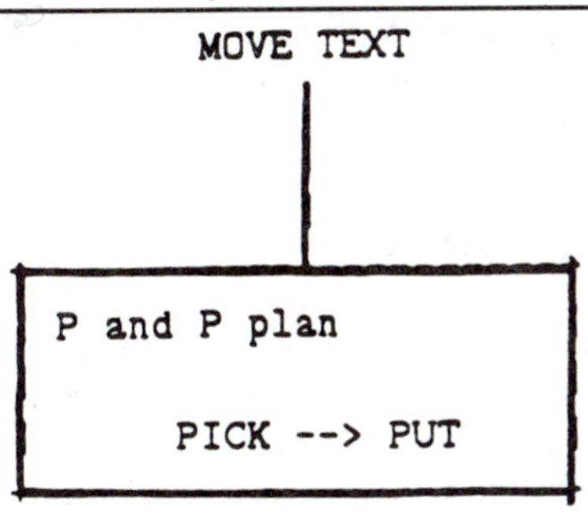

ganization. That is, the perceived similarity between the command pairs was based upon the use of the commands in plans to accomplish goals. For example, the expert group clustered Pick and Put together, because these two commands are used together when a user wants to accomplish the goal of moving a piece of text. This example, when contrasted with the Phase Two knowledge representation example in which novice users clustered Put with Insert and Replace because each of individual commands are used to accomplish the goal of adding information, illustrates the change in the representation of the commands from goal/action links to goal/plan links. Figure 9b presents a graphic representation for the knowledge structure that underlies this result.

In Phase Two we found that, in addition to having specific goal/action knowledge structures, users also organized the commands in a goal space. Given the knowledge structure changes that occur from Phase Two to Phase Three, it is interesting to look at whether or not these changes influence the goal space of the users. The multidimensional scaling results for the expert group in the present study suggest that there is only one primary change in the goal space. Recall that, in the initial learning phase, the dimensions of the goal space are editor/system, formatting/nonformatting, and begin/end sequence. Of these three dimensions the former two were also present in the multidimensional scaling for the experienced users. However, the begin/end sequence dimension is no longer used. This result suggests that experienced users continue to organize the commands in a goal space, but now they realize that interactions with the system don't occur in as rigid an order as they originally thought, so the begin/end dimension disappears. They still have sequence knowledge, but it is at the more local plan level instead of the global begin/end sequence level.

Because users begin to develop plan representations for their editing knowledge, we believe that it is during this phase in learning that users begin to mold the system to suit their own editing style by implementing these plans as macros. This molding makes their performance more efficient, because, using plan-macros, they can accomplish a goal with one command rather than several. We are currently testing this hypothesis by examining the number and content of macros created by users at various levels of expertise.

Phase Four: Increasing Expertise

Although the formation of simple plans results in some expertise in text editing, it is not until compound plans are formed that one can accomplish more advanced tasks. Phase Four of our model accounts for this ability and represents the completion of the acquisition that results in knowledge representations similar to those proposed in the GOMS model of expert performance. In this phase, users (a) combine simple plans into more compound plans to accomplish major goals, and (b) develop rules for selecting the best plan to achieve a given goal in a given situation. Once again, we observe a reorganization of the knowledge that results in the development of new links between the components of the representation. In this phase, we see a change from one-to-one correspondence between goal and plan to a one-to-several correspondence: that is, in Phase Three each plan or sequence of actions is directly linked to a specific goal (e.g., move text), while, in Phase Four, there are multiple plans that can be linked to each goal. However, because there are multiple plans, the links that connect these plans must have selection rules that tell the user under what conditions to access the plan to accomplish the goal.

Figure 10a shows the type of knowledge structures possessed by users in Phase Four. In this type of structure, there are several plans (sequences of actions) that may be instantiated to achieve a given goal. To be sure that the correct plan is chosen from the set of applicable plans, the links

Figure 10a. Phase Four—Increased Expertise.

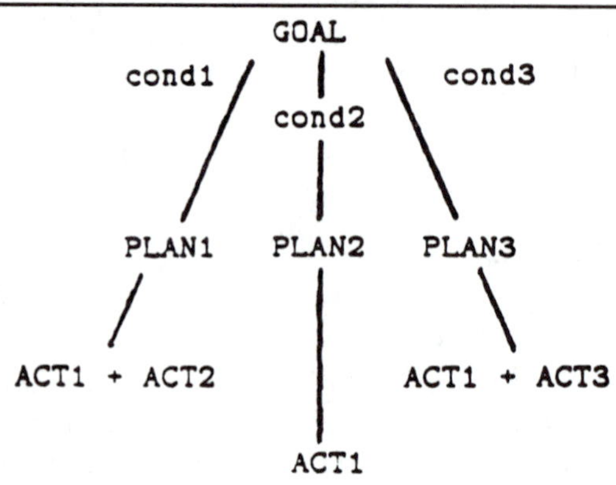

that connect these plans to the goal are conditions or selection rules that must be met before a given plan is chosen. These conditions can be based upon any distinguishing feature of the plans. Thus, at the highest level of expertise, goals are linked to plans using the conditions under which these plans are invoked, whereas goals were linked to simple plans in Phase Three and actions in Phase Two.

Robertson and Black (1983) present evidence for the evolution of compound plans. With increased experience, the time spent pausing between simple plans decreased, suggesting that users had combined the simple plans that they had learned to form compound plans. For example, the simple plan for moving the cursor to point to the item to be changed was combined with the command for changing the item (e.g., Delete). In addition, the decision time to initiate a compound plan during which subjects choose what they think is the most appropriate plan decreased as the editing session progressed, suggesting the acquisition of selection conditions that facilitate accessing the most appropriate plan.

Figure 10b presents a concrete example of a Phase Four knowledge structure. In the editor Robertson and Black used, to change one word to another word, there are three possible plans (sequences of actions) that may be invoked. The links connecting these three plans to the goal are conditions that must be met before the plan is selected. In the structure depicted, the conditions are based upon the relationship in length between the old word and the new word. In Figure 10b, we present a representation of this knowledge. The GOAL is to change word1 to word2. If word1 is longer than word2, then the Type Over and Erase plan would be used in which word2 is typed and then the extra letters from word1 are deleted. Similarly, if the two words are equal in length, then the Type Over plan is used and if word1 is shorter than word2, the Type Over and Insert plan is used.

Figure 10b. Phase Four—Example Structure.

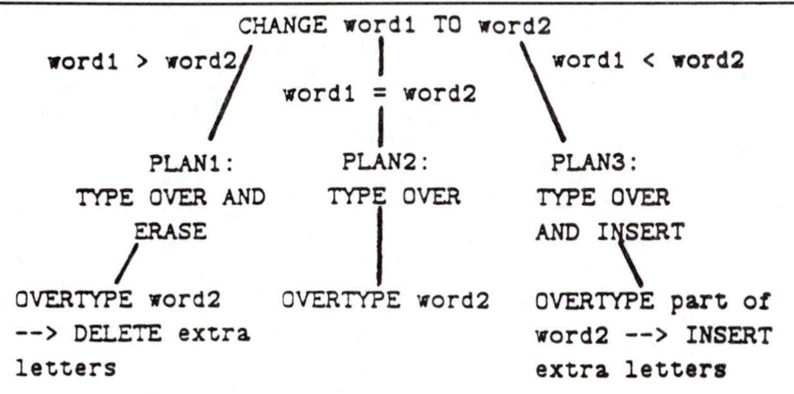

In addition to providing evidence for the compound plans of Phase Four, Robertson and Black also found evidence for the development of selection conditions. In the beginning of the training trials, when users had to change one word to another word, the majority of the users (approximately two thirds) would Delete the old word then Insert the new word. However, as experience increased, these same users stopped using the Delete/Insert plan and began to Overtype the old word with the new word. In this example, users appeared to develop a selection rule that can be stated as *If you want to change one word to another, use the Overtype plan.* This type of selection rule organizes the plans according to the priority that they have in efficiently accomplishing the goal.

THE LEARNING PROCESS

Our four learning phases describe the evolution in user behavior from problem solving to plan instantiation. That is, initially users must interpret each action in the editing task and monitor the success or failure of the action in terms of the final goal state. However, once the user is familiar with text-editing tasks, the commands are combined into plans that are applied when appropriate. It is the latter process that accounts for the plan-boundry pauses found in the Robertson and Black study, because deciding which plan to use takes longer than the transition from one plan action to another plan action when applying the plan.

Throughout this discussion of our four-phase model, we have made references to the GOMS model and the order in which the components of this model are acquired. Although we are not able to distinguish whether the Goals are acquired before, after, or simultaneously with the Operators, we are able to conclude that the Goals and Operators are acquired before the plans that are acquired before the Selection rules. In addition, we noted that, as the user becomes more experienced, the components build upon one another so that the chunks of knowledge represented using these components increase in size. Thus, we were able to trace the evolution of text-editing knowledge structures and extract an acquisition process that, in the end, results in knowledge representations similar to those found in the GOMS model.

Why do these learning phases occur: that is, what is the learning process that causes the user of a system to progress through these phases of learning? The initial, preconception phase is clearly necessary, because new users have only their prior knowledge to help them understand and use a system in the beginning. The final, increasing-expertise phase is also clearly necessary, because compound plans for accomplishing major goals and the selection rules for choosing plans at the most appropriate times are both necessary for skilled performance (Card, Moran, & Newell, 1983). But why are the two intermediate phases necessary?

Users cannot progress directly from phase one to phase four representations, because transforming phase-one representations to phase-four representations requires bringing more information together at one time than human working memory is capable of holding simultaneously. Thus, the two intermediate knowledge representations allow the user to progress from phase one to phase four in bite-size chunks that correspond to the limits of human working memory. For example, to progress from phase one to phase two, the users need only learn the links between the commands and the goals that can be accomplished, they do not need to simultaneously learn the sequencing links between the commands that are needed to combine the commands into plans. Then once the phase two command-goal links are mastered, the users can progress to phase three by learning the sequencing links that combine the relevant commands into simple plans for accomplishing elementary goals. Finally, once the simple plans are well learned, the user can progress to phase four by combining these simple-plan chunks into more compound plans and learning the conditions that determine when the various plans are most appropriately selected for accomplishing the goals.

However, while an intermediate step between phases one and four is necessary, phases two and three may not both be necessary. The simple plans of phase three are clearly a necessary precondition for the compound plans and selection rules of phase four, but phase two might be an artifact of the way that systems are currently taught to new users. In particular, current instruction manuals emphasize descriptions of individual commands at the expense of describing how these commands are combined into plans. We are currently investigating whether plan-based instruction manuals will allow new users to skip phase two and progress directly to phase three, thus significantly speeding up the process of learning new systems.

We are also pursuing the model's generality and testing its predictions. In particular, we are testing the generalization of our model to other domains. The most immediate extension from the text-editing domain is to computer programming. We have begun some preliminary work applying our model to this domain and plan to carry out a full scale longitudinal study in the near future. However, in the more distant future, we plan to try extending the model to noncomputer domains (e.g., learning physics, algebra, and the operation of devices).

To test the model's predictions, we are using it to guide training in text editing. Since our model was extrapolated from our observations of the natural evolution of knowledge representations for text-editing information, we believe that using this model to train users will facilitate the learning process. We are currently designing a study to test this hypothesis by comparing the performance of users whose learning was guided by the phases in our model to the performance of users whose learning was guided by commercial training materials in which the emphasis is on the kinds of commands that can be used in the system. That is, we plan to design tutorials that guide users through the transitions between the four phases of learning. We will first introduce the commands in terms of possible preconceptions that might exist for the commands and then show how these commands can be categorized by goals. Once the commands have been introduced, users will be guided in forming plans that consist of sets of commands that can be used to accomplish the goals previously introduced. Finally, we will show that the plans themselves can be combined into more complex plans with selection rules that allow for accessing the most appropriate plan.

CONCLUDING REMARKS

Our research fills a gap in the current literature by showing how we can use GOMS (Card, Moran, & Newell, 1983) and GAP tree (Spohrer et al., 1985) types of representations to examine the acquisition of computer knowledge. Our model not only illustrates how knowledge representations change to accommodate newly acquired information, but also addresses the order in which different types of knowledge (e.g., goals and plans) are acquired. In addition, we have examined the learning process for users who have naturally acquired text-editing experience, and thus believe that we have tapped learning mechanisms that can be applied to the design of computer-training materials.

REFERENCES

Adelson, B. (1981). Problem solving and the development of abstract categories in programming languages. *Memory and Cognition*, *9*, 422–433.

Anderson, J. R. (1983). *The architecture of cognition.* Cambridge, MA: Harvard University Press.

Anderson, J. R., Farrell, R., & Sauers, R. (1984) Learning to program in LISP. *Cognitive Science*, *8*, 87–129.

Arabie, P., & Carroll, J. (1980). MAPCLUS: A mathematical programming approach to fitting the ADCLUS model. *Psychometrica*, *45*, 211–235.

Card, S. K., Moran, T. P., & Newell, A. (1983). *The psychology of human computer interaction.* Hillsdale, NJ: Erlbaum.

Ehrlich, K., & Soloway, E. (1984). An empirical investigation of the tacit plan knowledge in programming. In J. Thomas & M. Schneider (Eds.), *Human factors in computer systems*, Norwood, NJ: Ablex Publishing Corp.

Johnson, S. C. (1967). Hierarchical clustering schemes. *Psychometrika*, *32*, 241–254.

Kruskal, J. B. (1964). Multidimensional scaling by optimizing goodness to fit to a nonmetric hypothesis. *Psychometrika*, *29*, 1–27.

McKeithen, K. B., Reitman, J. S., Rueter, H. H., & Hirtle, S. C. (1981). Knowledge organization and skill differences in computer programmers. *Cognitive Psychology*, *13*, 307–325.

Robertson, S. P. (1983). Goal, plan, and outcome tracking in computer text-editing performance (Report No. 25). New Haven, CT: Cognitive Science Program, Yale University.

Robertson, S. P., & Black, J. B. (1983). Planning units in text-editing behavior. *Proceeding of the Chi '83 Conference on Human Factors in Computing Systems* (pp. 217–221). Boston, MA: Association for Computing Machinery.

Schank, R. C., & Abelson, R. P. (1977). *Scripts, plans, goals, and understanding: An inquiry into human knowledge structures*. Hillsdale, NJ: Erlbaum.

Sebrechts, M. M., Black J. B., Galambos. J. A., Wagner, R. K., Deck, J. A., & Wikler, E. A. (1983). *The effects of diagrams on learning to use a system*. (Report No. 2). New Haven, CT: Learning and Using Systems, Yale University.

Shepard, R. N. (1962). The analysis of proximities: Multidimensional scaling with an unknown distance function. Part I. *Psychometrika*, *27*, 125–140.

Shepard, R., & Arabie, P. (1979). Additive clustering: Representations of similarities as combinations of discrete overlapping properties. *Psychological Review*, *86*, 87–123.

Spohrer, J. C., Soloway, E., & Pope, E. (1985). A goal/plan analysis of buggy Pascal programs. *Human Computer Interaction*, *1*, 163–207.

Takane, Y., Young, F. W., & deLeeuw, J. (1977). Nonmetric individual differences multidimensional scaling: An alternating least squares method with optimal scaling features. *Psychometrika*, *42*, 7–67.

11

Understanding and Learning Text-Editing Skills: Observations on the Role of New User Expectations*

Robert Mack

User Interface Institute
IBM Thomas J. Watson Research Institute
Yorktown Heights, NY

INTRODUCTION

The claim that prior knowledge influences the acquisition of new skills is a well-known generalization in psychology (see Bransford & McCarrell, 1974, for a relatively recent statement in contemporary cognitive psychology). Identifying this knowledge in particular domains like interactive computing, however, and showing its relevance to learning, has only begun to be carried out. An example is Bott's (1979) early study of college students learning to use a line editor. He reported examples of how prior experience with televisions, typewriters, and paper print technology could influence learners' interpretation of terminology, the operation of editor commands, and feedback. More recent studies at the IBM Watson Research Center have documented the role of typewriting knowledge in learning text editors for novices in realistic learning situations.

In these studies, office temporaries tried to learn commercially available word processing systems using self-study instruction (see Lewis & Mack, 1982a; Mack, Lewis, & Carroll, 1983; Carroll & Mack, 1984a,b). Learners often tried to generalize what they knew about typewriting to the word

*This chapter is an expanded version of a presentation given at the American Association for the Advancement of Science by the author and Mary Beth Rosson (Mack & Rosson, 1984). I would like to thank Mary Beth, John Carroll, Clayton Lewis, and John Gould for useful comments on the ideas presented in that talk, and developed further in this chapter. I also thank the editors for suggestions that improved the chapter.

processing system, but these expectations were often violated by the operation of the word processor. A common example is trying to use familiar typewriter operations like Carrier Return, Backspace, or Space to move the typing point, only to discover that these operations typically do not simply move the typing point, but actually can change what is typed. For example, pressing Return on many word processors ends the current line and inserts a new one.

Other studies have made similar observations about the role of typewriter knowledge in text editing (e.g., Douglas & Moran, 1983; Galambos, Wikler, & Black, 1983), as well as still other types of prior knowledge that have analogies to text-editing concepts and procedures (e.g., Foss, Rosson, & Smith, 1982; Rumelhart & Norman, 1978). These studies have included theoretical analyses of possible mechanisms by which prior knowledge influences acquisition of computing skills (see Bott, 1979; Barnard, Hammond, Morton, & Long, 1981; Carroll & Thomas, 1982; Carroll & Mack, 1985; Douglas & Moran, 1983; for more general discussion of metaphor and analogy in learning, see Gentner, 1985; Halasz & Moran, 1982; Holyoak, 1983; and Rumelhart & Norman, 1978).

This chapter summarizes selected data and conclusions about the knowledge new users, in this case secretaries, have about text editing, based on two further lines of research. One line of work tried to directly elicit expectations from new users about the operation of a word processor. A second has evaluated an editor prototype design that tried to accommodate these expectations. We review selected results from both studies, in order to provide further insight into how this knowledge, expressed in expectations, can influence new users' initial learning experience with text editors even beyond typewriting expectations. We also discuss implications for designing more intuitive editing interfaces.

PREDICTING TEXT-EDITING TASKS

In one line of research, reported more fully in Mack (1984), 12 office temporaries who were not experienced with computers were asked to predict how to do text-editing tasks like getting rid of a word in a line, or inserting a space between two words, just before observing a live demonstration of how to actually do the task with the text editor. Each task was indicated in a marked-up copy of the document, and observers were asked to make predictions in whatever terms they thought reasonable (but they were encouraged to refer to functions on the keyboard). The 12 participants were run individually, and each observed only one of three commercially available word processing systems. That is, there were four participants per system.

The results can be summarized in three generalizations. First, where text-editor operations resemble typewriter actions, participants assumed, at least initially, that the operations would act similarly. Second, participants tend to focus on the surface appearance of text-editing actions and outcomes. Participants seem to take at face value certain features of the editor that, in fact, can have other interpretations in terms of the editor program. Examples are line break symbols or blank areas of the screen that cannot be typed into. Finally, participants tend to have a simple view of operations. For example, participants seem to analyze operations like insert or delete, into constituent elements, and assume that each element is associated with an action. That is, they assume that each action has one outcome. This also seems to entail the assumption that one must exercise more explicit control over operations than is often required.

Typewriter Generalizations

Participants referred to typewriting methods in direct predictions for the first couple of demonstrations that involved cursoring, and less directly in terms of surprise at unexpected side-effects of superficially familiar operations involving Space or Carrier Return. For example, the first task demonstrated a revision operation which required moving the typing point to the revision point. Participants expected that the latter could be accomplished by using functions like Space, Backspace, or Return, as one would on a typewriter. In fact, Return inserted blank lines on all three systems, while Space and Backspace only moved the typing point on one system, inserting or deleting, respectively, on the other two systems.

Once demonstrated, participants were able to consider alternative uses for Space, Backspace and Return, and a couple of participants made comparisons that suggest an explicit recognition of these different possibilities:

> *So the directional keys* (Cursor functions) *only move the cursor. They do not move a sentence over as I thought they did. What I saw was the effect of space being inserted.*

Or this participant's comments when asked for a summary of how Return works after seeing a simulated typewriter operation using Cursor functions (the *E* in parentheses refers to an experimenter prompt):

> (E: Could Return work like that?) *Maybe not. Because if that* (i.e., the left and down cursor functions) *does that, this* (Return) *i can't do it.*

Typewriter expectations were also in evidence in predictions about making revisions. These predictions were initially limited to typewriter methods of erasing and retyping. For example, the first revision participants were asked to consider involved correcting two words that had been typed without a space separating them (*requestfor*). Eleven participants predicted that one would erase the rest of the line from *f* and then retype one position over. This limited expectation about text revision is quickly supplanted once text-editing methods are demonstrated, but examples persisted for most participants in subsequent predictions or descriptions. Here is a particularly clear statement from a participant on a system where, contrary to expectations, a user cannot simply type over material. The task is to replace a lower case letter with an upper:

> *I would try shift key and try to put a capital "c" over it* (i.e., the lower case *c*).

Four demonstrations later, the same participant expressed surprise at how a phrase was inserted (in protocol excerpts to follow "Mary" or pronominal equivalents, refer to a hypothetical "real user" the participant was to imagine she was observing, rather than the experimenter):

> *So she* (i.e., the hypothetical "Mary") *just inserted "text-processing". So you cannot type over an existing character. So this automatically inserts letters by moving over, pushing over the existing thing to the right margin.*

Still later, even after making this observation, the same participant described the replacement of a period with a comma (actually done with delete and insert) the following way:

> *Mary placed the cursor under the period and just seemed to type over it with the comma. . . . So you can just type over a period and make it a comma?* (E: Do you think so?) *Huh. Huh.*

Another participant stated her *preference* for being able to simply type over material even where she understood inserting to be a more powerful alternative. The task is to replace a phrase with a longer phrase at the beginning of a paragraph:

> (I would) *retype all this* (referring to the entire first line of the paragraph). (E: So you would be retyping over a lot of stuff?) *Or else you could just put it in insert.* (E: What would you do?) *I would feel most comfortable just typing over what is there.*

Most participants did not provide so explicit expressions of alternative typewriter-like expectations or preferences. However, anecdotal observation in real work situations reinforces the conclusion that typing by replacing is more readily understood, and preferred by novices. Many word processors allow users the choice of typing by replacement, or automatic insert.

Expectations motivated by typewriting are short-lived in this task. This is undoubtedly due in large part to the methodology where novel text-editing operations were demonstrated early on, and the questions asked in this context undoubtedly stimulated participants to modify their expectations, or adopt new ones. It is also plausible that, in the case of moving the typing point, it is simply not difficult to understand that Cursor functions are an alternative to Space, Backspace, and Return. We will see below that other participants from the same population of users also refer to typewriting expectations spontaneously, in a more realistic learning situation.

Surface Appearance

Typewriter expectations result in specific expectations about how certain text-editing operations would be done. A related characteristic of new users' understanding of text editing is their seeming *inability* to understand more than the surface appearance of what they see and experience. This is reflected in their descriptions of text-editing procedures, and in difficulty understanding certain relatively abstract features of editor design and document representation.

These descriptions tend to use terms that refer to familiar "objects" like words, lines, or text, and especially the surface appearance of blank nontext areas of the screen. Descriptions of adjusting, or reflowing text that accompanies inserting or deleting, tend to focus on the spatial rearrangement of text objects: *making space*, or *to give myself an open space*, in the case of inserting, *the line erasing itself*, to describe auto word wrap, *bring the sentence together*, or *move the rest of the sentence forward*, to describe reflowing lines after erasing material.

These examples may seem obvious, but certain operations require a deeper understanding of other features visible on the display. This understanding would reflect more accurately how the editor program works. Two examples are the interpretation of the *nontyping area* of the editor, and using *Carrier Returns* to end lines, where Return enters a nonprinting line break character.

The nontyping area is a characteristic of many editors in which blank space to the right of line ends is unavailable for entering text unless the user inserts characters or spaces out to that point. Typically, it is possible to move the typing point into such blank areas, but attempts to type in them results in an error message like "Nontyping Area. Move Cursor." Participants presented with demonstrations of this constraint could not give interpretations of what it meant. The inability to enter typing in a blank area seemed to conflict with a more concrete and familiar notion that blank areas within margins are available for typing.

Participants also seem to have trouble understanding that lines can be represented in some sense by characters, and that the space to the right of the character is, again, a nontyping area that is not really part of the document. Their alternative interpretation is that Return somehow creates a blank space that is part of the document. Here are some examples. Remember, "Mary" refers to a hypothetical user the participant was to imagine she was observing:

> *Oh. She* ("Mary") *pressed Return and made a space. When she pressed Return she created a space there. Oh. She didn't want a space there.*

Or

> (E: What should Return do?) *Bring you back to the left margin, but there is a space there.* (E: How would you fix it?) (There must be) *some way to delete the return. The last action that caused the space is . . .* (maybe) *line adjust?*

One participant described the effect of deleting a line end symbol, the effect of which is to join the next line to the end of the line with the typing point, in the following way: "("Mary") *deleted the nothing that was there*." In yet another case, "Mary" tried to add a phrase to the end of a line, beginning with typing over the period ending the original sentence with a comma. When one of the new characters replaced the line end symbol in this line, the next line was joined to it. The participant wondered why "Mary" was not able to just continue typing in the blank space that had been there prior to joining: "*I don't know why the machine did not use the space that was there*," instead of appending to it the next line. In all these cases, the effect of Return is seen as resulting in blank space to the right of the last character, space that is presumably available for typing.

The difficulties participants had with these two features, in the prediction and description task, reinforce anecdotal observations of real learners dealing with these features. These observations (e.g., Mack et al., 1983) suggest that the nontyping area is puzzling, especially coupled with the fact that Carrier Return can alter the form of the document in a way that is difficult to understand and undo. These difficulties suggest that new users' understanding of text objects on a text editor is limited to familiar elements of the document and does not spontaneously or readily extend to more abstract or unfamiliar features, at least not without explanation.

Typewriter expectations and the inability to see beyond the surface appearance of what is visible pertain to only a relatively small domain of text-editing operations. What happens when these participants begin to experience examples of novel text-editing methods?

Simple Actions

There is evidence that participants in this prediction and description task tend to expect actions to be simple and direct expressions of the goal underlying a given operation. One consequence is that, where there is more than one outcome of an action, participants seem to assume that these outcomes must each be associated with some explicit action, as if actions could only have one, univocal outcome. Another consequence, at least initially, is that participants seem to expect actions to be more directly related to the ultimate goal or outcome than is often the case for operations that may, in fact, require several steps.

Consider the following example, pertaining to inserting text. In one of the three word processors used, inserting is automatic—one simply positions the typing point and begins typing. However, all four participants observing this seemed to believe that some kind of explicit action was involved. Here is an example:

> (She—the hypothetical "Mary"—tried to) *separate* (the words) *'request' and 'for', to add a space between the words . . . how she got the thing to do it is beyond me.* (E: Did "Mary" do anything extra?) (I) *don't know. That I didn't catch? Maybe. . . . Some sort of button that would say you are going to do some kind of correction.*

The participant thought she had missed something even though she had not. Two thought that this action was somehow "illogical" without, presumably, first doing something else. The participant referred to in the immediately preceding protocol is an example. Here is what she said after viewing another demonstration of inserting:

> (It) *seems illogical that she* ("Mary") *could do that, to stick something in when there is no room for it . . . I could do that with any word, just move it over?*

Another example illustrates the extent to which a participant tried to construe what she observed to accommodate this assumption of simple action in the case of inserting. In this case, a participant had seen Backspace used as part of the procedure for inserting the letter *a* in the string *compny* (although in a way irrelevant to inserting the letter), but nevertheless interpreted its use as somehow allowing "Mary" to initiate inserting:

(E: Why did "Mary" use backspace?) *It must go back a space, to let you add something . . . the* "a" *between the two letters . . .* (It) *won't do anything.* (It) *must kind of put a hold on the machine so you can add something.* (Episode repeated. E: Did "Mary" have to do something to get "a" in there? Other than type "a"?) *Yeah. Definitely.* (E: You are confident?) *Oh yeah.* (You) *definitely have to go to the letter after where you want it, then backspace and put it* (the to-be-inserted letter) *in.*

The participant seemed committed to the view that inserting new text required some kind of additional action, and this led her to (mis)construe the irrelevant Backspace action as part of inserting.

Even some participants who observed the other two systems which do require an explicit action, namely, pressing an Insert function, seemed to expect a still simpler, more explicit procedure. Here is an example for the same task, inserting the letter *a* in the mistyped word *compny*:

(You will) *move arrow down, at 'n' hit 'insert', move it over,* (i.e., insert space) *backspace and put 'a' in and execute* (i.e., end inserting) (E: Do you have to do all that?) *Yeah. From what I've seen you do.*

In fact, while Insert on this system does open an insert window, one does not have to set up the space needed beforehand but simply begin typing. The procedure suggested by this participant's description of getting an "*open space*," and backspacing to the beginning to start typing, is more explicit than needed, or than demonstrated by the experimenter.

Another example involves replacing the word *Previously* with the phrase *Incidentally, in the past*. The participant predicted correctly that one would typeover the old word up to the material to be retained, and then insert the rest. But the participant actually counts out the number of blank spaces she thinks needs to be created, in order to insert the new material (*hold* refers to the name of the Insert function on this system):

Unfortunately, there isn't the same amount of letters in 'Incidentally, in the past' as in 'Previously,' . . . (I would) *put the arrow* (cursor) *to 'P', push 'Word' and 'Out'* (an invented sequence of keys for deleting which does not, in fact, delete) *let's say thirteen times. Then arrow* (cursor) *to 'we'* (first word of material to be retained) (press) *'Hold', move two spaces,* (press) *hold, go back, backspace all the way back, write* (i.e., type) *'Incidentally, in the past'.* (E: Can you think of an easier way?) *If I could I would say so.*

This participant had correctly predicted and described three previous cases of inserting, but in this case, and in an immediately preceding task, the participant reverted to a more explicit procedure of making an open space, repositioning the typing point at the beginning of the space, and typing the to-be-inserted material. This is all in contrast to getting into an insert mode and simply tpying.

The expectation of simple actions also arises in predictions about how to delete strings. Deletion typically involves one or more actions which have the effect of both erasing a string, and then automatically reflowing the line to eliminate the space occupied by the string. In contrast, half of the participants in the prediction tasks suggested that these two outcomes are associated with two separate actions: erasing and "closing up the gap." Here is an example:

> *I know she* ("Mary") *can remove letters using the Backspace bar but how she would close off the large space in between the words 'new' and 'Apex'* (resulting from the erased word) *I'm not sure.*

Or this example for a different participant and word processor:

> (I would) *erase with the Space bar and there would be a special button . . . to bring that sentence together to delete. Oh maybe you could just press delete* (notices delete key) *maybe . . . some sort of delete and then press a button to fill in, fill the space. There may be just one or two buttons to press to do that.*

We should point out that the other six participants did not refer to explicitly closing gaps: two gave ambiguous answers about what to do after erasing, while four thought that the gap would automaticaly be eliminated, as is the case. Insert and delete are the simplest cases of explicit control, but there are other examples involving predictions and descriptions of reflowing paragraph lines to right justify, or reflow to new margin settings. These operations proved complex, and the protocol data fragmentary. However, data from four participants using a word procesor where reflowing was not automatic suggest a more complex expression of explicit control of simple actions. Three of the participants on this system were presented with a case in which inserting material resulted in text that extended beyond the right margin and needed to be adjusted. Here is one participant's prediction for how to rearrange the paragraph:

> (E: How would you solve the problem?) (I would) *distribute the words in two lines.* (I would put) *'with', then 'a gentleman' on the next line* (i.e., shift words from the end of one line to the beginning of the next) (E: How would you do this?) (I would) *hit hold* (i.e., the insert key) *progress the line ahead from 'White Plains'* (E: Progress?) *Hit the space eleven times* (insert eleven spaces) *and . . . Put in "a gentlemen" . . .* (E: But would this solve the problem?) *You would have to do this one line at a time.*

The participant essentially suggests erasing material at the end of one long line, retyping or actually inserting it at the beginning of the next line, and iterating this process for all the lines of the paragraph. This prediction is a more explicit, and, by hypothesis, *conceptually* simpler approach, in contrast to reflowing automatically, or by some more global step. Notice also that the procedure involves making space to insert. Insert mode does not, of course, require this extra step. Two other participants suggested procedures that similarly involved some action applied to each line (a fourth was unable to make a prediction). In all cases, participants seem to analyze the procedure into specific effects and assume that each must be the result of some explicit action on their part.

Following each of these opportunities for making predictions, I was interested in coaxing participants into seeing the possibility of a more global approach. To this end, I gave participants the hint that an adjust key could be used to fix the problem and asked them how they might use it. Two participants now suggested positioning the typing point under the first character of the material to be adjusted and initiating an adjust procedure. The actual procedure happens to be not as direct as these modifed predictions suggest, but the latter are close.

Two other participants had no idea how one might adjust, even with considerable prompting:

> (E: Let your imagination go) *I can't visualize it.* (I can't) *see how you could do* (it) *without doing what we have used so far, erasing and* (retyping. I) *wouldn't know what to expect." Or this participant: "I'm sorry. I can't . . .* (it) *just doesn't make sense. Maybe it* (the label "adjust") *describes what you are doing, but it's not obvious, like 'erase' is obvious. 'Adjust' isn't.*

We can conclude from these examples that conceptualizing the reflowing of paragraphs can be difficult. One reason may be the competing view based on a conceptually simpler, and more explicit procedure of explicitly rearranging the lines of paragraph by erasing the ends of long lines, and then inserting the erased text at the beginning of subsequent lines.

The expectation of simple actions seems to affect the interpretation of a procedure after it is demonstrated. In one case, when a participant had a second chance to predict how to use adjust, in this case to effect a margin change, her predicted procedure misconstrued elements of the demonstration in a way consistent with a more explicit procedure. The participant suggested pressing Adjust at the beginning of the document file, which is roughly correct for a global page reformatting, and then pressing Return repeatedly to propagate adjusting for each line, which is not correct. She had seen Return used in the earlier demonstration as part of the paragraph adjust procedure. We infer that she simply assimilated its use there to her current expectation that some explicit action was needed in this bigger task, to specify adjusting of each part of the document.

Contrary Evidence for Expectation of Simple Actions

Not all the evidence is consistent with the hypothesis of simple actions. As we noted, four participants clearly predicted that, in getting rid of a word, the gap would automatically disappear. Two others were less clear but did seem to consider that reflowing might be done for them. On another task, no participant predicted that words would automatically wrap to a new line when typing went beyond the right margin. However, all but two were able to correctly interpret the feature. Automatic adjusting can also occur when margins are changed for a document that has already been typed. Two participants suggested that one would need to execute some kind of explicit action to effect the change in margins, but two others expected the letter to conform to the new margins as an outcome of simply changing the margin values (not all participants got to this demonstration):

> *But what I said would happen automatically didn't happen. The whole body of the letter is not reorganized because "Mary" changed the body to the right.*

These examples suggest that some participants can consider adjusting to be an automatic accompaniment of another action in certain situations.

Summary and Discussion

The prediction and description task provides insight into how people new to text editing try to understand text-editing operations. Operations like moving the typing point or making simple revisions are construed to some extent as they are on a typewriter. This conflicts with the fact that these operations typically change text, and do not simply move the typing point. Douglas and Moran (1983) make the same point by contrasting the strictly *locative* properties of such functions in typewriting, to the typically additional *mutative* properties of these functions unique to text editing. However, these expectations are short-lived, and in any case do not help to interpret novel text-editing operations. Here, a further expectation comes into play.

This expectation is that actions are simple in their effects and, as a corollary, that multiple effects of complex procedures must be the result of separate or explicit actions. For example, inserting or deleting involve on this view, respectively, "making space", or erasing and "getting rid of gaps." Other examples involve reflowing lines or paragraphs.

These predictions, of course, leave unanswered several questions about understanding the nature and impact of expectations on learning. For example, we cannot assess in a clean way how long these expectations last, because participants' understanding undoubtedly was influenced by the questioning, or how quickly experience with the procedures, in this case that of observing the "correct" procedure, results in "correct" understanding.

Finally, while we alluded to some evidence contrary to the expectations that actions like reflowing occur automatically, the overall hypothesis of simple actions does nonetheless suggest a coherent view of text manipulation. In particular, it suggests an editor design in which typing is replacing, Carrier Return does not insert, there are no invisible "surprises" like formatting symbols or nontyping areas, and users can explicitly manage reflowing rather than have it occur automatically. Such features are of course embodied in a piecemeal way in some existing systems. Many systems, for example, require some explicit action before inserting, with the default typing mode that of replacing (typeover). At least one word processing system allows a certain amount of control over reflowing by users. These considerations motivated a second line of research, whose relevance for understanding new user expectations we now discuss.

EVALUATING A NO SURPRISE EDITOR PROTOTYPE

The second line of research aimed at trying to accommodate expectations about typewriting, and simple actions in a prototype text-editor interface. This editor prototype was intended to be more concrete and familiar in appearance and operation for computer-naive users. Editor operations were designed around a small set of primitive objects implemented with function keys, like "Word," "Line," "Paragraph," and "Page," and actions like "Delete," "Insert," "Adjust," and "Format," whose designations and behavior were intended to better match new users' preoccupation with the familiar and concrete appearance of documents and text manipulation.

Put another way, the functions were intended to provide an object-action "language" which referred more directly to familiar text objects (e.g., "Word" or "Line") and actions applicable to them. For example, blank lines could be inserted and deleted by referring directly to the object of interest through the "Line" key, followed by "Insert." A combination like "Line" and "Delete" would delete a line, while "Paragraph" and "Adjust" would accomplish the reflowing of paragraphs with lines of uneven lengths. In particular, the form and content of documents was explicitly under participants' control, rather than rearranging itself automatically in unexpected ways. For example, inserting is an explicit action accomplished by pressing an "Insert" function key to make space (vs. initiating an open-ended insert mode). A "Delete" key erases nonblanks to the right of the typing point, which remains fixed at the beginning of the resulting gap. A key labeled "Adjust" closes up the gap by reflowing text leftward to the typing point within a line.

Two evaluations have been carried out to date, each involving six computer-naive office temporaries (from the same population as participants in the prediction study). Participants were given a half day to complete as many of eight typing tasks as possible, given only cursory reference materials in addition to the typing tasks. The tasks included typing, printing, and storing new documents, and retrieving and revising old ones (all one-page documents). We should emphasize that the goal of this prototype is to be learned with essentially no explicit instruction, and in neither evaluation was there explicit, step-by-step instruction. The overview materials referred to general text-editing concepts, specifics about powering on to the system, and brief, one- or two-line definitions of basic operations involving function keys and menu options.

A full account of the results to date beyond the scope of this paper. But several results reinforce and extend our conclusions about novice expectations in a more realistic learning situation.

Typewriting Expectations

The only operations which could literally resemble typewriting would be (a) operations that move the typing point such as Space, Backspace, Return, and (b) the typing mode of the editor (typeover). The prototype resembled typewriting in only one clear case. Carrier Return simply repositioned the typing point and had no other consequence, like inserting blank lines or ending a line at the typing point. Space and Backspace differed from a typewriter in that they were characters and could be used to erase text. Typing mode was also not literal overstrike, but typeover. Nonetheless, there is anecdotal evidence for expectations influenced by typewriter experience.

First, consider moving the typing point. Table 1 summarizes the total instances of positioning the typing point using the Cursor keys, Space, Backspace, or Return. It is clear that participants had little difficulty discovering and using cursor keys, since they constitute a large proportion of the total cases of cursoring: 92% of the total within-line, and 76% of the total between-line, cases. However, it is also clear that participants found

Table 1. Total Instances, Across Participants, of Transactions Involving the Typing Point Using Cursor Functions (Left, Right, Up, and Down), vs. Space, Backspace, and Carrier Return.

	Evaluation 1	Evaluation 2
Total within line	158	229
by left or right	145 (.92)	212 (.93)
by Space or Backspace	13 (.08)	17 (.07)
Total between line	234	294
by up or down	183 (.78)	219 (.75)
by Return	51 (.22)	75 (.25)

the Return function useful operation for positioning the typing point between lines in 25% of the cases. This may seem to be a small concession to typewriting, but, as we discussed earlier, on most word processors Carrier Return could not be used for this purpose, and indeed would result in an unexpected side-effect of changing the content of the document. Apart from this, there are a few cases (8% of the total within-line instances) where participants tried to move right or left using Space or Backspace, respectively. This behavior appeared to result from simple typewriter habit, and with a couple of exceptions, discussed below, caused no particular difficulty.

Reference to typewriter-like methods can also be inferred from attempts to make simple revisions. In typewriting, a typist corrects errors by erasing and retyping. In the editor prototype, correcting simple one-character typos, or replacing one string with another of equal length, can be accomplished by typing over the erroneous material. However, in about 33% of 61 cases participants did not use typeover directly, but first erased, either with Space or with Delete, and then retyped, as a typist would. There is nothing wrong with this method, but it suggests a typewriter-like approach, like whiting out and retyping, that contrasts with simply typing over.

Making corrections by typing over, or erasing first and then retyping, are both alternatives to text editing operations like delete and adjust, or insert. These operations are not inappropriate, but, when used in lieu of text editing methods, it seems reasonable to classify them more generally as a "fall-back" to simpler, more typewriter-like methods, in contrast to text editing operations proper.

The extent of fall-back can be seen in the summary data in Table 2. We have separated data from the two evaluations and indicated what proportion of the total within-line revisions were accomplished by text-editing methods involving Delete, Adjust, and/or Insert, and what proportion were

Table 2. Total Instances, Summed Across 6 Participants, of Transactions Involving Fall-back Alternatives to Text-Editing Operations for Within-Line Editing Changes (Range of Values Shown in Parentheses). Fall-back Includes Erasing and Retyping, and Typing Over Text.

	Evaluation 1	Evaluation 2
Total Within-Line revisions	96 (5–24)	139 (8–35)
Fail and Fallback	11 (.12) (0–3)	13 (.09) (0–12)
Immediate Fallback	14 (.15) (0–5)	10 (.07) (0–3)
Total fallback	25 (.26) (1–8)	23 (.17) (0–14)
Delete, Insert Adjust functions	44 (2–15)	112 (5–32)
Total immediate typos	32 (0–16)	62 (1–24)
Immediate typos by erase & retype	2 (0–1)	10 (0–9)

accomplished by fall-back methods as we have defined them. Within this latter category, we have further distinguished between immediate fall-back and that associated with a failed attempt to first employ a text-editing method. We can see that about 20% of the total 235 within-line revisions were completed by a fall-back method: About 10% of these are immediate, while 10% follow a failed attempt to use a text-editing method.

Table 3 makes the fall-back phenomena more concrete by describing summary protocols of four examples of fall-back. The first two excerpts are examples of immediate fall-back without first attempting some text-editing action. The last two of the examples illustrate fall-back following an attempt to use Adjust to reflow a gap in a line. A problem with the implementation of Adjust in the first evaluation is that, if the typing point is located at the beginning of gap (following erasure of a word or string), Adjust will concatenate the words bracketing the gap. This is easily solved by pressing Insert, but, as the excerpts indicate, many participants simply gave up and typed over the line to restore the space.

These are relatively systematic examples of typewriter-like expectations. There are also more idiosyncratic cases which, while affecting at most one or two participants, nevertheless indicate how troublesome can be the violation of even simple expectations. One learner evolved a complex procedure for correcting simple typos (i.e., those detected immediately after being made) by actually quitting the entire task and starting over.

Table 3. Four Examples of Fall-back to Simpler Editing Methods.

1. (Evaluation 1, Learner 1)
   Task: Delete single word in line

   Space over entire line, retype line without to-be-deleted word (vs. Delete specific word and adjust remaining spaces).

2. (Evaluation 1, Learner 3)
   Task: Replace single mistyped letter, immediately after typing it

   Cursor left once, Delete once and retype (vs. Backspace and retype)

3. (Evaluation 1, Learner 5)
   Task: Replace short word with longer, tries to insert a couple of extra spaces:

   Space to erase short word, types part of new word, then Adjust to insert more space. This has the side-effect of concatenating unfinished word with rest of typed text. Cursor left to beginning of unfinished word and types over entire line with new word in place of old.

4. (Evaluation 1, Learner 6)
   Task: Delete "s" of "pencils"

   Deletes correctly and presses Adjust to eliminate extra space, but concatenates "pencil" to next word. Types over entire line to eliminate the extra space (vs. Insert to insert one space to separate the words)

She did this repeatedly, even after having composed part of her document. The reason for this behavior is that she had tried to use the Delete key to correct the typo, but failed because the typing point was at the end of the string. Having evolved this alternative and more complex procedure, she persisted in it rather than exploring alternative procedures for Delete or alternative operations like Backspace. It seemed obvious that she was assimilating the Delete operation to Backspace as on a typewriter, and, sometime later, when asked about this, she admitted that she had been thinking about typewriters. A second participant initially interpreted Delete in the same way but quickly discovered the correct alternative.

Simple Actions

Now consider evidence reinforcing the claim that new users expect actions to be simple. Of course, the prototype was designed to accommodate this expectation to a large extent, so the design makes it difficult to detect evidence for this specific expectation, at least in terms of surprises. Nonetheless there were behaviors consistent with the hypothesis, and these are summarized in Table 4.

Participants in the first evaluation of the editor prototype gave evidence of difficulty initially understanding the concept of pointing and selecting

Table 4. Summary of Behaviors Consistent with Expectations that Actions are Simple in the Editor Prototype Evaluation.

- Confusing pointing and selecting
- "Correcting" word wrap
- References to, and behavior implying "making space"
- Revising by erasing and retyping
  (versus typing over)
- No complaints about having to adjust
  (even though complained about other "effortful" activities)
- Reflowing paragraphs: The "Eight Tile Puzzle" Metaphor of text manipulation

menu options by using Cursor functions to point, and a Select function to select the option. All six participants in the first evaluation seemed to expect that actions would more directly produce an outcome than was the case: e.g., that moving the typing point to a menu option should initiate that option by itself. Examples include participants who selected the option to "Work on a Document" and immediately tried to type, as opposed to filling in the name of the document, and then execute the filled-in menu to get a blank typing page. One participant was so surprised that selecting the option "Work on Document" did not immediately let her type that she abandoned that typing task entirely and moved on to another that involved retrieving a document. Similarly, some participants pointed to a menu option that required typing in some parameter value, like a document name, and immediately tried to fill in the value of the option, as opposed to first selecting that option, in order to move the pointer into the field where the value could be then entered.

These behaviors are consistent with the expectation that actions are simple: moving the typing point is the action, and the outcome should immediately follow. Participants did not immediately understand that their first action might instead be the first step of a procedure with more than one step leading to the intended goal. Pointing and selecting were simplified in the second evaluation, virtually eliminating these initial problems.

The second item of evidence refers to automatic word wrap. Only 5 of the 12 participants in the two evaluations used it, even after noticing that words wrapped automatically when the typing point moved past the right margins. The preferred alternative was explicitly pressing Carrier Return to end lines and start new ones, consistent with the notion of explicit control: that is, particpants expected that ending lines was an outcome that required an explicit (vs. automatic) action. Indeed, four of the six participants in the first evaluation spent time trying to "correct" word wrap, presumably believing that it was some kind of problem.

The third item of evidence in Table 4 refers to evidence for the notion of making space to insert, in contrast to automatically inserting or initiating an insert mode. There are 18 cases (for 6 participants) out of a total of 129 tasks involving insertion of new material (including blank lines) where there are direct or inferable references to the idea that inserting new text involves "making space" and then typing. These include cases where inserting would be the appropriate operation but participants tried the "incorrect" block Move alternative. An example is the participant who needed to restore a deleted line: "*I have to figure out how to make space in there.*" Or the participant who wanted to insert a space between two words: "*I wanted to make a space without typing it all over.*" Unable to solve how to do this, the participant ultimately fell back to typing over the entire line.

One might question whether or not particpants are just expressing themselves casually in these references, but there were two cases where the notion of making space clearly seemed to influence the *behavior* of participants. An example is the participant who wanted to replace a phrase (*computer*) with the longer phrase (*text-editor*) at the end of the line "about the new Apex 9000 computer system" To do this, the participant first used Space to erase the end of the sentence (i.e., *computer system*). Then she pressed Carrier Return to put the typing point at the beginning of the next, nonblank line where she also inserted a few extra spaces to move the text on the line further to the right. Her goal seemed to be to set up enough space to type the new phrase, retyping the word *system* and allowing additional material to spill-over from the first, to the second line. Further, she seemed to believe that this space had to be set up as a separate step in actually inserting new material.

There is nothing wrong with this approach but a quicker alternative is available by deleting *computer* and inserting a few extra spaces to accommodate *text-processing*. The participant would have found that text to the right of the typing point which exceeded the right margin would have wrapped automatically, and the new short line reflowed with the rest of the paragraph. The alternative behavior observed is, incidentally, similar to that for the participants in the prediction task who explicitly counted out the number of spaces needed to insert new words.

A fourth expression of simple actions may involve erasing and retyping to correct simple typos, or to replace strings of equal length, in contrast to simply typing over the erroneous letter or string. We described examples in Table 2 when discussing typewriter expectations. The connection to simple actions is that replacing a string with another can be analyzed into two actions: getting rid of the wrong string, and then typing in the correct one. The expectation of simple actions would predict that participants would associate an explicit action with both elements, even though, again, there is an alternative involving the single action typing over.

Operations of reflowing text, either within lines, or for paragraphs, provide yet another expression of the expectation of simple actions. Between the two evaluations there were 54 cases of adjusting gaps in lines. Of these, 58% were successful (summed over the two evaluations) immediately, and the remaining 42% successful with problems. The latter included cases of fall-back to simpler methods of erasing and retyping, or typing over. We will discuss the reasons for failure later, but for now we simply observe that in all these cases there was only one instance where a participant spontaneously complained about having to adjust. The others were not asked directly, but their lack of complaints in this connection contrasts with their spontaneous complaints about other operations which seemed difficult or unnecessary, for example, participants' unanimous but grudging belief that they could only move the typing point one position at a time.

A second type of adjusting involved entire paragraphs when insertions or deletions resulted in very uneven lines, or when participants created short lines by splitting lines one way or another. In the first case, a global procedure for reflowing contrasted with more piecemeal alternatives expressing a more explicit approach of interest to us. The second case involved restoring a paragraph consisting of short lines after it had been inadvertently reflowed, where there is no global action to solve this, such as "undo" reflowing.

There were six cases between the two evaluations where participants were faced with having to reflow a paragraph. Other cases were resolved in ways not relevant to the case for explicit management of reflowing, for example retyping an entire paragraph as part of another activity could have the same effect as reflowing to right justify margins. In four cases, participants tried a global operation, pressing the Paragraph key followed by Adjust, which is the preferred method in the prototype. However, in two cases, participants adopted a more explicit procedure similar to that we discovered in the prediction task data. Table 5 presents a protocol summary, including a schematic diagram of the paragraph in various stages of the participant's activity.

The episode was initiated by an inadvertent outcome: in the course of deleting a word at the end of a line, the participant inadvertently joined the next line to the end of the line with the typing point, or as much of the next line as would fit within the right margin. Unfortunately, the participant had not intended to join these two lines, the effect of which was to cause the residual words of the joined line to form a short line, requiring the participant to reflow the paragraph. This could have been accomplished by simply pressing the Paragraph key in combination with Adjust. The protocol summary describes what the participant tried to do instead.

The participant's first goal is to restore the situation by inserting the joined material on line L1 at the beginning of line L2 (the short line in

Table 5. Summary Protocol Excerpt Expressing the "Eight Tile Puzzle" Metaphor of Text Manipulation (In the Case of Paragraph Adjusting).

(1) Participant deleted end of line L1. Joined beginning L2 to L1, causing L2 to be shortened. Here is how participant describes her goal:

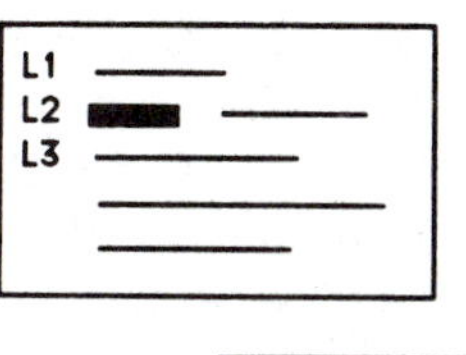

"*I have to move that sentence down* (i.e., the last segment of L1). (I) *erased that sentence* (i.e., the end of the original line L1)(I) *don't know how that happened.* (I) *need to move that* (i.e., segment remaining on L2) *over.*"

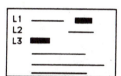

(2) Participant inserts blanks at beginning of L2 to move segment over to right. Inserts too much and last word of L2 short line wraps onto new blank line L3.

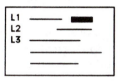

(3) Retypes end of L1 at beginning of newly formed space on L2. Has extra spaces between retyped words and original L2 text. Uses Adjust to close gap but concatenates. Uses Insert to separate segments. L2 is still too short because of single word that wrapped to (new) L3.

(4) Participant wants to put single word on L3 back on end of L2. Tries Move several times but does not work (Move is for block moves). Finally, uses Insert to move the word on L3 under space at end of L2. Tries Adjust to move it up into space on L2 but this left justifies word at left margin of L3. Insert again to move it back to end of L3. Cursor up to beginning of L2 and Adjust but this has no effect. Comments: "*I want to pick up the whole damn line is what I want to do.*" *In my head I am saying move that line back, insert* (i.e., restore) *words at the beginning of the line.* Eventually puts cursor at end of L2 and Adjust which joins single word on L3 to end of L2 and restores paragraph.

question). To do this, she uses Insert to make space, but in the process inserts too much space and causes the last word of L2 to wrap to a new inserted blank line, creating as a side-effect yet another short line L3. The solution remains the same: Press Paragraph followed by Adjust. But here again, the participant pursues her original strategy. She retypes the joined material on L1 at the beginning of L2. A subgoal here is closing up the extra space between the retyped material and the original material on L2. This she accomplishes using Adjust, which, unmodified by an object function like line or paragraph, simply eliminates gaps in lines.

Her next subgoal is to restore the orphaned word that had wrapped to the new L3. Here she tries several actions. First, she uses Insert to move the word on L3 to the end of the blank line. We can infer that she hopes to somehow pop the word in the space available at the end of L2. She tries Adjust to do this, but this left-justifies the word to the typing point which had been positioned at the beginning of the line L3. The participant inserts space again to move the word back under the space in L2. At this point she tries other operations, such as Move. Eventually, the participant tries Adjust at the end of L2, which joins the word on L3 to the end of L2, in effect deleting the extra blank line and restoring the paragraph to its original form.

The point of this example is that this participant did not consider a more global approach to reflowing the paragraph. Except for the unsuccessful sidetracks using inappropriate functions, the participant's strategy was not wrong—joining lines is possible in the prototype—and she ultimately succeeded. But her reasoning seemed constrained by a view of text manipulation in which segments of text are explicitly shuffled within lines or between lines to make line lengths more even.

As we noted at the outset, the evidence for simple actions, and its expression in the expectation of explicit control of reflowing, is sketchy in the prototype evaluation data, because the editor was designed to accommodate such expectations. Moreover, in some cases there were global methods for accomplishing tasks that make it difficult to observe more vivid alternatives of the expectation, represented, for example, in paragraph reflowing. Nonetheless, the evidence reviewed is consistent with the hypothesis that participants tend to expect simpler, more explicit forms of text manipulation than are typically available in text-editing systems. Before discussing the implications of this conclusion, we need to comment on evidence not entirely consistent with it.

Contrary Evidence for Expectation of Simple Actions

Five participants said or did things in the course of learning which imply that inserting or adjusting should be more automatic (or less explicit) than the prototype design seemed to allow.

One participant, for example, was surprised that Space erased text rather than simply moved the cursor. Rather than alter her expectation about how Space behaved, she constructed a more complicated account in which the effect of replacing letters with spaces was interpreted as "inserting space." She believed, in turn, that this resulted from some kind of insert mode that had persisted from an earlier use of the Insert function. The effect of Space was consistent with the *appearance* of blank spaces associated with the Insert operation, even though Space was actually erasing letters here.

Two participants thought that pressing Insert once should allow them to simply type at will—i.e., initiate an insert mode—and were surprised when this did not happen. Both interpretations imply the notion of a mode that, once initiated by an explicit action, allows subsequent actions like typing at will to occur automatically. In neither case, incidentally, did participants complain about having to do something to initiate inserting.

A third case was a participant who complained about having to adjust gaps in lines after deleting because she thought gaps should be automatically removed. It turned out that this participant, while having no direct experience with computers (as she had indicated in an initial interview), had nonetheless casually observed someone else and noticed how delete had worked on that system. The final cases reflect more on the scope of explicit action rather than the general notion itself. As we already noted, in four of the six cases of reflowing paragraphs, participants did attempt a more global action. Two other participants had deleted more than one word without reflowing the corresponding lines, so that, after several revisions, they had several corresponding gaps. With the typing point at the end of the document, they pressed the Adjust function, which had no effect. Their comments implied that they expected reflowing to occur globally. While they did not seem surprised at having to remove individual gaps, they did evidently expect that a more global action would have accomplished this goal.

These examples do not actually contradict the expectation of simple actions (with the exception of the participant who specifically complained about the lack of reflowing when erasing), because they all do refer to some explicit action. But they also suggest that people may differ about the extent or scope of simple actions.

Summary and Discussion

We have reviewed two lines of research from which we have inferred characteristics of new user expectations about text-editing activities. Typewriting knowledge is one component. The expectation that actions are simple is another, of special interest because it influences new users' understanding of text editing beyond those operations that are similar to typewriting. The expectation takes various forms, but, in tasks involving inserting, deleting, and reflowing text, a common expectation seems to be that text manipulation can involve the direct, explicit rearranging of segments of text.

This expectation suggests attributing to participants a metaphor which influences their understanding of text manipulation. Participants' understanding of text manipulation seems analogous to the Eight-Tile Puzzle game, in which one shuffles eight numbered tiles in a 3 × 3 frame until they are numbered consecutively. That is, making space or closing up gaps or rearranging the line lengths of paragraphs by nonglobal operations all suggest attributing to participants the expectation that text is manipulated by directly referring to, and rearranging segments of text. As the participant in the episode summarized in Table 5 put it, referring to her desire to return a joined segment to its original position at the beginning of the following (now shortened) line:

> *I want to pick up the whole damn line is what I want to do. In my head I am saying move that line back, insert* (i.e., restore) *words at the beginning of the line.*

The participant seems to be expressing a desire to directly manipulate a segment of text in order to rearrange its appearance.

The prototype evaluation was a short-term learning study, and it remains to discover how much, or what kind of, experience would be needed by participants before they would begin considering more global, or even automatic, alternatives to the Eight-Tile Puzzle metaphor. But it seems fair to conclude that the metaphor influences some participants' initial reasoning about key aspects of text editing activity.

What can we say about the origin of expectations about typewriting, or simple actions, as expressed in the Eight-Tile Puzzle metaphor? Typewriter expectations presumably come from experience using typewriters. The source of expectations about simple actions is less clear-cut. One might argue that the expectation of simple actions is just another expression of typewriter experience—or at least of the office tasks that involve paper documents, and typewriting, or manual approaches to manipulating their content. Examples include retyping large segments of text to make more localized changes, or rearranging document form and content by cutting and pasting parts, or even replacing one letter with another by erasing and retyping. The latter can be viewed both as a typewriter method and as an expression of simple actions (associating each logical element of the operation with an action).

To the extent that the expectation of simple actions does express a more direct approach to text manipulation in the novice's mind, explicit management of reflow can also be viewed as an example of direct manipulation. Direct manipulation is a principle of interface design in which actions and objects are designed to correspond in some sense to the analogous physical or manual operation in the domain of interest. (See Shneiderman, 1982; and Hutchins, Hollan, & Norman, 1986, for further discussion; also Smith, Irby, Kimbal, Verplank, & Harslem, 1982, for examples in an advanced workstation.) For example, instead of typing a command to delete a document, a user might *drag* a pictorial representation of the document to a pictorial representation of a wastebasket (as in Apple LISA); the action relates directly to what happens on the screen, and is analogous to physically tossing out a sheaf of papers.

Many of the obvious examples of direct manipulation pertain to manipulation of files and applications in the operating environment. Direct manipulation techniques for manipulating text content *within applications*, like word processing, include actions like *cut and paste*, which are intended to suggest manual counterparts. Explicit management of inserting, deleting and rearranging paragraphs operations in the No Surprise Editor prototype can also be characterized as a form of direct manipulation of text content, because these operations allow explicit and manual rearrangement of document content within and between lines.

The preceding analysis of the origin of expectations we have discussed seems plausible, but it is worth highlighting two aspects of behavior that may go beyond the metaphor of the paper office. First, operations which cause reflowing within lines or paragraphs on a video screen obviously represent metaphorical extension of the underlying knowledge of activities in the paper office. That is, literally manipulating segments of text is, of course, very difficult with paper: e.g., strings of characters do not actually reflow within and between lines as the Adjust operation permits. The same can be said for operations like cut and paste, for that matter, because the corresponding physical actions are obviously quite different, and the fact that document content is automatically reflowed does not have a counterpart. Certainly, more direct expressions of the underlying Eight-Tile Puzzle metaphor are possible: for example, actually dragging specified segments of a document from one point to another in a still more direct analogy to shuffling tiles in the puzzle, as the participant tried to do in Table 5.

A second point is that the expectation of simple actions, and explicit control is consistent with current views of how people acquire skills in general (e.g., Fitts & Posner, 1967; Anderson, 1982, for recent cognitive formulation). The initial phase of any skill acquisition seems characterized by deliberate, conscious attention to the elements of the target skill. Performance of the skill in this phase involves close attention to these component elements, and their relation to each other, and hence lacks the smooth execution of practiced skill. The expectation of simple actions, and its expression in the Eight-Tile Metaphor, similarly suggests an analogous deliberate, conscious attention to the constituent elements of text-editing operations. Moreover, participants seem to not only attend to each element, but assimilate each to a notion of simple action, according to which an explicit action is needed, on their part, to bring about that aspect of the procedure. Schneider (1982) has also suggested that novices have trouble chunking, treating as a whole the elements of commands, instead treating each as a separate procedure to be mastered.

How Expectations Affect Learning

Expectations can have multiple origins. They can derive from prior knowledge outside computing, like typewriting, or from participants' evolving knowledge of how the interface operates. Barnard, Hammand, Morton, and Long (1981) have distinguished between *compatibility* of methods for doing things with prior knowledge of methods for doing those things outside the computer domain, and *consistency* of those methods within the computer domain. For example, participants acquiring command languages expect consistency in the syntax of multielement commands (e.g., Carroll, 1982; Shneiderman, 1980, for general discussion). Whatever the origin of user expectations, it is a reasonable assumption that it is easier to learn interface operations when they work in accordance with user expectations.

The difficulty, of course, is that expectations can rarely match in full how an interface operates. Relevant expectations may simply be lacking. For example, knowing about typewriters does not really carry a participant very far into text editing. Expectations can only provide an approximation of what participants need to know. Mismatches are a major fact of learning computer systems. The effect of these mismatches is to induce users to try actions that do not match how the interface operates. This can result in failure at best, and, at worst in side-effects to which the user must respond in order to continue the task of interest.

One response to problems attending mismatched expectations is to change one's method for accomplishing a goal. An example associated with the prototype evaluation involves "fall-back" to a simpler method of accomplishing a goal. Examples include erasing and retyping, or simply typing over segments of text to correct a more localized problem that could be solved using text-editing operations like insert or delete (recall the data summarized in Table 2). "Fall-back" is relevant to learning, because it can be an obstacle to acquiring new skills. That is, to the extent that participants may give up trying novel, and potentially more productive, methods, and fall back on less productive, but simpler to understand, methods, participants may not benefit from the full capability of the computer.

Novice and more expert users of computers are similar with respect to this particular behavior. It is often claimed (see Rosson, 1984) that even experienced users have a limited repertoire of skills, relative to what is available, and potentially useful. Lack of breadth is often attributed, in turn, to difficulty learning new skills, particularly when a more familiar alternative procedure is available. A similar characterization can be made for novices in the evaluation study. Fall-back occurred in part as a result of problems experienced while trying a new operation, and also, presumably, because the simpler methods were just more familiar to participants.

Another response to problems is to try to recover from the side-effects associated with them. Side-effects often result in what Lewis and Mack (1982a) have called "tangled problems." The previous discussion provided examples of how expectations led to erroneous actions, and side-effects that led to recovery activity. The effect of mismatches on learning is unfortunately not easy to predict. To do so requires an understanding of the consequence, or side-effect the inappropriate action has, what actions are available to recover from the side-effects, and, more important, how all these actions and outcomes are interpreted by the user.

Using Space or Backspace to move the typing point in a line of text is an example of a generally simple and inconsequential mismatch. In the editor prototype, these operations, in effect, erase text by replacing with blanks. Participants were surprised at this, but, with two exceptions, understood the outcome and how to recover (i.e., simply reposition the typing point if necessary and retype). The two exceptions are participants who reasoned more creatively about the surprising effect of using Space to try to move through text. One participant, referred to earlier, attributed the appearance of spaces as an effect of an insert mode persisting from an earlier use of the Insert function. A second participant similarly attributed the erasing of text as the result of a delete mode persisting from an earlier use of the Delete function. These examples of creative reasoning make it hard to predict, in a simple way, the impact of violated expectations, because participants' interpretations can assimilate other, often wide-ranging, experiences (see Lewis & Mack, 1982b, for further discussion).

Other mismatches were more consistently difficult to interpret and recover from correctly. Consider the difficulties alluded to regarding the way line reflowing was implemented. The latter was designed according to a designers' model, such that text to the right of the typing point, beginning with the first nonblank, was shifted left to the typing point. This model presupposed that the typing point was at the beginning of the gap left by erasing when Adjust was initiated. In fact, the delete function was designed to insure that that text was erased "at a distance": The typing point remained fixed at the first character of the string being erased. This sensitivity of the operation to the position of the typing point conflicted, however, with alternative methods of erasing and alternative interpretations of how to recover from the side-effects of these methods.

For example, an alternative method of erasing used by some participants is to use Space to replace strings with blanks. In this case the typing point would be at the end of the gap, and adjust would seem not to reflow. This is easily solved by repositioning the typing point, but some participants gave up and fell back to typing over. On the other hand, participants who did try to reposition the typing point in this situation often put it at the space right after the leftmost word of the gap. Adjust in this case concatenated the words on either side of the gap. This is also easily solved by inserting a blank space. But here again, participants did not always have so clear-cut an interpretation about the side-effect, or how to recover. Rather than insert, for example, participants would try other actions like repeating Adjust or operations like Format or Move, the latter presumably seemed related to the notion of manipulating how lines looked and/or directly moving segments of lines.

It is worth noting here that these difficulties do not necessarily contradict the conclusion that participants better understand the explicit control of reflowing. Rather they can be viewed as relating more to problems with the *implementation* of Adjust than with the underlying notion of having to do so. Participants' expectations simply did not include sensitivity of adjust to position of the typing point. These mismatches are not hard to solve in principle. For example, the prototype was modified in the second evaluation so that line reflowing was insensitive to typing point position. But it was difficult to anticipate the difficulty participants would have with the operation.

We might argue that mismatches could be anticipated by a thorough task analysis and simulation of users' behavior in all possible contexts. This is an interesting possibility, certainly in line with theoretical or formal attempts to model user knowledge, predict misconceptions, and procedural complexity of interfaces (e.g., Card, Moran, & Newell, 1983; Douglas & Moran, 1983; Moran 1981; Kieras & Polson, 1985; Reisner, 1984; among others). However, the same examples discussed above suggest the difficulty with a predictive approach. The impact of procedural errors depends, not only on understanding how activities in the task domain are represented in the interface design, but also on how the latter interacts with what participants know, and how they reason in the face of difficulties. We can lay out logical possibilities for mismatches and misconceptions. But it is doubtful that we can know which ones will be pursued, or whether we really can anticipate how users will reason in the face of difficulties. It is just this interaction which is hard to predict and which suggests the need for empirical approaches to identifying and resolving problems with interfaces (e.g., Gould & Lewis, 1985).

To summarize, the impact of mismatches resulting from inappropriate expectations, depends on an interaction of the interface design with what users expect, and how they reason in the face of novel experiences. In some cases, mismatches can perpetuate simpler fallback methods. In other cases, mismatches can complicate error recovery in ways that are difficult to anticipate.

IMPLICATIONS FOR DESIGNING LEARNABLE INTERFACES

One response to evidence for the role of expectations on user performance, might be to nonetheless discount the need to take these expectations too seriously, on the grounds that learning problems related to them, like those associated with Carrier Return, or how reflowing occurs, are transient and/or can be resolved by training. This is fair enough where one can rely on the availability and effectiveness of training to convey novel concepts and procedures. However, our observations of learning (e.g, Mack, Lewis, & Carroll, 1983), as well as our understanding of industry views on the role and effectiveness of training (e.g., Seybold, 1981), suggest that training methods adequate to solving these kinds of problems are increasingly difficult to justify, either in terms of cost-effectiveness, or the burden placed on participants. Unaided self-instruction certainly does not seem adequate for avoiding problems users have trying to follow instructions (Mack, Lewis, & Carroll, 1983). These considerations have led us to conclude that expectations should be taken quite seriously in attempts to achieve key usability goals like helping learners get started on real work quickly, and insuring that they can progress quickly and easily to more advanced capabilities.

Accommodating expectations means, in most general terms, designing interfaces so that what users see, and what users do, better matches objects and actions they are familiar with, perhaps outside of the domain of computing. Examples are "Desktop" metaphors, and direct manipulation interface styles, found in advanced workstations like the STAR or LISA (Smith, Irby, Kimball, Verplank, & Harslem, 1982; Williams, 1983; for extended discussion, see Hutchins, Hollan, & Norman, 1986). The editor prototype discussed in this paper tried to accommodate expectations related primarily to the manipulation of text. For present purposes we have discussed this research only insofar as it provides insights into the nature and role of prior knowledge on learning text editing. Before closing, we need to comment on design implications of using novice expectations in the design of computer interfaces, specifically: (a) Is the prototype easier to learn? (b) What implications, if any, does such a design have for experienced users? (c) What role do expectations play in the overall design process?

It is actually too early to answer the first two questions, but the preliminary behavioral evaluation suggests the following comments. The prototype design does avoid key features which cause problems in other editors we have studied, particularly related to embedded formatting characters and nontyping areas (Mack, Lewis, & Carroll, 1983; see also Galambos, Wikler, & Black, 1983). We are encouraged that users in the second evaluation were able to complete a simple document creation task in a half hour, with no explicit training. But it is also clear that problems remain with the specific *implementation* of operations, like explicit management of reflowing. That is, users still had problems with the prototype.

We *can* also conclude from the preliminary evaluation that implementation details of operations are critically important, even where, as we have hypothesized, these operations may better match user expectations in general. Thus, while there was only one complaint about having to adjust in all the cases of having to reflow lines or paragraphs, participants nevertheless had difficulty mastering these operations. The second version of the editor prototype simplified how lines were reflowed by making reflowing within lines insensitive to typing point location: with one less distinction to master, participants did better. Further modifications can be anticipated along the same lines; namely, if the program can fill-in a procedural detail, let it do so.

With respect to experienced users, the implication of designing an interface with novice expectations in mind is the possibility that the resulting interface may not be useful for experienced users. Explicit management of operations, for example, implies more activity, hence more effort and opportunity for mistakes. This fact may trade off favorably with the understanding of reflowing that explicit control may, by hypothesis, provide novices, but such an editor could be cumbersome for experienced users. Should this be the case, the editor may be best regarded as a training vehicle which allows novice users to manipulate text in a way that conforms more closely to initial expectations, but provides some mechanism for graduating to more powerful editing functions (e.g., Mozeiko, 1982, or Reisner, 1977, for discussion of staging interfaces; Carroll & Carrithers, 1984, for empirical evaluation of one). For example, such a mechanism might allow users to tailor the interface so that reflowing occurs automatically once users are comfortable with the operation. Such a facility would also have to solve problems like fall-back to simpler typewriter-like methods.

It may also be necessary to embed novice-driven interface design principals in a more intelligent tutorial or help system that can take a more active role in helping users make the transition to more powerful, but potentially less immediately comprehensible functionality. This situation may be analogous to that faced by experienced users who are unwilling, or unable, to broaden their computer skills (Rosson, 1984). Rosson has proposed the voluntary monitoring of usage, with the intent of helping users discover more efficient methods of using computers. Additional discussion of still more active, or "intelligent," approaches to developing advice-giving systems can be found in Jackson and Lefrere (1984).

It is not clear, however, that the novice motivated implementations are going to be cumbersome for experienced users. To return to the example immediately above, explicit control of reflowing may be useful even for experienced users, at least for tasks where the user does not want the document on the screen to vary with respect to a hard-copy markup from which editing changes are being made (see, e.g., Gould & Alfaro, 1984).

With respect to the third question, perhaps the best way to express the role of novice expectations in the overall context of designing computers is to simply state that: *prediction data are only one, limited source of information, relevant to design.* Inferring and accounting for expectations are only part of a broader design *process* that involves many other considerations.

Methodologically, the strengths and weaknesses of predication data need to be assessed in their own right. Eliciting predictions seems, in my judgment, to be a variation of verbal protocol methods. In terms of Ericsson and Simon's (1980, 1984) seminal analysis, prediction data provide verbal protocols that are much more structured and constrained than simply thinking aloud. Predictions are constrained by the task context, including the specific features of the computer systems used to elicit predictions, by the follow-up, clarifying questions the experimenter asks, and undoubtedly by the cumulative concept of the interface the participant gradually evolves over the course of the study. Eliciting predictions undoubtedly shares some of the limitations of Think Aloud protocols in that they can be fragmentary, and participants in the task likely would not spontaneously articulate real expectations so systematically or overtly. The present study simply did not try to understand the impact on generating user predictions of this complex interaction of factors. It does seem reasonable to hypothesize, however, pending further study, that such data expose an immediate, online, and representative experience on the part of the person involved. Humans are cognitively active learners (Lewis & Mack, 1982a, b; Carroll & Mack, 1984a,b) for whom generating, and acting on expectations is entirely a natural activity.

More important to considerations of computer design is the observation that the process of designing the editor prototype involved much more than simply trying to make editing operations try to match the literal predictions participants generated in the first study. *Predictions and descriptions are not functional specifications.* The initial design of the prototype involved an informed, albeit intuitive, integration of several kinds of considerations, some of them already alluded to.

First, design designs mutually constrain each other, leading to compromises, trade-offs, and new possibilities. For example, using a limited set of function keys provided constraints that had nothing to do with the predictions. The terms used for primitive operations (e.g., *delete*, *adjust*, or *insert*) did not necessarily derive from participant's protocols. I found nothing in the protocols to contradict Landauer, Galotti, and Hartwell's (1983) conclusions about the difficulty of identifying consistent user-generated command descriptions. Consequently, I did not try to match surface descriptions of operations but tried to abstract from them an underlying semantic for the operation.

Implementation details of functions are also not typically articulated in the prediction or description data. It is not hard to make bad guesses about these details. We have already discussed an example of how the first evaluation of the prototype led to modification in how text reflowing was implemented in the second version of the prototype. As Whiteside, Jones, Levin, and Wixon (1985) have shown, getting the implementation details right is as important as choosing a particular style of interface.

Another consideration is that we do not want design decisions to necessarily exclude new possibilities for exploiting what computers can do. These new possibilities may simply not match anything about which new users can have meaningful expectations. For example, I never considered excluding text reflowing because it had no typewriter counterpart. Closely related to this consideration is the desire to balance interface design that accommodates novices with the needs experienced users have for powerful operations, as I suggested above.

All of these considerations lead to the observation that the prediction data underdetermine the set of design possibilities. Consequently, whatever emerges from this initial design process is itself a guess, or hypothesis, that must be tested. The only way to respond to this circumstance is to engage in behavioral evaluation and design modification in an iterative cycle that is completed when the design satisfies behavioral goals established at the outset. A full discussion of iterative design can be found elsewhere, most notably in Gould and Lewis (1985), where general principles are discussed, and several case studies cited. In this context, eliciting predictions is one of several ways to generate initial design decisions. As such, prediction data can be regarded as a form of task analysis, pointing to possible design features, but ultimately predictions are useful only in a broader design context in which behavioral evaluation and modification are inescapable.

CONCLUSION

We have discussed two lines of evidence for new user expectations about basic text-editing operations. One elicited predictions from computer-naive office temporaries about how to accomplish editing operations, like delete a word. The other observed learning behavior for an editor prototype intended to conform more closely to these expectations. Expectations in both studies made reference to how typewriters worked, regarding, for example, how the typing point is moved, or how revisions are made. Expectations also revealed that new users expect actions to be simple in that one action has one outcome. This expectation has different expressions for inserting, deleting, or reflowing lines and paragraphs. Expectations are relevant to learning, because they result in mismatches between what a learner wants to do, and what action he or she believes will accomplish the goal. The learning data also demonstrates how initial mismatches interact with interface design, and learners' reasoning to create more involved problem situations. Finally, we discussed implications for improving interface design, reinforcing the trend towards designing to accommodate user expectations.

REFERENCES

Anderson, J. (1982). The acquisition of cognitive skill. *Psychological Review, 89,* 369–406.

Barnard, P.J., Hammond, N.V., Morton, J., & Long, J.B. (1981). Consistency and compatibility in human-computer dialogue. *International Journal of Man-Machine Studies, 15*, 87–134.

Bott, R. (1979, March). *A study of complex learning: Theory and methodologies* (CHIP Report 82). LaJolla, CA: University of California, San Diego.

Bransford, J., & McCarrell, N. (1974). A sketch of a cognitive approach to comprehension: Some thoughts about understanding what it means to comprehend. In W. Weimer & D. Palermo (Eds.), *Cognition and the symbolic processes*. Hillsdale, NJ: Erlbaum.

Card, S., Moran T., & Newell, A. (1983). *The psychology of human-computer interaction*. Hillsdale, NJ: Erlbaum.

Carroll, J. (1982). Learning, using and designing command paradigms. *Human Learning: Journal of Practical Research and Applications*, *1*, 31–62.

Carroll, J., & Carrithers, C. (1984, August). Blocking learner error states in a training-wheels system. *Human Factors, 26*, 377–389.

Carroll, J., & Mack, R. (1984a). Learning to use a word processor: By doing, by knowing and by thinking. In J. Thomas & M. Schneider (Eds.), *Human factors in computer systems*. Norwood, NJ: Ablex Publishing Corp.

Carroll, J., & Mack, R. (1984b). Actively learning to use a word processor. In W. Cooper (Ed.), *Cognitive aspects of skilled typewriting*. New York: Springer-Verlag.

Carroll, J., & Mack, R. (1985). Metaphor, computing systems and active learning. *International Journal of Man-Machine Studies, 22,* 39–57.

Carroll, J., & Thomas, J. (1982, March/April). Metaphor and the cognitive representation of computing systems. *IEEE Transactions on Systems, Man and Cybernetics, SMC-12*, 107–116.

Douglas, S., & Moran, T. (1983, December 12–15). *Learning text-editor semantics by analogy*. In *Proceedings CHI'83 Human Factors in Computer Systems* (pp. 207–211). New York: ACM.

Ericsson, K., & Simon, H. (1980). Verbal reports as data. *Psychological Review, 87*, 215–251.

Ericsson, K., & Simon H. (1984). *Protocol analysis: Verbal reports as data*. Cambridge, MA: Bradford Books.

Fitts, P., & Posner, M. (1967). *Human performance*. Monterey CA: Brooks/Cole.

Foss, D., Rosson, M.B., & Smith, P. (1982, March 15–17). Reducing manual labor: An experimental analysis of learning aids for a text editor. *Proceedings CHI'82 Human Factors in Computer Systems*. New York: ACM.

Galambos, J., Wikler, E., & Black, J. (1983, December 12–15). How to tell your computer what you mean: Ostension in interactive systems. In *Proceedings CHI'83 Human Factors in Computing Systems* (pp. 207–211). New York: ACM.

Gentner, D. (1985). Structure mapping: A theoretical framework for analogy. *Cognitive Science, 7*, 155–170.

Gould, J., & Alfaro, L. (1984, August). Revising documents with text editors, handwriting recognition systems, and speech recognition systems. *Human Factors, 26*, 391–406.

Gould, J., & Lewis, C. (1985). Designing for usability: Key principles and what designers think. *Communications of the ACM, 28*, 300–311.

Halasz, F., & Moran, T. (1982, March 15–17). Analogy considered harmful. In *Proceedings CHI'82 Human Factors in Computer Systems* (pp. 207–211). New York: ACM.

Hutchins, E., Hollan, J., & Norman, D. (1986). Direct manipulation interfaces. In D. Norman & S. Draper (Eds.), *User centered system design: New perspectives in human-computer interaction*. Hillsdale, NJ: Erlbaum.

Holyoak, K.J. (1983). Analogical thinking and human intelligence. In. R.J. Sternberg (Ed.), *Advances in the psychology of human intelligence (Vol. 2)*. Hillsdale, NJ: Erlbaum.

Jackson, P., & Lefrere, P. (1984). On the application of rule-based techniques to the design of advice-giving systems. *International Journal of Man-Machine Studies, 20*, 63–86.

Kieras, D., & Polson, P., (1985). An approach to the formal analysis of user complexity. *International Journal of Man-Machine Studies, 22*, 365–394.

Landauer, T., Galotti, K., & Hartwell, S. (1983). Natural command names and initial learning: A study of text-editing terms. *Communications of the ACM, 26*, 495–503.

Lewis, C. (1982, February). *Using the "Thinking Aloud" method in cognitive interface design* (Research Report RC 9265). Yorktown Heights, NY: IBM Thomas J. Watson Research Center.

Lewis, C., & Mack, R. (1982a, March 15–17). Learning to use a text-editor: Evidence from thinking aloud protocols. In *Proceedings CHI'82 Human Factors in Computer Systems*. New York: ACM.

Lewis, C., & Mack, R. (1982b, March 19–24). *The role of abduction in learning to use text-processing systems*. Presented at the annual meeting of the American Educational Research Association. New York.

Mack, R. (1984, January). *Understanding text-editing: Evidence from predictions and descriptions given by computer-naive people* (Research Report RC 10333). Yorktown Heights, NY: IBM Thomas J. Watson Research Center.

Mack, R., Lewis, C., & Carroll, J. (1983, July). Learning to use a word processor: Problems and prospects. *ACM Transactions in Office Information Systems, 1*, 254–271.

Mack, R., & Rosson, M.B. (1984, May 24–29). *Observations on computer skill acquisition and implications for design*. Presentation at the Annual Meeting of the American Association for the Advancement of Science, New York.

Moran, T. (1981). The command language grammar: A representation for the user interface of interactive computer systems. *International Journal of Man-Machine Studies, 15*, 3–50.

Mozeico, H. (1982). A human/computer interface to accommodate user learning stages. *Communications of the ACM, 25*, 100–104.

Reisner, P. (1977). User of psychological experimentation as aid to development of a query language. *IEEE Transactions on Software Engineering, SE-3*, 218–229.

Reisner, P. (1984). Formal grammar as a tool for analyzing ease of user: Some fundamental concepts. In J.C. Thomas & M.L. Schneider (Eds.), *Human factors in computer systems*. Norwood, NJ: Ablex Publishing Corp.

Rosson, M.B. (1984). Effects of experience on learning, using, and evaluation a text editor. *Human Factors, 26*, 463–475.

Rumelhart, D., & Norman D. (1978). Accretion, tuning and restructuring: Three modes of learning. In J. Cotton & R. Klatzky (Eds.), *Semantic factors in cognition*. Hillsdale, NJ: Erlbaum.

Seybold, J. (1981). Training and support: Shifting the responsibility. *Seybold Report on Word Processing 4*. Media, PA: Seybold Publications, Inc.

Schneider, M. (1982). Models for the design of static software user assistance. In A. Badre & B. Shneiderman (Eds.), *Directions in Human/Computer interaction*. Norwood NJ: Ablex Publishing Corp.

Shneiderman, B. (1980). *Software psychology: Human factors in computer and information systems*. Cambridge, MA: Winthrop Publishers.

Shneiderman, B. (1982). The future of interactive systems and the emergence of direct manipulation. *Behavior and Information Technology, 1*, 237–256.

Smith, D., Irby, C., Kimball, R., Verplank, B., & Harslem, E. (1982, April). Designing the STAR interface. *Byte*, pp. 242–282.

Whiteside, J., Jones, S., Levin, P., & Wixon, D. (1985, April 14–18). User performance with command, menu, and iconic interfaces. In *Proceedings CHI'85 Human Factors in Computing Systems* (pp. 185–191). New York: ACM.

Williams, G. (1983, February). The LISA computer system: Apple designs a new kind of machine. *Byte, 18*, 33–50.

12

Integrative Modeling: Changes in Mental Models During Learning*

Marc M. Sebrechts

Psychology Department
The Catholic University of America
Washington, DC

Richard L. Marsh
Charlotte T. Furstenburg

Department of Psychology
Wesleyan University
Middletown, CT

According to an information-processing perspective, one of the central aspects of learning to use a complex device is the formation of an appropriate mental representation (Newell, 1981; Simon, 1981; Newell & Simon, 1972). The general argument of this approach is that, during learning, device characteristics are mapped onto a representation of the device, and that that representation, in turn, mediates subsequent performance. Given this model, an integral task of instruction is to improve the quality of the learner's representation or the rate at which it is acquired.

In this chapter we attempt to analyze several specific factors that could reasonably be expected to improve the learner's representation of a computer operating system. In particular we are concerned with the extent to which an "explicit model" presented as part of the instructional materials can be used to improve the learner's "mental model."

---

*The work reported here was supported by The National Science Foundation (grant IST-82-17572), Digital Equipment Corporation, and Wesleyan University. The views expressed are those of the authors and are not necessarily endorsed by the sponsors. Additional details of methodology and results are contained in Sebrechts, Marsh, and Furstenberg (1987).

Although our research is designed to assess the nature of representation during learning, we believe that it may also provide an analytic approach to some of the issues of cooperation that constitute the central theme of this book. *Cooperation* covers a very wide variety of activities, ranging from detailed assistance on specific tasks to encouragement or motivation. One key contribution of cooperation to learning consists in providing explanations as needed. Explanation, however, can take many forms, including specific procedures, analogies, descriptions of underlying mechanisms and models, or examples. Although there is agreement that cooperation can help learning in general, it is not clear which information components are critical to specific aspects of learning. This chapter begins to examine that issue. Of course, the context in which information is delivered, by machine or by another person, is likely to influence learning, and we do not directly address the role of that interaction. However, by examining the utility of specific types of information on skill acquisition, we should be able to provide an initial sense of the relative importance of these types of information in isolation from their interactive delivery.

PRIOR KNOWLEDGE AND LEARNING

The idea that learning depends on prior knowledge dates back at least to the *Meno*, in which Socrates argued that what appears to be learning is actually a remembering of something we previously knew. In current terminology, we would say that the information is already *stored*, but it is not accessible to the normal *retrieval processes*. Although there is little support for the strong nativist character of this view, many current theories place a strong emphasis on previous knowledge as mediating the way in which new information is represented. Some of these views emphasize the structure of knowledge called *schemata* (Bartlett, 1932; Head, 1920); others emphasize the episodic components of memory, suggesting that a new event reminds us of a related previous event (Schank, 1982; Ross, 1984); still others emphasize the process or *runnability* of a mental model (Gentner & Stevens, 1983).

There is no single view concerning the nature of mental representation in these various theories. A *schema* is usually taken to refer to a specific organized structure in memory and is generally associated with declarative knowledge (for alternative definitions, see Alba & Hasher, 1983). A *mental model*, in contrast, usually implies a representation that captures the workings of some device; it is generally associated with procedural knowledge. The boundaries between these and other forms of representation are not well defined. In addition, it is likely that the learning of procedures includes multiple representations, since declarative knowledge constitutes a central element of early skill acquisition (Fitts, 1964; Anderson, 1982).

In this chapter we will use the term *mental schemata* for those cases in which we wish to emphasize the structure of concepts in memory. Otherwise we will use the term *mental model* in its general sense to refer to the representation that people employ to guide their use of a device (Norman, 1983; see also Sowa, 1984). Finally, *explicit models* will be used for instructional information provided to the learner describing how or why a system works as it does. The explicit models are meant to convey directly the kinds of information that are assumed to be necessary for useful mental models. As a working hypothesis, we will consider any relevant instructional elaboration as a potential explicit model.

Schemata and Mental Models

Many of the current conceptions of the way in which a mental representation mediates learning can be traced back to Leibnitz's philosophical stance in the 17th century that "Nothing is in the intellect that was not first in the senses, except the intellect itself." Based on this general concept, Piaget (1954) argued that there are functional characteristics of the intellect, the *ipse intellectus*, that constrain the way in which cognitive structures can be modified. The two central functions of the intellect which serve to mediate learning are assimilation and accommodation. New information is either assimilated to current mental schemata, or schemata are modified to accommodate new information as it arrives.

This point was made rather effectively by Bartlett (1932), who demonstrated that the way people remembered a story depended on their previous knowledge. A recent resurgence of interest in Bartlett's results has lead to a number of related empirical demonstrations. When presented with information about a particular topic, people tend to recall relevant material that had not been presented but which fits with the topic schemata (e.g., Bower, Black, & Turner, 1979; Owens, Bower, & Black, 1979; Sulin & Dooling, 1974). Experts, who presumably already have appropriate schemata for a domain, can acquire information about that domain more quickly (e.g, Chase & Simon, 1973; Chi, Feltovich, & Glaser, 1981; Chiesi, Spilich, & Voss, 1979). Likewise, evoking appropriate schemata by elaborations (Schallert, 1976; Weinstein, 1978) or even titles (Bransford & Johnson, 1972) can increase understanding and retention of the gist of a story or description.

Recently, Rumelhart and Norman (1978, 1981) have developed a model that provides a more detailed description of three aspects of schema development: *Accretion* is the process by which new information is interpreted in terms of previously existing schemata. *Tuning* is a gradual change of the schemata to achieve a better fit to the information. *Restructuring* involves the development or creation of new schemata. These mechanisms provide a useful way to account for learning by analogy; new concepts can be understood in terms of previously stored concepts that have a similar structure. This model can explain a substantial number of confusions in various domains. An extensive analysis of text editing, for example, showed how this model can account for the specific types of errors in simple skill acquisition (Bott, 1979).

The example of text editing illustrates a close link between schemata and mental models. In both cases, the mental representation is taken as mediating behavior. In general, however, schemata have been emphasized more within the context of learning, whereas mental models have been emphasized as representations that guide performance (Gentner & Stevens, 1983). Mental models have been used to characterize performance on physics problems (Larkin, 1983), the operation of a calculator (Young, 1983), and strategies in navigation (Hutchins, 1983). Whereas schemata have come to be associated with well-defined knowledge, the structure of mental models has remained vague (Norman, 1983; but see Johnson-Laird, 1983, for an alternative view). In general, however, it is assumed that it is the presence of these representations that mediates learning, memory, and performance.

Explicit Models

If this general description of learning is correct, then providing subjects with instructional elaborations that constitute an explicit model of a domain may help them to develop better mental models. That is, it may be easier to develop an appropriate representation if the relevant structure is provided explicitly through instruction. The fact that explicit organization can affect even a simple memory task was demonstrated by Bower, Clark, Lesgold, and Winzenz (1969). They showed that, if lists of words were explicitly organized into categories, recall was substantially better than when the same lists were presented in random order.

The import of such explicit structures has also been shown for more complex skill acquisition. Mayer (1975a) asked subjects to learn a BASIC-like computer programming language. Some of the subjects were provided with a model that described major functional units of the computer in terms of more familiar domains, whereas other subjects simply received the general instructions about the language. Those subjects who received the model before training showed better performance on a transfer task than those who had received no model or those who received the model after training.

Similar effects have been found for other tasks. Halasz and Moran (1983) analyzed problem solving using a calculator and found that giving subjects an explicit model of the calculator stack-group resulted in more efficient solutions to novel problems. Kieras and Bovair (1984) found that a "phaser bank" model of a control panel produced faster learning and better retention than rote learning. Smith and Goodman (1984) examined performance of subjects on the assembly of a basic electrical circuit. They found that using structural or functional explanatory schemata as part of the instructions improved recall of procedural steps and performance on a transfer task. In each of these cases, the explicit model presumably helped subjects to organize the training materials (Ausubel's (1968) advance organizer); it provided subjects with a schema to which they could assimilate the new information.

It is not the case, however, that instructional elaborations always help learning. In fact, in at least some cases, elaborations of material can actually decrease performance. For example, Reder and Anderson (1980, 1982) found that, when tested on summary information, subjects receiving detailed texts did worse than those receiving summaries alone, even if study time for the two types of material was equated. These results have been used to argue that author-provided elaborations have limited utility in improving declarative knowledge.

Different results have been found for the domain of skill acquisition. Reder, Charney, and Morgan (1986) tested the role of elaborations for subjects learning how to use a personal computer. They found that both subject-generated and author-provided elaborations could improve performance. In addition, they found that whereas syntactic elaborations (i.e., examples) improved performance on basic computer file manipulation tasks, conceptual elaborations did not. A study by Pepper (1981) confirmed the utility of examples for learning simple computer-programming tasks.

EXPLICIT MODELS AND LEARNING

Previous empirical results make it clear that prior knowledge is critical to learning. In addition, there is some empirical evidence that the status of prior knowledge can be manipulated by providing explicit model information which can then be used to structure subsequent learning. At the same time, however, arguments about the importance of schemata and mental models have perhaps overemphasized the coherence of learning. A number of our previous studies on skill acquisition (Sebrechts & Dumont, 1986; Sebrechts, Furstenberg, & Shelton, 1986; Wagner, Sebrechts, & Black, 1985; Sebrechts, Deck, Wagner, & Black, 1984; Sebrechts, Galambos, Black, Deck, Wikler, & Wagner, 1984; Deck & Sebrechts, 1984; Sebrechts, 1983) have suggested that the mental models which subjects develop have limited coherence, multiple components, occasional conflicts, and that they are continually changing during initial learning.

This chapter therefore has two objectives. First, it describes some of our empirical data on the effects of explicit models (in the form of different types of elaboration) on the formation of mental models. More specifically, the studies examine the role of (a) *subject-generated elaboration* resulting from verbalization during learning, and (b) different types of *experimenter-provided elaboration*. In addition, our analyses emphasize the way in which these elaborations influence performance on different types of tasks.

Second, the results are analyzed to determine the character of the changing mental models of the learner. Performance measures, types of errors, and verbal protocols are used to construct a description of the subject's evolving representation of the task domain. We are especially concerned with trying to assess the adequacy of a schema-like description of mental models in this context. It should perhaps be emphasized that, although the domain of interest is computer skill acquisition, early learning in this domain appears to be largely a function of declarative knowledge. The representations of novice users therefore are likely to be quite different from those reported for experts (e.g., Card, Moran, & Newell, 1980; but see Landauer, 1987, for limitations on those results).

METHOD: EXPERIMENT 1

Subjects

Eight university students served as subjects and provided both performance and verbal protocol data. All of these subjects had some computer experience, but no prior experience with UNIX. Computer experience was assessed by a questionnaire prior to the beginning of the study and was used as a covariate in our analyses. Verbal protocols were also collected from three additional subjects who did not constitute part of the balanced design on elaboration condition. For those subjects, only protocol data is considered in the analysis.

Materials

Instructions. Descriptions of 16 frequently used UNIX commands (Hanson, Kraut, & Farber, 1984) were separated into four *command subsets* (booklets) covering the following topics, as indicated in Table 1: (a) organization of directories and files, (b) methods for manipulating files, (c) options for executing commands and moving information through the system, and (d) the VI (visual) editor.

Table 1. Commands Presented in Each Command Subset.

<table>
<tr><th>Command Subset (Booklet)</th><th>Commands</th></tr>
<tr><td>File Organization</td><td>change directory (cd)
list (ls)
make directory (mkdir)
remove directory (rmdir)</td></tr>
<tr><td>File Manipulation</td><td>link (ln)
copy (cp)
remove (rm)
move (mv)</td></tr>
<tr><td>Command Execution</td><td>redirection operators (⟨,⟩)
concatenate (cat)
pipeline operator (|)
tee (tee)</td></tr>
<tr><td>Visual Editor</td><td>insert (i)
delete (x,d)
change (r,c)
put (p)</td></tr>
</table>

For each command subset, eight versions of instructional materials were written. Each version included one of four types of *conceptual-elaboration* and one of two types of *syntactic-elaboration*. The different versions of the study materials are illustrated for the make directory (mkdir) command in Table 2.

The four conceptual-elaboration conditions consisted of the following: (a) The *Simple* version stated the relevant concepts, with minimal elaboration concerning their meaning. For example, subjects were told that files had associated names and addresses, but they were not told how those names and addresses were represented in the system; (b) The *Redundant* version included an extended restatement of the simple concepts while avoiding new explanatory material. This condition was roughly matched in length to the model and analogy conditions; (c) The *Model* version provided a functional model describing the reason for certain properties of the commands and their function within the system. Thus, for example, the remove command was explained in terms of changes in file names and addresses within a directory; (d) The *Analogy* version provided an analogy from a familiar domain which described the underlying functional model and its relationship to the commands. The directory structure was compared to a branching tree, the manipulation of files was described in terms of a library card catalog system, the flow of information during command execution was described in terms of water flowing through pipes, and the editor was compared to a multifunction tape recorder.

Table 2. Examples of a Portion of the Changing Files Booklet for Each of the Elaboration Conditions Using the "mkdir" Command.

Elaboration (Example Type)	Command Description
Simple (Abstract)	The make directory command adds a new directory name to a directory. To add a new directory name, type mkdir DIRNAME ⟨cr⟩ where 'DIRNAME' is the name you want to give to the newly added directory.
Simple (Concrete)	The make directory command adds a new directory name to a directory. Suppose, for example, that your working directory contains a directory named 'personal' and two files named 'george' and 'susan'. To make a new directory named 'jobs' in your working directory, you would type mkdir jobs ⟨cr⟩ Your working directory would then contain the files 'george', and 'susan' and the directories 'personal' and 'jobs'.
Redundant (Abstract)	The make directory command adds a new directory name to a directory. Whenever you wish to add another directory to those you already have, you can use the make directory command. To add a new directory name, type mkdir DIRNAME ⟨cr⟩ where DIRNAME is the name you want to give to the newly added directory. After you have used this command, there will be a new directory name, 'DIRNAME', added to a directory.
Redundant (Concrete)	The make directory command adds a new directory name to a directory. Whenever you wish to add another directory to those you already have, you can use the make directory command. Suppose, for example, that your working directory contains a directory named 'personal' and two files named 'george' and 'susan'. To make a new directory named 'jobs' in your working directory, you would type mkdir jobs ⟨cr⟩ The computer would then add a new directory name to your working directory so that it would contain the files 'george' and 'susan' and the directories 'personal' and 'jobs'.
Model (Abstract)	The make directory command creates a new directory within the hierarchy of directories on the system by adding a new directory name to an existing directory. The new directory is identified as a special file which can contain the names and addresses of other files and directories you wish to add. To make a new directory, type mkdir DIRNAME ⟨cr⟩

(Continued)

Table 2. Continued.

Elaboration (Example Type)	Command Description
	where 'DIRNAME' is the name you want to give the newly created directory. The computer starts at the root directory and searches through the directories listed in the pathname. The next to last name must be an existing directory. The last name will be the name of the new directory which is to be added.
Model (Concrete)	The make directory command creates a new directory within the hierarchy of directories on the system by adding a new directory name to an existing directory. The new directory is identified as a special file which can contain the names and addresses of other files and directories you wish to add. Suppose, for example, that your working directory contains a directory named 'personal' and two files named 'george' and 'susan'. To make a new directory named 'jobs' in your working directory, you would type mkdir jobs ⟨cr⟩ The computer would then add the name 'jobs' to the names already in your working directory so that it would contain the names of two files, 'george' and 'susan', and the names of two special files or directories, 'personal' and 'jobs'.
Analogy (Abstract)	The make directory command adds a new branch to the directory tree by adding a new directory name to an existing directory. To make a new directory, type mkdir DIRNAME ⟨cr⟩ where 'DIRNAME' is the name you want to give the newly created directory. The computer starts at the base of the directory tree and climbs over the branches listed in the pathname until it gets to the next to the last branch. Then it grows a new branch whose name is the last name in the pathname.
Analogy (Concrete)	The make directory command adds a new branch to the directory tree by adding a new directory name to an existing directory. Suppose, for example, that your working directory contains a directory named 'personal' and two files named 'george' and 'susan'. To make a new directory named 'jobs' in your working directory, you would type mkdir jobs ⟨cr⟩ The computer would then add a branch named 'jobs' to your working directory. The leaves 'george' and 'susan' and the branches 'personal' and 'jobs' would all be connected to your working directory.

To some extent all of these elaboration conditions contain metaphoric and analogical components, as will any abstraction. The relevance of that fact and some of its consequences for computer interface design have been noted by Hutchins (1987). The differentiation among types of elaboration is therefore not absolute; rather it reflects the most explicit and distinctive aspects of each elaboration type.

There were two types of syntactic-elaboration in the form of examples used to display command format. One set of materials used an *abstract example* that provided the general form of the command (e.g., "mv NAME1 NAME2<cr>"). The other set of materials used a *concrete example*, with a specific description of an individual instance of command use (e.g., "mv george harry<cr>").

Questions. Three types of questions were developed for each set of concepts and commands (each booklet): *Procedure* questions, testing knowledge of specific procedures; *Command Use* questions, testing knowledge of the ways in which commands could be used without requiring knowledge of specific procedure syntax; and *System* questions, testing knowledge of overall system structure or function. (We use the term *Procedure* rather than *Procedural* questions to highlight the fact that, although the questions require knowledge of specific procedures, they do not distinguish between "procedural" and "declarative" knowledge.) Examples of the different question types are shown in Table 3.

Procedure

Subjects were run on 3 consecutive days. Verbal protocols, which served as a measure of subject-generated elaborations, were collected throughout the experiment. Subjects were asked to read all materials aloud and to say whatever came to mind while reading or answering questions. Conceptual-elaboration condition, command subset, and question type were varied within subject. Syntactic-elaboration condition was varied between subjects.

On the first day the subjects were presented with a General Introduction to the UNIX operating system and the experimental procedure. In addition, they received basic information about special characters used in the instructions and questions (e.g., <cr> for carriage return). On each of the first 2 days, subjects were presented with two command subsets, each followed by a related set of questions that provided initial performance measures. On the third day they were presented with all of the questions from the previous 2 days in quasirandom order; responses to these questions yielded final performance measures.

Table 3. Examples of the Different Types of Questions.

Procedure Questions from the Changing Files Command Subset:

Suppose you are in the '/usr/yourname/math' directory which contains a file named homework.
Suppose also that the 'yourname' directory contains a file named 'termpaper'.
What would you type to give the file 'homework' the additional name 'assignment'?
What would you type to transfer the file named 'termpaper' to the 'math' directory?
What would you type to sever a link to a file?

Command Use Questions from the Visual Editor Command Subset:

	True	False
The delete command erases text from the computer's memory.	[ ]	[ ]
You can use the change command to change the position of a line of text in your file.	[ ]	[ ]
You can use the 'dd' command together with successive 'p' commands to copy a line of text.	[ ]	[ ]

System Questions from the File Organization Command Subset:

In what directory is the file '/usr/research/exp1'?
Suppose you specify the pathname '/usr/lab/name/assign/num1' when you give a command. What does this pathname tell you about the organization of directories in the system?
Why is the concept of a working directory useful?
That is, why does UNIX bother to keep track of your working directory?

METHOD: EXPERIMENT 2

Eight university students, none of whom had participated in Experiment 1, served as subjects in Experiment 2. The method was the same as in Experiment 1, with two exceptions. First, no verbal protocols were collected. This change was designed to indicate whether or not subject-elaboration through verbalization had influenced performance. Second, those questions from Experiment 1 which produced problems of interpretation for subjects, or which showed no variance in response, were slightly modified.

GENERAL RESULTS

Before describing the specific results, we will provide a general overview of the dependent measures used. This will be followed by an analysis of the results first in terms of experimenter-elaboration and then in terms of subject-elaboration. Finally, we will analyze specific measures of learning together with three component processes that characterize the gradual integration of information that constitutes learning in these studies. In general, we will jointly analyze the effects for the two experiments whenever appropriate.

The three types of data analyzed are performance measures, error types, and verbal protocols. *Performance measures* consisted of answers to questions about individual procedures, command usage, and system function. The effects of syntactic elaboration on performance as evidenced by differences between Concrete and Abstract Examples are shown in Figure 1. The effects of conceptual elaboration conditions on initial performance are shown separately for Procedure, Command Use, and System questions in Figure 2. Changes in performance between initial and final testing are shown in Figures 3 and 4. Figure 3 provides relative measures of "learning" and "forgetting" for different types of questions. Figure 4 provides a more detailed breakdown of changes in true and false responses for Command Use questions.

Error analyses consisted primarily of a detailed breakdown of the types of errors made on Procedure questions. A detailed classification of these errors is provided in Appendix A. Even with the very restricted set of tasks and commands presented, subjects made a tremendous number of different types of procedure specification errors (approximately 500).

For purposes of analysis, errors are collapsed into two broad classes: *conceptual* and *implementation*. This rough classification is intended to capture some of the most important differences among errors. Conceptual errors are those that appear to reflect a general misconception about the way in which commands are used. Implementation errors, in contrast, reflect difficulties with formulating the command correctly, including problems with command names, pathname specification, and the sequencing of arguments.

Protocol analyses were carried out for the data of Experiment 1. These analyses consisted of a classification of subjects' conceptions of the system based on their verbal protocols and related responses to questions. This classification, together with associated frequencies, is presented in Appendix B.

EXPERIMENTER-PROVIDED ELABORATION

Previous research by Reder, Charney, and Morgan (1986) has shown that performance on a series of computer tasks was enhanced by the presentation of syntactic elaborations (examples) during learning, but not by more conceptual elaborations. Our data extend those results by suggesting that both syntactic and conceptual elaborations can influence the learning of a "cognitive skill" (Card, Moran, & Newell, 1980), although such effects depend upon the relation between elaboration type and the task to be performed.

Syntactic Elaboration: Concrete Examples

Our data replicate the effects of syntactic elaborations on performance. As can be seen in Figure 1, the primary effect of example type was in the Procedure case. In Experiment 1, the Concrete Example condition resulted in 12% more correct responses to Procedure questions than the Abstract Example condition. In Experiment 2, the Concrete Example condition resulted in an 18% difference in the same direction. Concrete examples did not improve performance on Command Use or System questions. Their utility appears to be confined to those instances in which specific syntactic information is needed.

Figure 1. Mean percentage of questions answered correctly for two types of examples, Concrete or Abstract. Effects are shown for three types of questions (Procedure, Command Use, System) during Initial and Final testing.

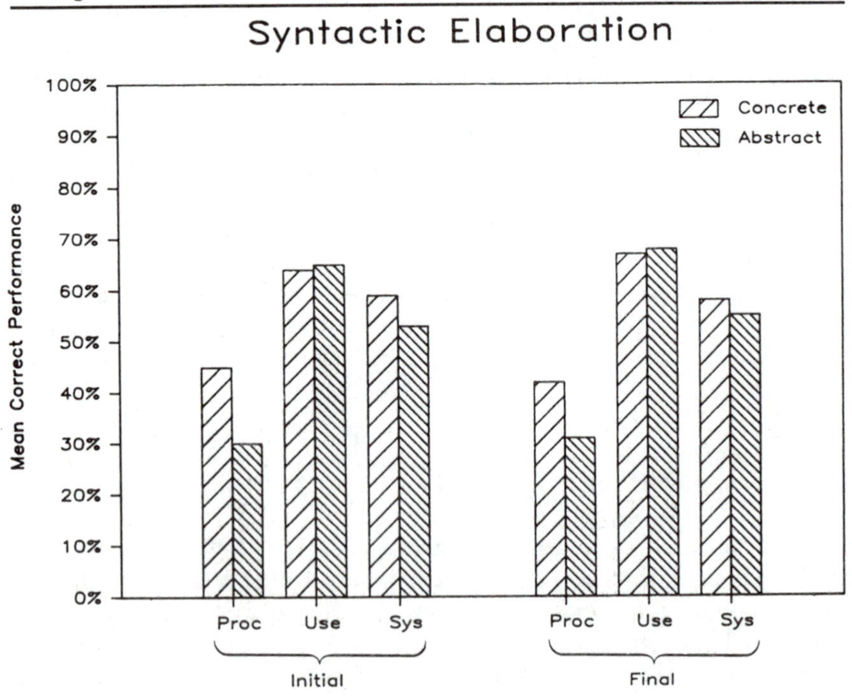

These syntactic effects appear to be based in part on an improved understanding of the commands. The Concrete Examples resulted in substantially fewer of the conceptual errors (App. A, I-III) that occurred on Procedure questions (42%) than did the Abstract Examples (58%). These errors reflect problems in general conceptions about what commands to use, and not just difficulties in specifying appropriate pathnames or arguments. Concrete examples may therefore be useful in clarifying conceptual distinctions that are important for correct command selection and formulation.

Other research we have conducted with very different instructions (Sebrechts, Deck, & Black, 1983; Sebrechts et al., 1984) also emphasizes the importance of highlighting distinctions among commands (see Murdock, 1960, for a general discussion of stimulus "distinctiveness"). In those studies, subjects were given either standard text instructions or the same instructions arranged in a spatial layout that isolated individual procedures specifying their location within a command hierarchy. The spatial layout group generated fewer procedural errors, suggesting that the conceptual clarity of that presentation helped decrease confusions in command formulation.

Conceptual Elaborations

Conceptual elaborations also affected performance, as shown in Figure 2. For the two experiments combined, the Model condition produced the best initial performance on Procedure questions. (Effects of the other elaboration conditions depended on the experiment as described below under "Subject-Elaboration".) As with syntactic elaboration, conceptual elaboration influenced the number of conceptual errors on Procedure questions. The Model and Analogy conditions resulted in roughly 10% fewer conceptual errors across the two experiments. There was, however, relatively little effect of elaboration condition on the frequency of implementation errors, which reflect problems with specific command formulation.

Conceptual elaborations had relatively little effect on Command Use questions. The author-provided models did not include explicit statements concerning command generalization, and they apparently did not contain sufficient information to enable subjects to infer those generalizations.

The greatest effects of conceptual elaboration occurred in the results for initial performance on System questions. The Model and Analogy conditions showed significantly better performance than the Simple and Redundant conditions. The model information did help subjects to understand general system concepts, although that understanding was not necessarily directly related to knowledge about specific commands.

Verbal protocols confirmed the importance of explicit models in de-

Figure 2. Mean percentage of questions answered correctly on initial testing for Procedure, Command Use, and System questions as a function of elaboration condition (Simple, Redundant, Model, Analogy).

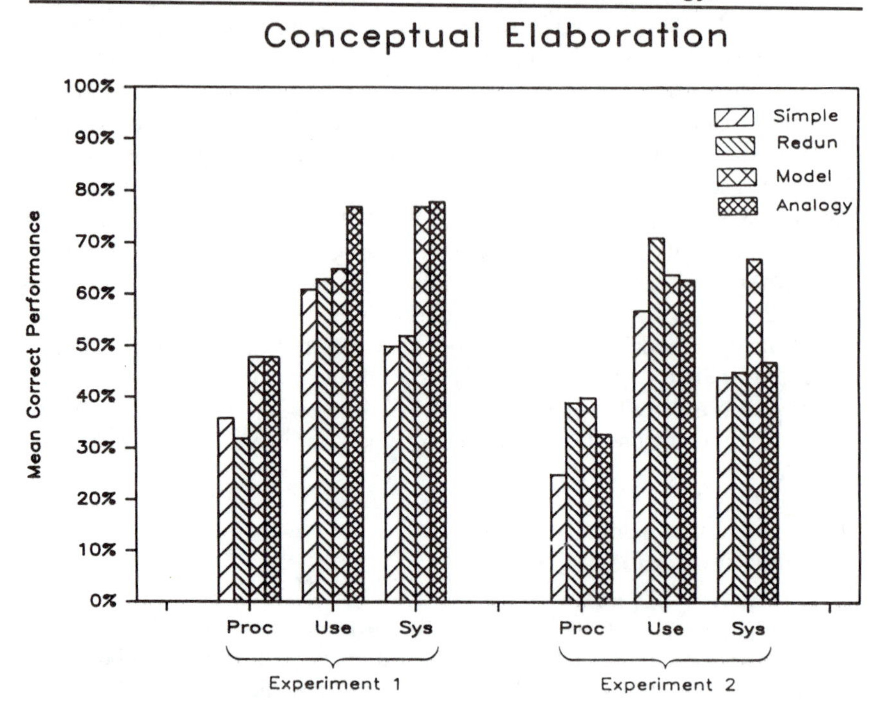

veloping a conception of the system. This was particularly evident in tasks that required subjects to understand system organization and to differentiate among a number of system components: directories; files; filenames; addresses; home, working, and root directories; complete and abbreviated pathnames. Subjects' protocols indicated numerous confusions about these concepts (App. B, II.A–II.O), which produced difficulties in answering a wide range of questions. Some subjects, for example, failed to differentiate among standard files, which can contain text, and directories, which in UNIX are special files that hierarchically organize the information about the names and locations of files (Appendix B, II.A). One subject, who said that he did not understand the distinction, had difficulty performing even the simple operation of changing to another directory. Another common difficulty concerned the specification of file locations. In UNIX, a file can be specified by a pathname indicating the sequence of directories leading to a specific file. The separation between each directory name in a pathname is specified by a '/'. Any path can start at the top of the hierarchy, specified by an initial '/', or it can start from a current directory, without the initial '/'. Many subjects had difficulty fully understanding these distinctions (App. B, II.E and II.F). One subject who misunderstood pathnames, for example, reinterpreted the effects of certain commands: In one case, he was presented with a command "rm thesis/outline," which instructs the system to remove the file 'outline' contained in the subdirectory 'thesis.' This subject apparently reinterpreted the slash as a file delimiter and suggested that two files "thesis" and "outline" were being removed: "maybe you're removing two files?"

In general, confusions among basic concepts were less evident in the protocols from the Model and Analogy conditions than from the Simple and Redundant conditions. There were 43 instances of such "specific confusions" (App. B, II) in the Simple condition, 30 in the Redundant condition, 23 in the Model condition, and 14 in the Analogy condition. Presumably, the elaborated information helped to provide a framework for distinguishing among the various concepts.

Experimenter-provided models. In order to determine if the subjects consciously invoked the experimenter-elaborations in problem-solving, verbal protocols were analyzed for use of our specific models and analogies. There were a limited number of references, indicating that the subjects were using the model information in at least some cases (App. B, V–VI). One subject for example, indicated that he understood the concept of linking with the following assertion: "I was just thinking that they aren't really separate files but have separate names." Likewise, his understanding of addresses appeared to help clarify the distinction between copying a file and moving a file; when asked if a file's address changed when the file is moved, he said: "cp [the copy command] makes a new address [and hence a new physical file], move maintains the same address [and hence keeps the same physical file]." Model concepts were also useful in determining certain system capabilities. For example, one subject, when asked whether or not deleted text can be recovered, said that it could because it is stored in a "buffer."

A number of references to explicit models also occurred during learning about the editor. The use of special characters such as a carriage return, which are not visible on the screen, are often problematic. Three subjects in the Model condition used the "string of characters" description to explain the use of special characters. One subject explained the deletion of a space by reference to the fact that the "delete word" command (dw) tells the machine to remove the characters including the space following the word.

The use of the models, however, is not always successful. One subject had particular difficulty with root, home, and working directories. When asked if the "ls" command (without any associated argument) would work, she said: "it'll do your home thing." In this case she appeared to have an erroneous model of the home directory as the default. That is true only if the home directory is also your working directory when the command is issued.

Experimenter-provided analogies. Protocol data also indicated that subjects used the analogies successfully in answering questions. For example, as one subject described the directory structure for a particular question, she said, "That is my branch, my working directory," connecting the tree analogy to the directory structure. Later she continued the analogy to a particular file: "That would be a leaf."

Another subject very explicitly employed the tape recorder analogy in reasoning about the editor. One of the questions asked what would happen if the "dw" command was given when the cursor was located in the middle of a word. This command is a mnemonic for "delete word." The two obvious choices are that the command will delete the whole word or that the command will delete a portion of the word, depending on the cursor location. The subject resolved the issue appropriately by using the analogy: "Does it delete from where you are to the end? If we go with the tape recorder model—from where you are to the end."

Another subject tried to answer questions by reasoning with our library analogy. This analogy is used to explain files (books) in terms of filenames (labels on call cards) and addresses (call numbers). When asked to describe the two kinds of information associated with a file, she said, "It's got a reference card and a place on the library shelf." The analogy was clearly influencing her reasoning about the problem. Interestingly, as the analogy broke down, so did her reasoning about the files. The move command is used to rename a file; technically it associates a new name with the old address. When asked whether a file's pathname and address would be changed by a move command, the subject said, "If you move it, it's not the same address. It's like moving a book, the call number would change." The subject appears to have assumed that moving was manipulating books instead of call cards. Oddly enough, moving a book would not normally change the book's call number, but in this case the subject assumed that it would and therefore made the wrong inference about files and addresses.

In general, performance in Experiment 1 tended to be best in the Analogy condition. It was therefore rather surprising that most of those subjects made at least one comment about how they disliked an analogy or found it confusing. This suggests two interesting issues for instructional design. First, subjects' evaluative comments may not be good predictors of learning. Second, the reaction to the analogies may in fact increase their utility. If subjects recognize difficulties in the analogies and therefore try to resolve those problems, they may in fact be thinking more deeply about the concepts. The deeper processing of the analogies may in turn increase their utility.

User-generated models. In the absence of explicit models provided by the experimenter, subjects tended to utilize their own models and analogies. This was most evident in responses to System questions which asked about more general functional properties. A detailed analysis of conceptual confusions in response to those questions revealed six principal models of the system: (a) a container model, which suggests that files are contained in directories; (b) a location model that describes the system in terms of an individual's location (usually in a directory); (c) a misconstrued hierarchy model, in which subjects conceive of a structured relationship of directories including erroneous distinctions between the "levels" at which root, home, and working directories are located; (d) a file reference model, which confuses the way in which files are accessed, including a confusion of a file with its name; (e) a default model, in which subjects assume that the system will follow a set of defaults, although the defaults sometimes do not correspond to those of the system; and (f) a black box model, in which the subject describes relations between input and output in terms of intermediate processes consisting of unknown operations. The most prominent models consisted of some form of spatial metaphor indicating that files are "contained" in directories or that the system could be thought of in terms of the user's location "in" a directory.

In addition to general models of system functioning, subjects developed individual "local" models to account for specific behaviors (e.g., App. B, III–E,K,Q). Thus, for example, in UNIX, the systems '>' and '<' are redirection operators that control the input and output of commands. Several subjects had seen those symbols used in other contexts as directional arrows, so they erroneously thought that they served as pointers or indicators of the locations of a directory. In another case, a subject was describing what happened when the end of a line was reached in the editor. He knew that other systems would break the line between words. He incorrectly used that as a model for what would happen with the VI editor. As in these examples, many of the models reflected the subjects' prior knowledge; they attempted to isolate some information from another system and to construct a related model for the new context.

On other occasions, subjects constructed models that showed no evidence of being based on any other system with which they were familiar. One subject developed a model of processing to account for the way in which multiple commands worked. UNIX allows for the output of one command to be passed to another command on a single command line (using "pipes"). One subject failed to understand how information was passed and therefore developed a parallel processing model in which all commands on a given line would be processed at the same time. Another subject had difficulty combining the files from several directories into a common directory. He eventually tried to solve this problem by renaming all of the directories so they would have a common name. He erroneously concluded that, "if they have the same name, they must have the same file."

In general, subjects in the Simple and Redundant conditions tended to use their own models of the system more frequently than in the Model and Analogy conditions. The use of such models appears to be an important component of learning, and the absence of adequate explicit models can actually encourage subjects to generate their own models.

Task Type and Elaboration

Both syntactic and conceptual elaboration can improve performance, but the effectiveness of those elaborations depends on the task requirements. Thus, syntactic elaboration improved performance on Procedure questions, presumably because the concrete examples provided specific syntactic information which helped to correctly specify a procedure. Conceptual elaborations, in contrast, were most useful for performance on System questions; the models provided information about the system that could be used to describe more general functional principles.

There was also some evidence that the utility of conceptual models was somewhat more general. The models improved performance on the Procedure questions during initial performance, and the analogies improved initial performance on both Procedure and Command Use questions. These effects suggest that functional or analogical information about the structure of the system may be useful in providing a framework for organizing commands. This is consistent with some of the literature on advance organizers (Ausubel, 1968; Mayer, 1975a, 1981; Bransford & Johnson, 1972). As previously noted, other research has suggested that a model designed to emphasize system organization through spatial layout can influence performance by helping to clarify distinctions among components of procedures (Sebrechts et al., 1984). The results here suggest a similar process. The presence of organizational information clarifies distinctions and helps to produce better performance.

SUBJECT-GENERATED ELABORATION

The effects of experimenter-elaboration interacted with whether or not subjects were induced to provide their own elaborations through verbal protocols (Figure 2). Overall performance was better in Experiment 1 than in Experiment 2. This may, however, reflect differences between the two groups of subjects in addition to effects of subject-elaboration. Of greater interest are the *relative* effects of the author-provided elaborations within each of the two experiments.

In Experiment 1, the Model and Analogy conditions resulted in the best performance for all three question types. For Experiment 2, Redundant and Model conditions were best for Procedure questions; the Redundant condition was best for Command Use questions, and the Model condition was best for System questions.

In brief, there are two major consequences of subject-generated elaboration in our data that need explanation: Subjects who were required to provide verbal protocols (Experiment 1) tended to show relatively better performance in the Analogy condition, and relatively worse performance in the Redundant condition, than those who were not required to do so (Experiment 2).

Analogies. If learning is dependent on prior knowledge, then analogies should be especially useful aids to learning, since they provide a way to map new information onto prior knowledge or schemata (Rumelhart & Norman, 1981). The importance of analogies has been recognized as part of a general theory of reasoning (Sternberg, 1977), as an effective learning technique (Collins, Warnock, Aiello, & Miller, 1976), and as a contributing factor in creativity (Poincare, 1913).

In order for the analogies to be useful, however, the learner must understand the nature of the mapping. In common analogies such as "an atom is like a solar system," we assume that a person will be able to identify the nature of the links between the two domains. In more complex analogies, however, subjects often fail to recognize the analogical relationship (Reed, Ernst, & Banerji, 1974). Finding the correct associations may be especially problematic when dealing with a computer operating system. Thus, there is no immediate, intuitive mapping produced by the phrase "the structure of files and directories is like that of a library card catalog." This is partially a product of the subjects' limited familiarity with the computer domain. Moreover, since subjects may not recognize what constitute the salient features of the analogous domain, they may actually generate the wrong mappings (see Halasz & Moran, 1982). Recognizing that such erroneous extrapolations are possible, we tried to make the links between the computer domain and the analogous domain explicit. When we introduced the concept of an address, for example, we would extend the library card catalog analogy by saying that "addresses are like call numbers." Nonetheless, understanding the series of connections implied by these analogies requires substantial mental effort. Presumably, by encouraging the subject to process the descriptions more fully, the verbal protocols in Experiment 1 clarified the mappings between the computer task domain and the analogous domain.

Redundancy. Even the Redundant condition showed some improvement relative to the Simple condition given the appropriate circumstances. This effect was evidenced in the Procedure and Command Use questions of Experiment 2. This is a surprising result, since the Redundant condition was designed to provide a rough control for instructional length of the model condition, while providing as little additional information as possible. So, for example, in the case of the list directory contents command (ls), the Simple condition stated that "The list command displays the contents of a directory on the screen of your terminal. To list the contents...." The Redundant condition for the same command embedded the same information in an expanded text: "The list command displays the contents of a directory on the screen of your terminal. It allows you to find out what is in a particular directory. To list the contents...."

A person with at least moderate computer experience would likely find the Redundant condition to be repetitious; the added Redundant statement would be informationally redundant with that provided in the Simple condition. Although the second statement does have some content, it provides no new reduction of uncertainty (Garner, 1962) beyond that provided by the person's knowledge of the situation and the statements in the Simple condition. In other words, given a person's prior knowledge about displaying information on a screen and the new information that directory contents can be displayed by the list command, the user already "knows" that the list command provides a means to "find out what is in a particular directory."

For more novice users like those in our studies, however, the statements do not appear to be informationally redundant. Interestingly, none of our subjects mentioned the repetitious nature of the statements in any of their verbal protocols. Presumably, added statements provided multiple ways to understand the basic concepts, thus increasing the probability of comprehension and retention. This is what has been described as "redundant elaboration" (Anderson & Reder, 1979; Reder, 1979). According to this view, the added statements provide alternative ways to structure and later retrieve information. For the novice, "find[ing] out what is in a particular directory" may provide a different way of thinking about "display[ing] the contents of a directory on the screen."

Redundancy must therefore be understood in terms of the prior knowledge of the user. What constitutes "informational redundancy" for the person with substantial prior knowledge may serve as "elaborative redundancy" for the novice with limited prior knowledge.

The effects of this elaborative redundancy, however, are still rather task specific. The redundancy did help to clarify commands, as evidenced by the fact that performance improved (in Experiment 2) for subjects on Procedure and Command Use questions, both of which dealt with the way commands were used. In contrast, it did not help with Systems questions; a better grasp of commands did not provide a more general understanding of the underlying principles described in the Model condition.

Likewise, the utility of "elaborative redundancy" depends upon the context. The Redundant condition did not produce the same performance advantages in Experiment 1. Presumably, the Redundant materials already contained most of the elaborations that could be made based on the given information. Attempts to extend that material through subject-elaboration were therefore likely to add little new correct information and could actually interfere by misdirecting subjects or by forcing them to concentrate excessively on irrelevant differences in phrasing.

Interaction of Types of Elaboration

Previous studies (Reder et al., 1986) have indicated that performance on a set of file manipulation tasks could be improved by subject-elaboration, but that such elaboration provided no added benefit beyond that of author-provided elaboration. Our data suggest an extension to these results. In the Reder et al. (1986) case, conceptual elaboration appears to have included several different types of elaboration. In our studies, by separating different types of conceptual elaboration and different types of questions, we found that subject-generated elaboration can interact with the type of author-provided elaboration. Subject-generated elaborations had little relative effect on functional models or examples, a positive effect on analogies, and a negative effect on redundant elaborations. It would, of course, be important to determine if the way in which subject-generated elaborations are induced also affects this interaction.

Verbal Protocols as Subject-Elaboration

Based on our analysis of the results, it appears that reading and thinking aloud serve to foster subject-elaboration, which in turn interacts with the experimenter-provided elaboration of the instructional materials. This is a modification of other views on the effects of verbalization. In a detailed review of verbalization, Ericsson and Simon (1980) suggested that, in general, concurrent verbalization that draws on information to which a subject is already attending, has relatively little effect on performance. As information load increases, the tendency to verbalize decreases, in some cases because of a decrease in accessibility.

In our data, it also appears to be the case that, when subjects are reading and thinking aloud, they are simply verbalizing the information that is accessible to them from their own representation (what Ericsson & Simon called "heeded information"). In addition, however, the process of verbalization may modify, redirect, or clarify the heeded information. That is, the verbalization appears to change the nature or relative depth of processing (Craik & Lockhart, 1972; Cermak & Craik, 1979). This change in processing, what we have called subject-elaboration, in turn modifies the effectiveness of instructional elaboration.

ELABORATION AND LEARNING

In the present studies, the persistence of training effects was measured by comparing final performance (day 3) to initial performance. Syntactic elaboration, in the form of Concrete Examples, continued to provide better responses to Procedure questions during the final testing, although there were no long term effects of example type on other questions (see Figure 1). This result reflects the fact that Procedure questions require specific syntactic knowledge that can be retrieved by reference to individual examples.

The effects of conceptual elaboration, in contrast, decreased by final performance. As during initial performance, the Model and Analogy conditions tended to produce better performance when verbal protocols were collected (Experiment 1), and the Redundant condition tended to produce better performance in the absence of verbal protocols (Experiment 2). Subjects who received model information still performed better on System questions. However, other effects of elaboration on System, Procedure, or Command Use questions were no longer reliable. This change is probably a consequence of the integration of pieces of information across the different commands resulting in a better (though still incomplete) model of the system. That general model influences performance on all commands regardless of the elaboration condition under which initial learning of the command occurred. This effect is most striking for the Simple group, who received basic descriptions for a set of commands with no elaboration. Since they are given very limited information about one set of commands, they must try to utilize the elaborations they are given for other sets of commands in order to make generalizations.

With the objective of providing a more fine-grained description of learning, we analyzed the amount of learning and forgetting for each type of question; the results of these analyses are shown in Figure 3. *Learning* was assessed by determining what percentage of questions that had received a wrong answer during initial performance received a correct answer during final performance. *Forgetting* was measured by determining what percentage of questions that had been initially correct were wrong on the final performance test. In general, subjects showed more forgetting than learning on Procedure questions, more learning than forgetting on Command Use questions, and roughly equivalent learning and forgetting on System questions.

Figure 3. Mean percentage change in performance from Initial to Final testing for three types of questions (Procedure, Command Use, System) in Experiments 1 and 2. *Learning* indicates changes from incorrect responses in initial testing to correct responses during final testing; *forgetting* indicates changes from correct to incorrect responses.

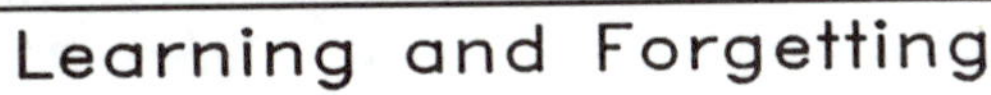

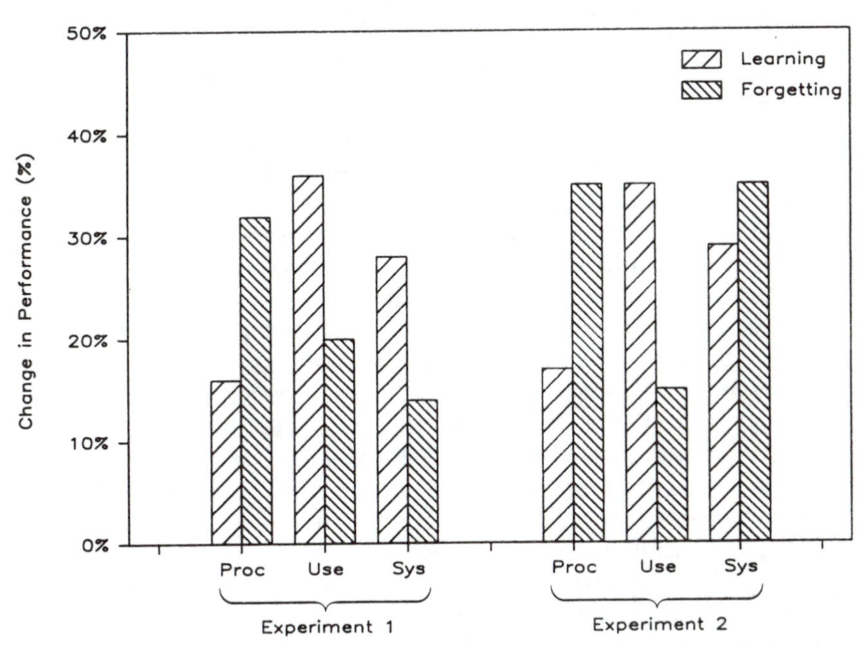

The decrease in performance on Procedure questions reflects the difficulty in retaining specific syntactic information. The models that were presented to the subjects provided generalizations about function; they did not provide generalizations about command structure. Subjects did, however, develop some of their own generalizations, which were often misleading. Thus, in their protocols, some subjects expressed the belief that commands must be two letters (App. B, I-A). This belief was evidenced by errors of substituting two letter abbreviations such as 'rd' for the remove directory ('rmdir') command and 'md' for the make directory ('mkdir') command (App. A, V-A, C).

If information is gradually being integrated into an improved representation of the system, then errors generated by model misconceptions should decrease, whereas those resulting from procedure specification should not. This is in fact what occurs. The number of *model misconceptions* (App. A, I) decreased substantially from initial to final performance. Presumably, as subjects learned more about the system, they were better able to construct a more integrated model of it as a whole.

In contrast, there was a substantial increase in the number of *command misconception* errors (App. A, II). Although the subjects were able to integrate information from the different command subsets into a more general model, they concurrently suffered from increasing confusion regarding specific commands. The emerging model apparently failed to provide access to individual commands, since functions were not linked to specific command syntax.

It is evident that, as time passes, information is also integrated with knowledge from other systems, leading to new errors. During initial performance subjects did not show signs of confusion with other systems. During final performance, however, subjects did show such *intrusion errors* from other systems, the most common being some form of 'delete' and 'directory' (App. A, VI).

In addition, subjects began making up commands which did not appear to come from any other system. These *fabricated commands* (App. A, VII) included combinations of previous knowledge and partial information about the new system. In trying to eliminate a directory, for example, a subject suggested 'deldir' (App. A, VII-D), a modification of the correct 'rmdir' (remove directory) using the new concept and associated command, but replacing remove ('rm') by the more familiar delete ('del').

The model which users are developing contains a number of generalizations and specific facts, both correct and erroneous. Understanding the functions of the system does help to constrain the range of possible commands, but it by no means guarantees procedural accuracy. For example, in trying to break a link, a number of subjects construed commands that seemed to incorporate the appropriate concepts, but with some combination of constructs. In fact, in UNIX, 'rm' serves to break the link between a filename and a file. Subjects, however, introduced a number of alternatives that suggested removing a link ('rmln' or 'rl'), severing a link ('svln'), unlinking ('unlink') as separate from removing a file ('rm') (App. A, VII-E).

As can be seen in Figure 3, the percentage of learning was substantially greater for Command Use than for Procedure questions. For Command Use, the same pattern was present in both experiments with substantially greater learning than forgetting. Unlike the Procedure questions, the Command Use questions did not require specific syntactic knowledge. General information about how to use commands can be helpful for this type of question; thus, studying one command can help in responding to a question about another command. As a consequence, to the extent that the material from the different command subsets was actively integrated, subjects appeared to be actually learning and improving their understanding of how commands are used.

At the same time, inappropriate integration can once again hurt performance. When the learner reads a true–false Command Use question,

Figure 4. **Mean percentage change in performance from initial to final testing for *true–false* Command *Use* questions in Experiment 1 and 2. *Learning* indicates changes from incorrect responses in initial testing to correct responses during final testing; *forgetting* indicates changes from correct to incorrect responses. Responses are divided into changes from *false* during initial testing to *true* during final testing (*False to True*) and from an initial true to a final false answer (*True to False*).**

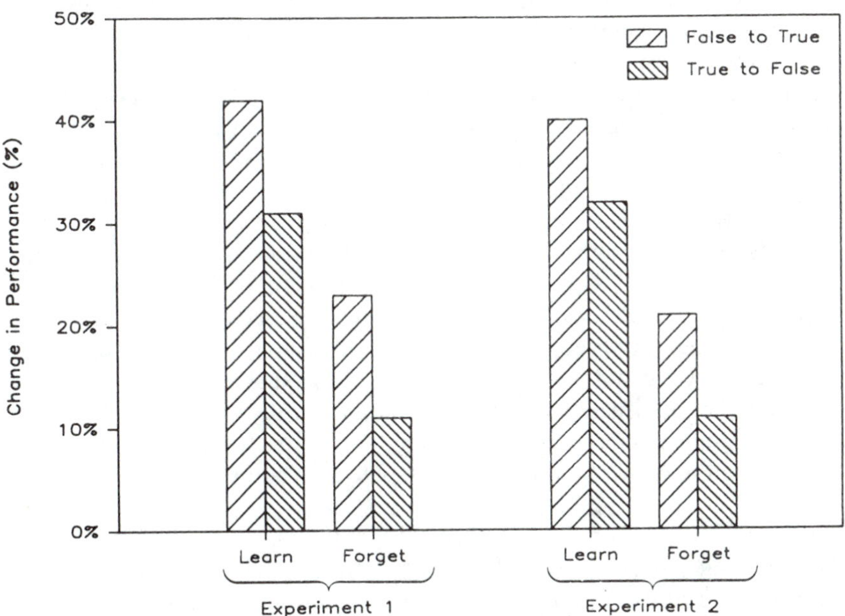

the information contained in that question has some probability of being integrated into his or her current knowledge base even if the statement is false. This is especially the case if the learner is unsure of the correctness of the statement. Thus, a statement such as "The mv command gives a file an additional filename" is false. However, reading the statement may in fact help develop a model that includes that statement. The fact that such problematic integration occurred is illustrated in Figure 4. Subjects tend to change their answers from false to true substantially more than the reverse, whether or not that change was correct (learning) or incorrect (forgetting).

For System questions there was slightly more learning than forgetting in Experiment 1, whereas in Experiment 2, learning and forgetting were roughly equivalent. Neither of these System question differences, however, reached statistical significance. As noted previously, the effects of conceptual elaboration were reduced during final performance. That is precisely the effect that would be expected if a more general integrated model were being used to direct responses.

INTEGRATIVE MODELING: A PROCESS DESCRIPTION OF EARLY COGNITIVE SKILL ACQUISITION

The effects of elaboration on learning can be summarized as follows: (a) author-provided elaborations can improve initial skill acquisition; (b) subject-generated elaborations can also improve skill acquisition; (c) the utility of author-provided and subject-generated elaborations interact; (d) the utility of elaboration depends upon the type of task; (e) there are gradual transitions in learning, with decreased effects of elaboration over time; and (f) changes in learning are task dependent.

The results can also be characterized in terms of the mental models that subjects are developing. First, subjects have confused mental representation of the domain, as indicated by the large number and types of errors (see Lewis & Mack, 1982, for a related result). Second, those representations can be improved by providing subjects with explicit model information about the system. Third, over time, the subjects' mental transformation of the information come to dominate explicitly provided information. Fourth, subjects' own elaborations can improve performance, presumably by helping the explicit models to be matched to their own previous mental models.

Performance, error, and protocol results indicate a gradual integration of information into a developing mental model. Much of the data can be described in terms of schematic knowledge; new information is structured in terms of prior knowledge. More generally, however, the acquisition data suggest a more complex phenomenon. The learning we observed cannot be described as well-defined discrete transitions. Instead, assessment of a learner's knowledge suggests that the state of knowledge at any point in time is ill defined. Subjects provide vague, incomplete, and frequently conflicting responses. Likewise, "states" tend to be continuous rather than discrete. Information is actively being structured even at the time of response. This is evidenced in the protocols in which subjects change their response as it is being made. These changes in representation over time constitute a central feature of the data we have studied here and elsewhere (Sebrechts & DuMont, 1986). These changes in mental models are what we describe as integrative modeling.

Integrative modeling is meant to emphasize that the way in which learning occurs is through integration of various components. The emphasis is on the process of dynamic modeling rather than on a static model. This learning model contrasts with more restrictive schematic views in the same way that a dynamic memory model which organizes information when needed (Schank's, 1982, *MOPs*) contrasts with a relatively stable memory structure (Schank & Abelson's, 1977, *scripts*).

Transitions in performance data between initial and final testing support the gradual integration of concepts. Effects of different kinds of elaboration are diminished over time. General concepts remain relatively strong, whereas individual procedures are forgotten. The consequences of these transitions are reflected in the error data we described. Subjects show a merging of prior knowledge and newly acquired information. Intrusions and constructions increase with time.

These changes are also evidenced in subjects' protocols in the first experiment. There are numerous forms of reasoning that constitute the integrative process, many of which are listed in Section VIII of Appendix B. Here we will illustrate five processes that occur rather consistently: concept differentiation, inferential completion, case-based generalization, assumption of causal similarity, and development of parallel structures.

Concept differentiation. One concept that is difficult for subjects is the notion of a *mode* or system state. Certain commands can be issued only in certain contexts. Thus, for example, 'x' can be used to delete a character in the editor only when in command mode. One subject expressed his confusion about these conditions: "Is x the delete command? How do I get into the delete command? How can I type x without destroying something?" At this point he fails to clearly differentiate the command mode and the insert mode in the editor. Later, when trying to remember an editor command, his emerging understanding of the mode concept is evidenced in his differentiation between working in the editor and giving other system commands: "Oh yeah, you're already in the editor." On rereading the material, he indicates that he has now recognized the "mode" distinction within the editor but cannot find it described: "It doesn't tell me anything about the command mode or the insert mode." As he reads on, he finally finds the explanation corresponding to his emerging model: "Oh yeah, this explains the modes." He apparently now differentiates between a command mode and an insert mode, resolving his earlier problem with deletion.

Inferential Completion. At any time during the course of learning, some information is missing. Subjects often actively construct components of their model as needed to provide inferential links between what they know (or believe they know) and a more complete description of the material they are learning. In our study, subjects received some instructions about directories before receiving similar instructions for files. Part of that material described the fact that "rmdir" could be used to remove directories, although all files had to be eliminated from that directory before it could be removed. One subject incorrectly assumed that since she had not been informed of a command to remove files, the "rmdir" must be able to do it: "It has to let you remove files so you can clear the directory." A number of other subjects followed the same strategy. Rather than waiting for the information about removing files, they tried to infer a more complete model.

Case-based generalization. In addition to making inferences about specific command use, subjects also took examples they encountered as the basis for forming more general models. For example, in the case of the editor, subjects learned how to delete some text which was saved in a buffer and then to put that text at a new location. This occasionally resulted in generalizing a single "copy location" model. Information could be moved from the text into a buffer and back again. This model suggests that information once moved out, no longer exists in the buffer. That is incorrect, since the information is copied from the buffer and not deleted. This erroneous generalization caused difficulty when subjects needed to copy information more than once, since they did not understand that it was still present in the buffer. Generalizing from individual cases is an important aspect of learning, although any form of inductive reasoning runs the risk of over-generalization. (See Soloway, 1986, for an interesting related application in machine learning.)

Assumption of causal similarity. Subjects also tried to understand the cause underlying particular system behavior. For example, in the editor it is necessary to use the <esc> key to end an insertion. It provides a means to move out of insert mode and back to command mode; in that state, typed information will be interpreted as a command rather than as information to be placed into the text. The <esc> key also needs to be pressed after the 'change' command. In trying to develop a reason for the use of escape, subjects assumed a causal relationship similar to that for insertion. While reading about the change command, for example, one subject said: "I have to end with <esc>, oh, anything with insert you need an <esc>," implying that the change command includes an insertion. When asked if the <esc> key needs to be pressed after the change command in the editor, another subject correctly noted that "It probably goes into insert mode and thus you need to press <esc>." Given the fact that the command needed to be terminated with an <esc> and that they previously had learned to use <esc> as a way to end insertion, they correctly reasoned that the new usage must be caused by a need to exit from an insertion.

Development of parallel structures. Another way in which subjects integrated specific aspects of their knowledge was by identifying parallel structures across commands. For example, subjects learned two ways to change characters in a file. The first method uses 'r' followed by a single new character to be inserted in place of the old one. The second method uses 'cw' followed by a new word. At first the subject finds this confusing, but then recognizes the parallel with the previously learned method for deletion: "R for what? Oh for replace! So replace just. . . (long pause) for letter to letter only. The change will do a word. This is like x for delete. To delete it, I would say d something and for changing, I would say c something." The subject recognizes that just as 'x' deletes a single character, 'r' replaces a single character. Likewise, just as 'c' specifies a change or substitution, 'd' specifies a deletion. However, both of these latter commands require an argument specifying what is to be deleted or changed. More specifically, the subject learned how to change or delete a word using the argument 'w'.

The foregoing examples are meant to illustrate integration rather than to provide an exhaustive description of the concept. In fact, however, it is quite evident that even these restricted examples cannot be described as reflecting a single unitary process. Rather, we see integrative modeling as incorporating two major component processes by which information is initially acquired: assimilative structuring and fragmented structuring.

Assimilative Structuring

As we indicated previously, learning is heavily dependent on prior knowledge. This fact was noted in Piaget's (1954) use of the term *assimilation*, which emphasizes the notion that new information was fit into (or *assimilated* into) the existing mental representation. Mayer (1975b) used the term *assimilative encoding* as a descriptor of how people used prior knowledge as a basis for encoding new information.

There are two ways in which our data can be described in these terms. First, many errors and verbal descriptions suggest that subjects are utilizing prior knowledge as the basis for understanding the current system. Secondly, the effects of elaboration can be seen as providing a model to which other information can be assimilated. We call the process underlying these effects *assimilative structuring*. It is assimilative insofar as it depends primarily on the organization of prior knowledge. We use the term *structuring* in contrast to Mayer's *encoding* for two reasons: First, the process does not appear to be confined to the time of initial encoding. Rather, the attempt to fit information together is an ongoing activity. Second, we mean to qualify assimilation in a way that suggests a dynamic state of change rather than a simple "lock and key" fit to a preexisting model. In other words, it is not pure assimilation.

Central to assimilative structuring is the notion that new information can be mapped to prior concepts. Figure 5 illustrates schematic representations for commands which may produce such mappings. Following a semantic network description used by Norman, Rumelhart, and the LNR Research Group (1975), the nodes in the figure represent procedures. Figure 5a represents the general procedure for changing a file; pressing a command followed by an old filename and a new filename causes the old file to be changed to the new file. Figures 5b, c, and d provide parallel procedural representations for renaming, copying, and changing files. Blank nodes indicate those procedures that are the same as those from similarly located nodes in the change-files diagram. These particular diagrams are only skeletal structures for the procedures without specific content.

This figure is, of course, an abstraction that does not capture the dynamic nature of concepts. However, it can be used to illustrate some of the examples in our data. One type of assimilation consists of a simple mapping of *names*. For example, one subject noted that "Move is just like renaming a file." This subject had used the 'rename' command on another system and his mental model of the command function is approximated by Figure 5b. By equating the new command name 'mv' with the old command name 'rename' in that diagram, the representation is kept intact. A simple replacement in the "rename-file" schema is all that is needed to learn about 'mv' in this case.

Not all name mappings, however, are as successful. The same subject later runs into difficulty when he tries to map 'directory' to his notion of 'account': "Well I assume that a directory is something similar to an account." The subject's conception of an *account* is apparently that of a single directory in which all of a user's files are kept. It does not, for example, allow for subdirectories or superior directories as in the UNIX examples. The subject is performing an assimilation, but in this case, he has carried over a number of misconceptions.

In several other cases, when subjects did not recall an appropriate command, they attempted to find a mapping between old command names and new ones. Thus, for example, when one subject was trying to combine files he said, "I would want a command that would append one to another." On the DEC-20 with which this person was familiar, "append" was the appropriate command. In reasoning about the present system, he was assuming that a similar process should exist, perhaps under a different name. However, in this case, the subject had not learned any such command, and so his mapping strategy failed.

Although the mapping of names was frequently successful in our data, such mappings work only if the procedures are already known by another name. More commonly, procedures are learned by drawing appropriate analogies which require mapping of *relations* (Gentner & Gentner, 1983). For example, our subjects were unfamiliar with any procedure that was the equivalent of "link." One way to learn this procedure is by building a

Figure 5. Schema-like representations of command conceptions evidenced by several subjects. (a) The general framework. (b) Rename-file schema with UNIX to DEC-20 name-mapping, mv = rename. (c) Copy-file schema. (d) Link-file schema. Confusion of file and filename introduces error into schemata.

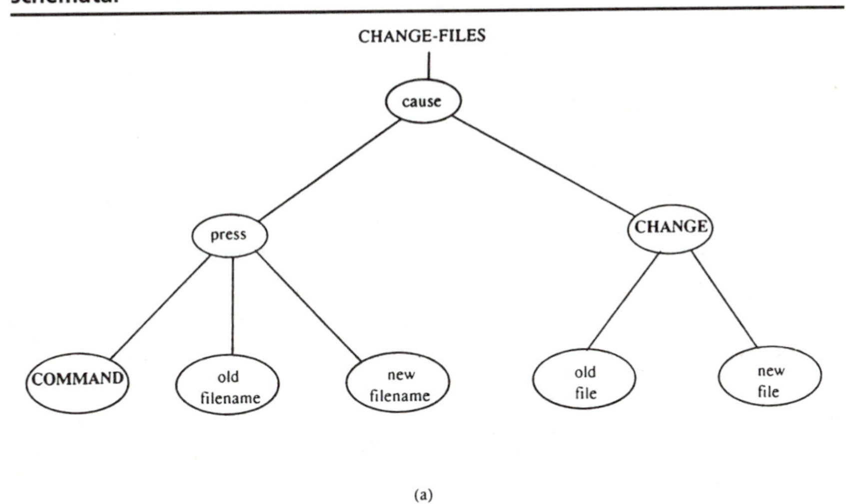

(a)

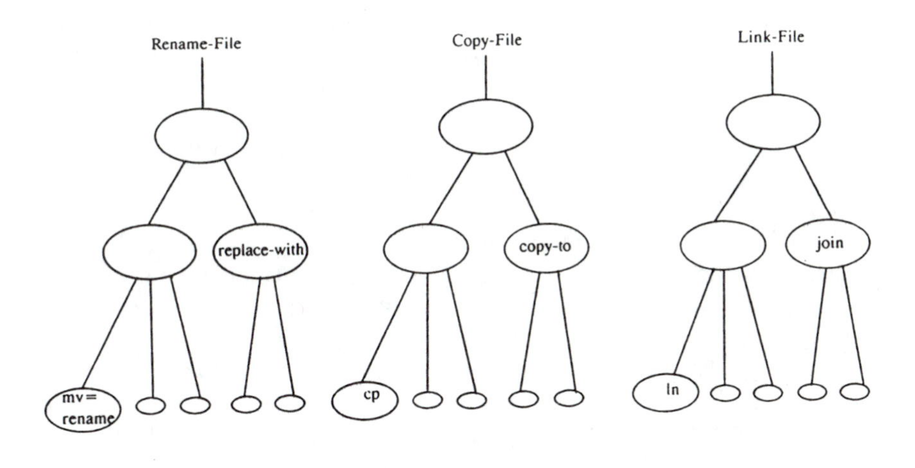

(b) (c) (d)

representation comparable to that for renaming, or copying. Schematically, Figure 5c can be built by making small modifications to Figures 5b or 5c. Building the representation in this way makes it easier to develop an appropriate way to formulate the command: command name, old filename, new filename. However, this approach is not without problems. By assuming that the three commands have the same structure, the subject fails to differentiate the action of changing a "file" from that of changing a "filename." In fact, copy creates a new "physical" file, whereas move and link do not. This results in a certain amount of confusion among the commands (App. A, III-F,G,H).

There were many examples of other types of conceptual mappings. Thus, for example, one subject tried to map the directory concept to a more familiar directory outside of the computer domain: "Directory. . .directory like a telephone." Another subject was asked what would happen if text was inserted beyond the end of the line. The subject tried to draw an analogy between the action of the VI editor in UNIX and the editor they had used previously. The previously used system included word wrap, a feature that avoids dividing a word at the end of the line. The subject assumed that the same functional properties must hold in VI: "I know it won't break at the middle of a word." In fact, the VI editor will divide a word when the right margin is reached.

Interestingly, subjects frequently try to assimilate functional characteristics to previously known systems even in the absence of appropriate context. The same subject, when asked if a file could be concatenated to itself (i.e., if the same file could be used as both the input and output for typing out the file), concluded that it could, based on the fact that another editor he knows changes the version number rather than the actual file. In the UNIX case there was no information provided that gave any indication of a comparable change in version number.

Even from these few examples, it is evident that assimilation can become rather complicated. There is some pure mapping of names, but most of the mappings we observed included functional relationships as the basis for more complicated inference. One subject provided a particularly interesting use of assimilation in the development of a mental model of certain procedures. This subject used the tree analogy as a way to specify the pathnames for commands. In reading about a particular set of files and directories she traces through the hierarchy: "That would be a leaf... oh....another branch...off 333 [the name of one of the directories]...a branch with two leaves....As I understand it, you take the pathway first and then the name of what you want." Interestingly, this subject has used the analogy to think about pathnames quite successfully. However, she also assumes that her previous knowledge of locating something and then performing an action at that location will work here. She thus incorrectly assumes that you first specify the pathname (which she refers to as "pathway" in several cases), and then give a command. This reflected an assumption made by several subjects, that specifying pathnames was equivalent to actually traversing the system, a metaphoric walk in the woods in this case.

This example illustrates the fact that assimilative structuring describes a component learning strategy which may itself subsume other processes. The subject is fitting the directory structure to a tree, and building the tree metaphor to deal with systematic movement through the tree. She then applies the revised analogy to her particular problem. The importance of such constituent processes has been noted in Rumelhart and Norman's (1981) description of learning by analogy (which includes accretion, tuning, and restructuring).

Our data likewise suggest that there is substantial variability in the role that prior knowledge plays in learning. In some instances, as in the case of name substitution, emphasis is placed on *assimilation*. In other cases, such as mappings between editors, the emphasis is on *structuring* the appropriate match between prior knowledge and new information. In general, assimilative structuring emphasizes those learning processes that depend primarily on the well-ordered schema-like aspects of knowledge. Whereas this is an important component of learning, we argue that any theory that relies exclusively on this approach would be too restrictive.

Fragmented Structuring

Although these mappings and subsequent integration are quite common, they, by no means, provide for an exhaustive account of what the subject is doing. In order for subjects to use assimilative structuring, they must have the appropriate knowledge base. In Greeno's (1983) terms, subjects must have the correct "ontology of a domain," which defines the types of entities with which they are familiar. One of the characteristics of a computer system, is that it is an abstract entity for which subjects have a very limited ontology. As a consequence, much of their learning appears rather disjointed, consisting of separate components of knowledge. We use the term *fragmented structuring* to emphasize the fact that much of what is learned consists of components of knowledge that are relatively isolated from other aspects of the domain. These components may be individual facts or small task-specific models that show relatively little connection to a more general model of the domain.

Many of the subjects' comments indicate that they have simply retained a set of individual facts. In distinguishing different pathnames, for example, one subject uses a single salient fact: "There's a slash if it goes from the root and not a slash if it goes from the working [directory]." Likewise, in determining if the command "sort people | people2" will work, the subject comments, "It [the pipeline operator, |] needs to be followed by another command." She appears to be referring to her memory of specific formatting descriptions and examples. Another subject makes the formatting problem more explicit: "I know it's [the 'tee' command] in conjunction with the pipeline, but I'm not sure what came first." The subject appears to have no general principle for argument order outside specific instances. Similar conflicts with retention of format examples were evidenced in using pathnames. When one subject saw a problem with a link command followed by a particular pathname, he was puzzled by the '/' in the pathname because it conflicted with the specific instance he had seen before: "I don't remember the slash being used." Later, when he saw the remove command with an abbreviated pathname, he commented, "This is not the way I remember the format." In fact, the formats for pathnames are generally independent of commands. However, in these cases, rather than remembering any general principle, the subjects seem to retrieve specific instances. Even when mnemonics are available, subjects occasionally fail to use them. One subject notes, for example, "I thought dw was to change a word." In fact, 'dw' is for deleting a word and 'cw' is for changing a word. If the subject had generalized the principle of mnemonics rather than trying to retain individual instances, this confusion would probably have been avoided.

This type of reasoning fits well with the utility of specific concrete examples. Part of this utility stems from the fact that examples can be used to construct more general descriptions. This is consistent with recent experiments reported by Ahn, Mooney, Brewer, and DeJong (1987) which indicate that subjects can extract schemata from single examples. To some extent this is evidenced in our data as well. The schemata we found, however, were often incomplete or mistaken as in our earlier description of inferential completion and case-based generalization.

More commonly, rather than trying to from a general schema, subjects frequently used specific instances with which they were familiar. This strategy of focusing on specific details, however, can be problematic. In one instance, bullets (•) were used to highlight commands, but the subject took the bullets as part of the command syntax. The importance of the individual examples for learning in the computer context was demonstrated by Ross (1984). He found that, when subjects were learning about computer text-editing procedures, the context tended to remind them of particular examples. As a consequence, they would often select a procedure based on superficial resemblances between a problem and a previous solution. The importance of individual instances or episodes has been emphasized in a more general theory of *reminding* as descriptive of memory (Schank, 1982).

Fragmented structuring, however, is not restricted to the use of individual examples. It is also evidenced by partial or incomplete models that seem poorly integrated with other concepts in the domain. Thus, for example, as we noted previously, several subjects inferred that all commands must be two letters long despite the fact that they had been given conflicting examples. One subject, when attempting to remove a superior directory, tried a list of related names and then used the two-letter rule to constrain the options: "Get rid of it...remove it...list, change? I know there has to be two letters that stand for it...maybe rm." (In this case the correct command is 'rmdir,' so the specific rule was misleading.)

Another example of such a fragmented model is evidenced by a subject who has difficulty understanding the role of "tee." This operator allows output to be sent to separate locations (much like a T-pipe in plumbing allows water to flow in two directions), and must be used in conjunction with another command. One subject apparently developed the following rule: letters make up commands (e.g., rm, mkdir, ls, tee), whereas symbols make up "functions" (<, >, |). When answering a question using the pipeline operator, he suddenly realizes a problem with this model: "I just realized that tee is a function, it's not a command like sort or list. It's in a different class than those." The subject apparently continues to use the partial model of two classes of operators, but reassigns 'tee' from "commands" to "functions."

The fragmentary nature of models during learning has also been noted for other domains. For example, diSessa (1986) examined students' models of LOGO programming and found that they consisted of a number of partial explanations. He described their representations as "distributed models," contrasting the students' "patchwork" of ideas with more unified descriptions that are frequently ascribed to experts. DiSessa (1983) has also provided a related account for some of the explanations offered for the solution of physics problems. Based on a series of protocols, he argued that physics-naive students have a collection of basic phenomena ("phenomenological primitives") in terms of which they see the world. These primitives can be used to provide self-contained explanations, until they are later abandoned or integrated into more complete abstract description.

Much of the knowledge displayed by our subjects appears to have this patchwork quality. Subjects exhibit certain primitives in their restricted ontology. Files, for example, are often difficult to understand in terms of names and addresses, because they are thought of as objects to be directly manipulated (see Shneiderman, 1982, 1983). As a consequence, subjects have difficulty reasoning about novel situations in which the underlying characteristics are important. For example, for many users it is difficult to understand how a deleted file can be recovered, since they believe that operations occur to the file itself, and therefore assume that "delete" physically removes the file.

Pathnames provided another example of alterative ways of thinking about basic concepts. As we have noted, a number of subjects use various models or analogies (e.g, a tree) as a way of decomposing the meaning of a pathname. This is a conceptual organization which we try to encourage with explicit models. Other subjects, however, appear to think of a pathname as a kind of proper name rather than as a description of the underlying hierarchy. It is an isolated fixed construct generated by a set of rules concerning the user's location and the presence or absence of slashes. This approach is much more complicated and increases the user's memory load.

There is a tendency to think that these and other examples of fragmented structuring are problematic, that they fail to provide the understanding offered by a more integrated schema, and that they will therefore lead to decreased performance. Although it is probably true that the lack of a coherent schema generally implies decreased understanding, it does not necessarily entail a decrement in observed performance. There are two major reasons for this. First, it appears to be the case that certain information is appropriately stored in fragmentary form. For example, remembering that a command must always come first on a command line does not require deep understanding, and it serves as a useful rule (recall that, in fact, some of our subjects developed a "deep" model that incorrectly "explained" why pathnames should precede commands). Second, on many routine tasks it may be easier to retain a somewhat restrictive distorted view (a "false" primitive, or what Rubinstein & Hersch, 1984, call an "external myth") rather than a more accurate complete view.

For instance, in our previous example of file deletion, it is often easier for a subject to simply assume that deletion ("rm" in the UNIX case) actually eliminates the file permanently than to construct the appropriate model of storage and reference. Many people seem to work efficiently with this flawed conception for some time. Oddly enough, our observations suggest that, in some cases, these people are better off than those with a better model. When they accidently delete a file, they moan and start redoing the work; frequently they will use a backup they have created. The person who understands the system somewhat better realizes that the information is still available on the system, but often does not know how to recover it. We have in fact informally observed people spending more time trying to find a lost file on the system than it would take to retype the relevant material. Fragmented structuring is meant to specify the character of a learning process. Its utility depends on the conditions for using the acquired knowledge.

Integrative Modeling Summary

Consistent with many recent arguments, we believe that prior knowledge is central to learning. People learn new information in terms of their current knowledge. However, we need to temper our Bartlettian enthusiasm for organized schemata with the harsh reality of our messy mental attics. Based on the observations of child development, the need for this balance should not come as a surprise. The "blooming, buzzing, confusion" (James, 1890) is ordered only gradually. Either we accept the notion that prior concepts are "given" (à la Plato or Fodor, 1981), or we need to solve some part of the problem of induction. And along the way, we start with isolated, fragmentary, incoherent pieces of knowledge. That perspective is certainly not without precedent. In a discussion of the development of scientific explanation over 100 years ago, Goethe (1970) argued that our common sense views include a number of disjoint special cases. It is only after substantial effort that we sift through these isolated instances and develop the more abstract models that serve as scientific explanation.

A key factor in integrative modeling is an emphasis on process rather than structure. Within the context of our study, there is little evidence of completed static models. Rather, the mental model is in a constant state of construction. To some extent, this is what we would expect in any learning context, and it is likely that an expert has a more stable representation. Even experts, however, demonstrate changes in their knowledge over time, and it is that dynamic character of memory that is emphasized by integrative modeling.

Integrative modeling also proposes that knowledge includes schematic and fragmented constituent elements. This concept is perhaps best visualized as a network with relatively coherent, highly structured regions (schemata) and relatively isolated, unstructured regions (isolated facts). However, since the emphasis is on process, we make no assumptions about the specific structure of memory. In fact, recent descriptions of memory have noted that script-like retrieval does not necessarily reflect script-like memory. Instead, it may be generated by appropriate control processes operating on an unorganized associative network with only local connections between concepts and propositions (e.g., Kintsch & Mannes, 1987). Likewise recent distributed models of memory have emphasized the separation of structural and conceptual organization in memory (Rumelhart, McClelland, & the PDP Research Group, 1986).

Our data do not provide any adequate criteria for distinguishing among various models of memory. Instead, they indicate the need for a representation that can accommodate schema-like as well as fragmented learning and retrieval, gradual transitions towards increasing integration, partial models, and occasional conflicts. Precompiled script models do not easily fit these constraints.

It is also important to note that our results deal with a restricted set of data. Despite the fact that subjects are learning a procedure, performance does not appear to have the automaticity normally associated with proceduralized knowledge. Proceduralization of knowledge often implies a change in the type of representation (see Anderson, 1982); a mental model may be used to develop the appropriate procedures, but it can be argued that the model is no longer needed once the procedure is fully automated. The classic example of this is the driver who no longer is able to describe how to drive a car with a stick shift once that skill is learned. Integrative modeling, in contrast, is meant to describe the transitions that occur during the phase of learning in which actions are not fully automated. Given estimates of 100 hours for cognitive skill acquisition (Anderson, 1982), that phase certainly covers much of the ground described in recent empirical studies. In addition, our observations of experts indicate that some of their most interesting behavior occurs when they are confronted with novel problems and return to a nonautomated problem-solving mode.

Finally, it should be noted that the component processes of assimilative and fragmented structuring are not meant to refer to different kinds of knowledge but to different strategies that may be used across types of information. In the computer domain, for example, Shneiderman and Mayer (1979) noted an important differentiation between system specific syntactic knowledge, and more general semantic knowledge applying across systems. Our study likewise includes both components, although they exist at multiple levels and are not easily separable. Syntactic aspects of a command can either be retained by reference to a specific instance (fragmented) or by using a model of command formulation from another system (assimilative). Likewise, an understanding of the system structure can be part of an underlying model (assimilative) or knowledge of the specific way in which an individual concept is used (fragmented). For example, in the case of a "buffer," some subjects clearly have a general conception of storage mechanisms and recognize that a buffer preserves a copy of information. Other subjects, we have observed, however, do not have a general model of this sort. Instead, they have learned that deleted text is stored in a buffer and that it is then inserted elsewhere by a 'put' command. When we have probed their knowledge of a buffer, however, their model seems confined to the command context in which they have used it with relatively little evidence of any generalization.

To some extent, integrative modeling goes against our traditional conceptions of understanding which we take to imply a unified, well-organized mental representation. For a *complete* understanding of any device or system, such a model would be required. But people rarely possess such exhaustive knowledge. Integration helps to structure and relate the components, but the mental model continues to be specified by both assimilated and fragmented components. Although our data only addresses early learning, we believe that this description is far more general. People differ greatly with respect to the degree of integration of their knowledge, but even experts continue to possess mental fragments that help drive the acquisition of knowledge.

IMPLICATIONS FOR DESIGN

Our research has been concerned primarily with improved understanding of the role of mental models in acquiring a cognitive skill. These results, however, suggest that certain types of information will be useful in instructional design as well as pointing to a number of limitations in our current knowledge.

Our data suggest that syntactic elaboration can improve the learning of procedures. Individual examples proved useful in learning how to formulate commands, consistent with reports by Reder et al. (1986). In general we would argue for designing instructions in a way which supports reasoning by reminding from specific instances (Schank, 1982; Ross, 1984). Analyses of errors and protocols suggested that, although examples provided fragmented knowledge about the domain, that form of knowledge can be very useful.

At the same time, care must be taken to avoid incorrect inference from examples. In certain cases, subjects in our study adopted irrelevant specifics of individual examples as part of their emerging domain knowledge. Multiple examples should be constructed to distinguish the relevant and irrelevant aspects of examples. Our results indicate that general abstract format statements are not adequate for that purpose.

The results also suggest that conceptual elaboration can be useful under appropriate conditions. Within the context of our studies, providing explicit functional descriptions was generally the most effective form of elaboration. Including model information as part of the instructions can help to provide subjects with an initial framework for understanding how commands work, a type of advance organizer (Ausubel, 1968). Providing a model can also help to encourage subjects to think more fully about the material and to develop a better understanding of the overall system. The long term benefits of these elaborations are not clear from our studies.

The results of Redundant conceptual elaboration indicate that the target user must be taken into account in designing materials. What appear to be repetitious statements may in fact provide useful support to novices.

It is also evident from our data that instructional materials should be designed in a way that encourages subject-elaboration. Design strategies that may achieve that objective include deliberately providing incomplete instructions (Carroll, Mack, Lewis, Grischkowsky, & Robertson, 1985) and informing subjects of errors, but requiring them to discover the correct answer on their own (Lewis & Anderson, 1985). Our observations suggest limits to both of these strategies. If subjects do not feel they have sufficient information to continue searching for a solution, they are often frustrated. Thus, while the instructions should encourage exploration, they should also include a way to find the solution if all else fails. Our suggestion would be to provide an exploratory instructional or tutorial manual and separate, heavily indexed reference materials or equivalent online help.

There are also some general restrictions to the principles we have described. First, in order to separate the role of the information content of the instructions from that of interaction with the system, we used paper-and-pencil tests. Naturally, this imposes restrictions on our conclusions, since we believe that the effectiveness of instructional materials can interact with the use of the system during learning. (Elsewhere, Sebrechts & Deck, 1986, we suggest some reasons why *interactive* learning is not always advantageous.) In addition, the importance of elaboration for manual design is constrained by the length of the resultant manual, the system to be used, and the users' strategies. We briefly describe each of these issues.

Manual length. One of the central constraints addressed here concerns the relationship between the length of instruction and the type of instruction. In descriptions of a "minimal manual," Carroll (1984; Black, Carroll, & McGuigan, 1987; Catrambone & Carroll, 1987) has emphasized the fact that discarding instructional material that is not directly useful to initial performance can speed learning. The difficulty, of course, is in establishing what is useful for whom. For the populations studied, it has been established, for example, that detailed material about disk usage (Carroll et al., 1985) or descriptive material explaining automatic carriage returns (Black et al., 1987) is not helpful. In contrast, our results showed that explanations of function, analogies, and even restatement of key ideas can improve initial performance. We view these results as complementary. The material in initial training manuals should be reduced, and our manuals were in fact much shorter than standard commercial manuals. Our results, however, focus on the kinds of materials that it may be useful to keep in the process of reducing the manuals. In brief, it appears that syntactic and conceptual elaboration may be useful within the context of a shortened manual. However, it is critical to try to assess the generality of these claims. Someone who has never used a typewriter may need some information about carriage returns, and a library analogy will only work for those who understand how a library is organized.

These qualifications should appear obvious in retrospect, but instructional guidelines often avoid specifying their context of application. In making any specific claims about how much information to provide during instruction, we need to address the problem of generality. Information that cannot be integrated will not be useful. Of course, in the computer context, we want the computer itself to help constrain, if not eliminate, many of the current instructional problems.

System type. The utility of particular kinds of instructional material may also depend on the type of system used. Several studies that suggest reducing elaboration have used menu-based systems (Carroll, 1984; Carroll et al., 1985; Black et al., 1987) or have used constrained training systems (Carroll & Carrithers, 1984; Catrambone & Carroll, 1987). In our case we examined a system, UNIX, that was not menu-based. This shifts a substantial portion of the task to one of recall rather than of recognition. It is likely that utility of elaboration interacts with system type. If the system allows for easy recovery or avoids errors, then less instruction is necessary. If, in contrast, the system does not provide those facilities, then the instructional material may provide a means to guide recovery. Unfortunately, it is usually difficult to provide the appropriate recovery information within the context of a standard instructional sequence. Material can either be designed to make such recovery more visually explicit (e.g., Sebrechts et al., 1984) or some form of indexing needs to be added to facilitate that task.

User strategy. Our subjects did not all use the different types of elaboration with equal success, despite the fact that we attempted to control for subject experience. In part this can be accounted for by the fact that we do not have precise tools for equating relevant experience. In addition, however, there are differences in the strategies that people favor in learning to use a device. Some people prefer a more exploratory mode, whereas others focus on presented material (Deck & Sebrechts, 1984). In terms of instructional manual design, it is probably necessary to have alternative materials that the user can select. Ideally, the user would be able to select, interactively, the information they preferred.

Integrative Modeling and Cooperative Learning

In this chapter we have argued that there are different types of explicit models, and that the utility of such models depends on their ability to facilitate integrative modeling. Explicit instructional information is useful insofar as it can provide an integrative learning bridge.

Interestingly, the specification of such bridge constructs constitutes the informational core of cooperative learning. And, from this perspective, it is the cooperative metaphor that should direct our construction of manuals. A cooperative context can be informationally defined as one in which the information provided is matched to the current mental representation of the user as needed. In the case of a manual, we try to isolate the most salient properties of the dynamic cooperative context and apply those in what is constrained to be a static context. Of course, given these constraints, the traditional manual cannot respond directly to different requirements of individuals. Part of the promise of new computer systems is that they will assume some of the dynamic properties of cooperation and thus enable instruction to be tailored on an individual basis.

Integrative modeling emphasizes that, in either scenario, we need to focus on the rules for optimal mapping between information presentation (explicit models) and the learner's representation (mental models). There are at least five psychologically important tasks in describing such rules. First, we need a theory of skill learning to characterize what constitutes relevant information and to define when and how that information should be made available. Second, we need a taxonomy of how such information is mapped onto different types of tasks. Third, individual differences in learning strategy will need to be described more completely to help determine the generality of instructional design principles. Fourth, a better description is needed of the techniques people use to acquire the information they need (Alty & Coombs, 1981; Aaronson & Carroll, 1986; Robertson & Swartz, 1987). Fifth, we need to analyze how informational utility interacts with the specific characteristics of a device and the manner in which the information is provided (manuals or online systems). In this chapter, we have provided some preliminary concepts concerning the first two of these issues.

REFERENCES

Aaronson, A., & Carroll, J.M. (1986). *The answer is in the question: A protocol study of intelligent help*. Yorktown Heights, NY: IBM Thomas J. Watson Research Center Research Report.

Ahn, W., Mooney, R.J., Brewer, W.F., & DeJong, G.F. (1987). Schema acquisition from one example: Psychological evidence for explanation-based learning. *Ninth Annual Conference of the Cognitive Science Society* (pp. 50–57). Hillsdale, NJ: Erlbaum.

Alba, J.W., & Hasher, L. (1983). Is memory schematic? *Psychological Bulletin 93*(2), 203–231.

Alty, J.L., & Coombs, M.J. (1981). Communicating with university computer users: A case study. In M.J. Coombs & J.L. Alty (Eds.), *Computing skills and the user interface* (pp. 7–71). London: Academic Press.

Anderson, J.R. (1982). Acquisition of cognitive skill. *Psychological Review, 89*, 369–406.

Anderson, J.R., & Reder, L.M. (1979). An elaborative processing explanation of depth of processing. In L.S. Cermak & F.I.M. Craik (Eds.), *Levels of processing in human memory*. Hillsdale, NJ: Earlbaum.

Ausubel, D.C. (1968). *Educational psychology: A cognitive view*. New York: Holt, Rinehart, & Winston.

Barlett, F.C. (1932). *Remembering: A study in experimental and social psychology*. Cambridge, England: Cambridge University Press.

Bott, R.A. (1979). *A study of complex learning: Theory and methodologies*. (Center for Human Information Processing Technical Report No. 82). San Diego: University of California, San Diego.

Black, J.B., Carroll, J.M., & McGuigan, S.M. (1987). What kind of minimal instruction manual is most effective? In *Proceeding of CHI + GI* (pp. 159–162). New York: The Association for Computing Machinery, Inc.

Bower, G.H., Black, J.B., & Turner, T.J. (1979). Scripts in memory for text. *Cognitive Psychology, 11*, 177–220.

Bower, G.H., Clark, M.C., Lesgold, A.M., & Winzenz, D. (1969). Hierarchical retrieval schemes in recall of categorical word lists. *Journal of Verbal Learning and Verbal Behavior, 8*, 323–343.

Bransford, J.D., & Johnson, M.K. (1972). Contextual prerequisites for understanding: Some investigations of comprehension and recall. *Journal of Verbal Learning and Verbal Behavior, 11*, 717–721.

Card, S.K., Moran, T.P., & Newell, A. (1980). Computer text-editing: An information processing analysis of a routine cognitive skill. *Cognitive Psychology, 12*, 32–74.

Carroll, J.M. (1984). Minimalistic training. *Datamation, 30*, 125–136.

Carroll, J.M., & Carrithers, C. (1984). Training wheels in a user interface. *Communications of the ACM, 27*(8), 800–806.

Carroll, J.M., Mack, R.L., Lewis, C.H., Grischkowsky, N.L., & Robertson, S.P. (1985). Exploring a word processor. *Human-Computer Interaction, 1*, 283–307.

Catrambone, R., & Carroll, J.M. (1978). Learning a word processing system with training wheels and guided exploration. In *Proceedings of CHI + GI* (pp. 169–174). New York: The Association for Computing Machinery, Inc.

Cermak, L.S., & Craik, F.I.M. (1979). *Levels of processing in human memory*. Hillsdale, NJ: Erlbaum.

Chase, W.G., & Simon, H.A. (1973). Perception in chess. *Cognitive Psychology, 4*, 55–81.

Chi, M.T.H., Feltovich, P.J., & Glaser, R. (1981). Categorization and representation of physics problems by experts and novices. *Cognitive Science, 5*, 121–152.

Chiesi, H.L., Spilich, G.J., & Voss, J.F. (1979). Acquisition of domain-related information in relation to high and low domain knowledge. *Journal of Verbal Learning and Verbal Behavior, 18*, 257–273.

Collins, A.M., Warnock, E.H., Aiello, N., & Miller, M.L. (1975). Reasoning from incomplete information. In D.B. Bobrow & A. Collins (Eds.), *Representation and understanding: Studies in cognitive science*. Orlando, FL: Academic Press.

Craik, F.I.M., & Lockhart, R.S. (1972). Levels of processing: A framework for memory research. *Journal of Verbal Learning and Verbal Behavior, 11*, 671–684.

Deck, J.G., & Sebrechts, M.M. (1984). Variations on active learning. *Behavior Research Methods, Instruments, and Computers, 16*, 238–241.

diSessa, A.A. (1983). Phenomenology and the evolution of intuition. In D. Gentner & A.L. Stevens (Eds.), *Mental Models*. Hillsdale, NJ: Earlbaum.

diSessa, A.A. (1986). Models of computation. In D.A. Norman & S.W. Draper (Eds.), *User centered system design: New perspectives on human-computer interaction*. Hillsdale, NJ: Erlbaum.

Ericsson, K.A., & Simon, H.A. (1980). Verbal reports as data. *Psychological Review, 87*(3), 215–251.

Fitts, P.M. (1964). Perceptual-motor skill learning. In A.W. Melton (Ed.), *Categories of human learning*. New York: Academic Press.

Fodor, J.A. (1981). *Representations*. Cambridge, MA: MIT Press.

Garner, W.R. (1962). *Uncertainty and structure as psychological concepts*. New York: Wiley and Sons.

Gentner, D., & Gentner, D.R. (1983). Flowing waters or teeming crowds: Mental models of electricity. In D. Gentner & A.L. Stevens (Eds.), *Mental models*. Hillsdale, NJ: Erlbaum.

Gentner, D., & Stevens, A.L. (1983). *Mental models*. Hillsdale, NJ: Erlbaum.

Goethe, J.W.V. (1970). *Theory of colours* (C.L. Eastlake, Trans.). Cambridge, MA: MIT Press. (Original work published 1810.)

Greeno, J.G. (1983). Conceptual entities. In D. Gentner & A.L. Stevens (Eds.), *Mental models*. Hillsdale, NJ: Erlbaum.

Halasz, F.G., & Moran, T.P. (1982). Analogy considered harmful. *Proceedings of conference on human factors in computer systems* (pp. 383–386). Gaithersburg, MD: National Bureau of Standards.

Halasz, F.G., & Moran, T.P. (1983). Mental models and problem solving in using a calculator. In A. Janda (Ed.), *Proceedings of the CHI '83 conference on human factors in computing systems* (pp. 212–216). New York: ACM.

Hanson, S.J., Kraut, R.E., & Farber, J.M. (1984). Interface design and multivariate analysis of UNIX use. *ACM Transactions on Office Information Systems, 2*(1), 42–57.

Head, H. (1920), *Studies in neurology*. New York: Oxford University Press.

Hutchins, E. (1983). Understanding micronesian navigation. In D. Gentner & A.L. Stevens (Eds.), *Mental models*. Hillsdale, NJ: Erlbaum.

Hutchins, E. (1987, September). *Metaphors for interface design*. Paper presented at NATO-sponsored workshop in Multimodal Dialogues Including Voice, Venaco, Corsica, France.

James, W. (1890). *The principles of psychology*. New York: Holt.

Johnson-Laird, P.N. (1983). *Mental models*. Cambridge, MA: Harvard University Press.

Kieras, D.E., & Bovair, S. (1984). The role of a mental model in learning to operate a device. *Cognitive Science, 8*, 255–273.

Kintsch, W., & Mannes, S.M. (1987). Generating scripts from memory. *Ninth Annual Conference of the Cognitive Science Society*. Hillsdale, NJ: Erlbaum.

Landauer, T.K. (1987). Relations between cognitive psychology and computer system design. In J.M. Carroll (Ed.), *Interfacing thought: Cognitive aspects of human-computer interaction*. Cambridge, MA: MIT Press.

Larkin, J.H. (1983). The role of problem representation in physics. In D. Gentner & A.L. Stevens (Eds.), *Mental models*. Hillsdale, NJ: Erlbaum.

Lewis, M.W., & Anderson, J.R. (1985). Discrimination of operator schemata in problem solving: Learning from examples. *Cognitive Psychology, 17*, 26–65.

Lewis, C., & Mack, R. (1982). Learning to use a text-processing system: Evidence from "thinking aloud" protocols. *Proceedings of conference on human factors in computer systems* (pp. 387–392). Gaithersburg, MD: National Bureau of Standards.

Mayer, R.E. (1975a). Different problem-solving competencies established in learning computer programming with and without meaningful models. *Journal of Educational Psychology, 67*, 725–734.

Mayer, R.E. (1975b). Information processing variables in learning to solve problems. *Review of Educational Research, 45*, 525–541.

Mayer, R.E. (1981). The psychology of how novices learn computer programming. *ACM Computing Survey, 13*, 121–141.

Murdock, B.B., Jr. (1960). The distinctiveness of stimuli. *Psychological Review, 67*, 16–31.

Newell, A. (1981). Physical symbol systems. In D.A. Norman (Ed.), *Perspectives on cognitive science*. Norwood, NJ: Ablex Publishing Corp.

Newell, A., & Simon, H.A. (1972). *Human problem solving*. Englewood Cliffs, NJ: Prentice-Hall.

Norman, D.A. (1983). Some observations on mental models. In D. Gentner & A.L. Stevens (Eds.), *Mental models*. Hillsdale, NJ: Erlbaum.

Norman, D.A., Rumelhart, D.E., & the LNR Research Group (1975). *Explorations in cognition*. San Francisco: W.H. Freeman & Co.

Owens, J., Bower, G.H., & Black, J.B. (1979). The "soap opera" effect in story recall. *Memory & Cognition, 7*, 185–191.

Pepper, J. (1981). Following students' suggestions for rewriting computer programming textbooks. *American Education Research Journal, 18*, 259–269.

Piaget, J. (1954). *The construction of reality in the child*. New York: Basic Books.

Poincare, H. (1913). The value of science. In *The foundations of science* (G.G. Halsted, trans.). New York: Science Press.

Reder, L.M. (1979). The role of elaborations in memory for prose. *Cognitive Psychology, 11*, 213–234.

Reder, L.M., & Anderson, J.R. (1980). A comparison of texts and their summaries: Memorial consequences. *Journal of Verbal Learning and Verbal Behavior, 19*, 121–134.

Reder, L.M., & Anderson, J.R. (1982). Effects of spacing and embellishment on memory for the main point of the text. *Memory and Cognition, 10*, 97–102.

Reder, L.M., Charney, D.H., & Morgan, K.I. (1986). The role of elaborations in learning a skill from an instructional text. *Memory and Cognition, 14*(1), 64–78.

Reed, S.K., Ernst, G., & Banerji, R. (1974). The role of analogy in transfer between similar problem states. *Cognitive Psychology, 6*, 436–450.

Robertson, S.P., & Swartz, M. (1987). Question asking during procedural learning: Strategies for acquiring knowledge in several domains. *Ninth Annual Conference of the Cognitive Science Society* (pp. 325–335). Hillsdale, NJ: Erlbaum.

Ross, B. (1984). Remindings and their effects in learning a cognitive skill. *Cognitive Psychology, 16*, 371–416.

Rubinstein, R., & Hersch, H. (1984). *The human factor: Designing computer systems for people*. Burlington, MA: Digital Press.

Rumelhart, D.E., McClelland, J.L., & the PDP Research Group. (1986). *Parallel distributed processing*. Cambridge, MA: MIT Press.

Rumelhart, D.E., & Norman, D.A. (1978). Accretion, tuning and restructuring: Three modes of learning. In J.W. Cotton & R. Klatzky (Eds.), *Semantic factors in cognition*. Hillsdale, NJ: Erlbaum.

Rumelhart, D.E., & Norman, D.A. (1981). Analogical processes in learning. In J.R. Anderson (Ed.), *Cognitive skills and their acquisition* (pp. 335–360). Hillsdale, NJ: Erlbaum.

Schallert, D.L. (1976). Improving memory for prose: The relationship between depth of processing and context. *Journal of Verbal Learning and Verbal Behavior, 15*, 621–632.

Schank, R.C. (1982). *Dynamic memory: A theory of reminding and learning in computers and people*. New York: Cambridge University Press.

Schank, R.C., & Abelson, R.P. (1977). *Scripts, plans, goals and understanding*. Hillsdale, NJ: Erlbaum.

Sebrechts, M.M. (1983). Cognitive guidelines for computer-based instruction: A navigational case study. *Proceedings of the human factors society - 27th annual meeting* (pp. 642–646). Santa Monica, CA: Human Factors Society.

Sebrechts, M.M., & Deck, J.G. (1986). Techniques for acquiring computer procedures: Some restriction on "interaction". In *Proceedings of the human factors society - 30th annual meeting* (pp. 275–279). Santa Monica, CA: Human Factors Society.

Sebrechts, M.M., Deck, J.G., & Black, J.B. (1983). A diagrammatic approach to computer instruction for the naive user. *Behavior Research Methods and Instrumentation, 15*(2), 200–207.

Sebrechts, M.M., Deck, J.G., Wagner, R.K., & Black, J.B. (1984). How human abilities affect component skills in word processing. *Behavior Research Methods, Instruments and Computers, 16*, 234–237.

Sebrechts, M.M., & Dumont, M.D. (1986). Mental models in computer skill acquisition: An empirical study. *Proceedings of the IEEE International Conference on Systems, Man and Cybernetics* (pp. 750–755). New York: The Institute of Electrical and Electronics Engineers.

Sebrechts, M.M., Furstenberg, C.T., & Shelton, R.M. (1986). Remembering computer command names: Effects of subject generation versus experimenter imposition. *Behavior Research Methods, Instruments and Computers, 18*(2), 129–134. (Also Wesleyan University, Cognitive Science Program, Research Report MMS-85-112.)

Sebrechts, M.M., Galambos, J.A., Black, J.B., Deck, J.G., Wikler, E., & Wagner, R.K. (1984). *The effects of diagrams on learning to use a system* (Tech. Rep. 2). New Haven, CT: Yale Psychology Department, Learning and Using Systems Project.

Sebrechts, M.M., Marsh, R.L., & Furstenberg, C.T. (1987). *Effects of instructional elaboration on mental models while learning a cognitive skill* (Research Report MMS-87-138). Middletown, CT: Wesleyan University, Cognitive Science Program.

Shneiderman, B. (1982). The future of interactive systems and the emergence of direct manipulation. *Behavior and Information Technology, 1*, 237–256.

Shneiderman, B. (1983). Direct manipulation: A step beyond programming languages. *IEEE Computer, 16*(8), 57–69.

Shneiderman, B., & Mayer, R.E. (1979). Syntactic/semantic interactions in programmer behavior: A model and experimental results. *International Journal of Computer and Information Sciences, 8*, 219–238.

Simon, H.A. (1981). Cognitive science: The newest science. In D.A. Norman (Ed.), *Perspectives on cognitive science*. Norwood, NJ: Ablex Publishing Corporation.

Smith, E.E., & Goodman, L. (1984). Understanding written instructions: The role of an explanatory schema. *Cognition and Instruction, 1*, 359–396.

Soloway, E. (1986). Baseball: An example of knowledge-directed machine learning. In G.H. Bower (Ed.), *The psychology of learning and motivation*. New York: Academic Press.

Sowa, J.F. (1984). *Conceptual structures: information processing in mind and machine*. Reading, MA: Addison-Wesley.

Sternberg, R.J. (1977). *Intelligence, information processing, and analogical reasoning: The componential analysis of human abilities*. Hillsdale, NJ: Erlbaum.

Sulin, R.A., & Dooling, D.J. (1974). Intrusion of a thematic idea in retention of prose. *Journal of Experimental Psychology, 103*, 255–262.

Wagner, R.K., Sebrechts, M.M., & Black, J.B. (1985). Tracing the evolution of knowledge structures. *Behavior Research Methods, Instruments, and Computers, 17*(2), 275–278.

Weinstein, C.E. (1978). Elaboration skills as a learning strategy. In H.F. O'Neil, Jr. (Ed.), *Learning strategies*. New York: Academic Press.

Young, R.M. (1983). Surrogates and mappings: Two kinds of conceptual models for interactive devices. In D. Gentner & A. L. Stevens (Eds.), *Mental models*. Hillsdale, NJ: Erlbaum.

APPENDIX A
CLASSIFICATION OF PROCEDURAL ERRORS

I. *Model Misconceptions*
   A. File/directory confusion
   B. File/filename confusion
   C. File/program (command) confusion
   D. Pathname/address confusion
   E. Address confusion
   F. Command/text confusion (in VI)
   G. Misconstrued hierarchy
   H. Confusion of root/home/working directories
   I. Root directory confusion (=/usr)
   J. Home directory confusion (=/usr)
   K. Working directory confusion
   L. Directory organization confusions:
      1. Directory contains everything nested beneath it
      2. Does not understand a directory's contents
      3. Does not realize file with same name already exists
   M. Default confusions:
      1. Does not understand concept (specifies arguments)
      2. Default incorrect
      3. Does not specify what default is
      4. Specifies "screen" for default output
   N. Pathname/command confusions:
      1. Thinks specifying pathname moves you to the location
      2. Thinks specifying multiname pathname (either complete or abbreviated) moves you to the location specified even if it follows a command
      3. Thinks any name enclosed in '/'s moves you to the location specified
   O. Hypothetical entities (assumes system knows what command or argument to use)
   P. Thinks must specify the pathname before command
   Q. Thinks 2 subdirectories with the same name can exist in the parent directory
   R. Scope operator/command confusion

II. *Command Misconceptions*
   A. cd:
      1. Must specify the directory from which to change
      2. Generally does not understand
   B. ls:
      1. Lists contents of files in directory specified
      2. Lists names of directories only

APPENDIX A Continued.

C. mkdir:
   1. Makes a directory in an existing directory
   2. Changes a file to a directory
   3. Prints
   4. Displays

D. rmdir:
   1. Redirects
   2. Remembers
   3. Moves directory to location specified

E. ln:
   1. Connects 2 existing files (or directories)
   2. Creates 2 separate files linked together
   3. Combines
   4. Links a file to a directory
   5. Copies
   6. Puts a file in a directory
   7. Links 2 files under same filename
   8. Creates a new file that contains 2 existing files
   9. Remembers
   10. Generally does not understand

F. mv:
   1. Moves the user to existing directory
   2. Combines
   3. Copies (original file or directory still exists)
   4. Generally does not understand

G. rm: Generally does not understand

H. cp:
   1. Combines
   2. Moves (transfers)
   3. Concatenates
   4. Copies but have to issue separate command to name copy
   5. Name copy
   6. Copies to a file in a directory
   7. Changes a file's name

I. cat:
   1. Consolidates
   2. Categorizes
   3. Puts (moves)
   4. Sends to existing file
   5. Renames
   6. Generally does not understand

J. ⟨,⟩:
   1. ⟨ must precede ⟩
   2. Can only use one or the other in command sequence

APPENDIX A Continued.

3. 〉 sends file to a directory
4. Used as a command
5. Used with incorrect command (e.g. mv, ln, etc.)
6. Sends a command to a file
7. 〉 sends to a new file on the screen
8. Generally does not understand

K. tee:
1. Sends to file only
2. Sends file to a directory
3. Used tee alone (without other command or pipe)
4. Must specify 2 destinations
5. Destination must already exist
6. Displays new file on screen (displays on screen under new file name)
7. Displays on screen only
8. Does not think is a command
9. Generally does not understand

L. pipe:
1. Connects 2 separate commands with different arguments
2. Separates arguments
3. Connects 2 files or directories
4. Connects command to a file (with single command)
5. Does not understand when to use
6. Puts file in a directory
7. Uses with inappropriate commands (e.g. cd)
8. Generally does not understand

M. pipeline/tee: Believes they cannot be used together

N. VI–i: replaces (replaces & inserts)

O. VI–c:
1. Changes the position of two words
2. Adds text on the line above the text
3. Erases
4. Does not realize 'c' puts text in the delete buffer

P. VI–d:
1. Erases
2. Changes text display, not the text itself
3. 'dd' moves line

Q. VI–x: Erases

R. VI–p:
1. Replaces
2. Specifies text which is to be changed
3. Moves text
4. Generally does not understand

S. VI–〈esc〉: "turns off" command

APPENDIX A Continued.

III. *Command Confusions*

A. cd: rm
ls
mkdir
cp
mv

B. ls: cat
mv
sort
cd
format

C. mkdir: cd
d
ls
⟩
cp

D. rmdir: rm
d
⟩
mkdir
dd

E. rm: dd
d
cd
ln
mv
⟨esc⟩
rmdir

F. mv: cat
ln
cp
⟩
mkdir
pipe
tee
⟨
rm
cd
c

G. ln: cp
mv
⟨
ls
cat
c
tee
⟩
mkdir

H. cp: mv
ln
ln,rm
c
cat
⟩
tee
mv,ln

I. cat: mv
ls
format
ln
tee
⟩
⟨
sort
d
cd
ln,mv
cp
cp,ln
p
who

J. tee: ⟨
⟩
⟨,⟩
cat
mv
cp
dis,cp
rm
ls
ln
pipe

K. ⟨,⟩: ln
⟨
⟩
tee
cat
mv

APPENDIX A Continued.

pipe
L. pipe: ⟩
⟨
/
M. VI–i: p
d
r
c
⟨
format
N. VI–dd: cd
O. VI–d: c
x
r
P. VI–x: r
Q. VI–c: i
d
R. VI–r: x
d
c
i
S. VI–p: i
mv
d
r
mv
T. /: pipe
⟩

IV. *No Command Given*

V. *Mispellings*
A. rmdir: rm
rd
rdir
rem
rmd
redir
re
B. rm: r
rem
C. mkdir: m
md
mk
ma
mdir
D. cd: ch
cdir
chdir
E. ls: ld
lsdir
F. ln: lk
G. cp: c
cop
H. mv: m

VI. *Intrusions* (from other systems)
A. mkdir: cr
B. cat: print
ty
dir
C. rmdir: del
delete
D. rm: del
delete
E. p: undel
F. mv: dir
G. tee: dir

VII. *Fabrications*
A. cd: dp
cn
wd
wdir
dis
mk
print
path
B. ls: a
ds
ldir
ty
display
C. mkdir: nd

APPENDIX A Continued.

add
a
ad
adir
nw
ndir
nf
sdir
make

D. rmdir: deldir
rv
er

E. rm: deldir
def
dc
dl
df
er
break
rmln
rl
svln
rdir
unlink
un
md
mk

F. mv: tr
mF
com
send
cb
cdir

G. ln: rn
ref
rl
mk

H. cat: view
com, combine
dis
combine & save
send, sd
ds
cdir

I. VI–dd: dl
bb

J. Indeterminates:
mk
er
erase
command
rename
show

VIII. *Pathname Errors*

A. Abbreviated when complete required
B. Single abbreviated when multiname required:
  1. Specifies final name only
  2. Specifies next to last name only
C. Multiname abbreviated when single required (working dir/dir . . .)
D. Working directory as single abbreviated
E. '/' precedes abbreviated name
F. No '/' before complete name
G. Initial '/' means 'change directory'
H. Complete name as '/name/'
I. Abbreviated name as '/name/'
J. Specifies each name in multiname separately with each enclosed in '/'s.
K. No '/' between names in multiname abbreviated
L. Interprets multiple sources as pathname
M. Believes '/' separates arguments
N. Does not understand what initial '/' means
O. Does not understand multiname abbreviated (what '/' means)

APPENDIX A Continued.

- P. Extra '/' at end of pathname
- Q. Incorrect sequence of names
  1. Order of names reversed
  2. Omitted a name
  3. Extra name at end
  4. Extra filename before final name
  5. Specifies wrong name

IX. *Argument/Format Errors*

- A. Format Errors:
  1. Command/argument order:
     - a) argument command argument
     - b) argument command
     - c) argument argument command
  2. Argument order
- B. Argument Specification Errors:
  1. Incorrect argument type:
     - a) Dirname when filename required
     - b) Filename when dirname required
     - c) Filename required when either acceptable
  2. Incorrect number of arguments:
     - a) Missing argument
       - (1) Source
       - (2) Destination
     - b) Extra argument
       - (1) Source
       - (2) Destination
     - c) Must be a command with each argument
     - d) Cannot specify multiple sources
  3. Source does not exist
  4. Uses [ ] around arguments
  5. Specifies output alone on separate command line

X. *Command Sequencing Errors*

- A. Missing commands:
  - tee
  - pipe
  - cat
  - ls
  - sort
  - format
  - dd
  - >
  - count
- B. Extra Commands:
  - tee
  - cat
  - ls
  - ln
  - rename
  - cp
  - mkdir
  - format
  - mk
  - cd
  - >
  - sort
  - who

APPENDIX A Continued.

- C. Input file specification errors:
  1. Incorrect position of input file (usually at end)
  2. Specifies input for each command in sequence
- D. Uses series of separate commands instead of piped sequence but does not specify an intermediate file
- E. Uses series of separate commands when a single command is appropriate; does not specify an intermediate file
- F. Thinks each command must be issued separately
- G. Thinks each command operates on original source file
- H. Incorrect order of commands (multiple commands required)

XI. *VI Cursor Position Errors*
- A. Character to left
- B. Sequential characters
- C. Must be at beginning of word
- D. Did not specify

XII. *VI Mode Errors*
- A. Exec/VI
  1. Use VI command in exec mode & vice versa
  2. Use ⟨esc⟩ after exec command
- B. Command/insert
  1. No ⟨esc⟩ when required
  2. after i
  3. after c
- C. Use ⟨esc⟩ in commmand mode
  1. after r (or c)
  2. after x
  3. after d
  4. after p
- D. Interprets command as text to be inserted (p) or replaced (c)
- E. Interprets text as command
- F. No concept of insert mode (to insert text must insert spaces and type in text)
- G. Assumes normally in insert mode
- H. Assumes normally in replace mode

XIII. *VI Scope Errors*
- A. Incorrect Scope:
  1. d as sentence
  2. d as line on screen
  3. d as from cursor to end of sentence
  4. d as from cursor to end of line
  5. d as word
  6. d as next 2 words
  7. d as character
  8. d as most recently inserted text
  9. d as everything except last word (word cursor is on)

APPENDIX A Continued.

10. d as character to right of cursor to end of word
11. w as character to left of cursor
12. w as entire word
13. w as character
14. w as from character to right of cursor to end of word
15. w affects spaces both before and after word
16. w as line on screen
17. w as to end of file
18. w as word excluding final space
19. w as character cursor is on to end of sentence

B. Extra d

C. Conceptual confusions:
1. Operator not specified (gives text to be affected)
2. Gives operator and text to be affected
3. Ignores operator
4. Operator as part of text
5. No operator specified
6. Operator specified when not appropriate (after p)

D. No concept of delete buffer

XIV. *VI Miscellaneous Errors*

A. Text did not wrap around (word wrap)
B. Extra space(s) inserted on screen
C. Text to be inserted not specified
D. Text incorrect
E. Inserted to right of cursor
F. Text put to left of cursor
G. Text put to left of the word cursor is on
H. Text put to right of the word cursor is on
I. Text on subsequent lines indented
J. Screen not specified
K. Extra space inserted
L. Text inserted in incorrect place
M. Spaces inserted instead of text
N. ⟨esc⟩ precedes command

XV. *General Miscellaneous Errors*

A. Misunderstood question
B. Misunderstood where s/he was in the hierarchy
C. Thought we meant 'c' by replace
D. No answer given
E. No explanation given
F. Vague explanation
G. "Don't understand"
H. "Nothing would happen"
I. "Is no command"

APPENDIX A Continued.

J. "Format incorrect"
K. Refuses to extend
L. Does not understand '*' wildcard
M. Directory contents not specified
N. Directory contents incorrect (missing files or directories)
O. Error message would be given
P. "Why an explanation?"
Q. Special Constraints on 'rmdir' command
   1. Tries to remove working directory
   2. Tries to remove higher-level directory
   3. Tries to remove non-empty directory

APPENDIX B
CLASSIFICATION OF PROTOCOLS IN EXPERIMENT 1

	Total Number of Occurences	
	Initial	Final
I. *General Confusions*		
A. Commands must be 2 letters.	2	2
B. Command format confusion: misunderstands what a command is; that most commands require input.	3	2
C. Command sequencing confusion.	1	1
D. EXEC/ VI confusion.	2	9
E. Command/argument order confusion.	2	2
F. Confusion w/ optional arguments.	1	—
II. *Specific Confusions*		
A. File/directory confusion.	14	1
B. Working directory/workspace confusion.	1	—
C. Root/home/working directory confusion.	4	—
D. File reference confusion (file/filename).	6	—
E. Pathname confusion.		
1. Complete.	6	3
2. Abbreviated.	2	1
3. Multiname abbreviated.	4	1
F. General '/' confusion.	7	3
G. Model confusion: why does the system do it this way?	1	—
H. Address confusion.	7	6

Appendix B Continued

	Total Number of Occurences	
	Initial	Final
I. Default confusion.	2	—
J. VI Mode confusion: INSERT/COMMAND.	5	—
K. Cursor position confusion.	3	3
L. Scope operator confusion.	6	3
M. Word wrap confusion.	1	—
N. Buffer confusion.	3	—
O. Command function confusion.		
1. Confused w/ another command.	7	5
2. Confused w/ command from another system.	7	3
3. Confused by English meaning of command name.	15	7
4. Confused about when to use command.	9	3
III. *Subjects' Own Models*		
A. Location.	21	7
B. Container	3	1
C. Misconstrued hierarchy.	7	3
D. Specify location before issuing command.	4	—
E. Ownership: root for everyone, home just for user.	1	—
F. Hierarchical model of addresses, filenames, dirnames: address nested beneath the filename which is nested beneath the dirname.	1	—
G. N-ary tree.	—	—
H. Information flow.	1	—
I. Default.	2	3
J. Black box.	2	2
K. Parallel processor vs. serial processor model: executes commands in command sequence simultaneously.	1	—
L. Personification of computer.	2	—
M. Tee as a *function*: different from other commands.	—	—
N. ⟩,⟨ as directory location indicators.	—	2
O. Redirection operators as arrows, pointers.	4	1
P. Buffer.	—	1
Q. Word wrap breaks between words (as in WPSIM).	1	—
R. In VI, space character as delimiter; separates command from text.	1	—

Appendix B Continued

	Total Number of Occurences	
	Initial	Final
S. DEL/INSERT for change.	1	—
IV. *Subjects' Own Analogies*		
A. CPU as a brain.	1	—
B. Tree.	—	1
V. *Subjects' Use of Our Models*		
A. File Organization	—	—
B. Manipulating Files	2	2
C. Command Execution	1	1
D. Visual Editor (VI)	—	—
VI. *Subjects' Use of Our Analogies*		
A. File Organization	4	—
B. Manipulating Files	—	—
C. Command Execution	—	—
D. Visual Editor (VI)	1	—
VII. *Difficulties with Analogy*		
A. File Organization	—	—
B. Manipulating Files	4	—
C. Command Execution	2	—
D. Visual Editor (VI)	2	—
VIII. *Subjects' Learning & Problem Solving Strategies*		
A. Learning.		
1. Summarizes.	21	—
2. Relates to other systems.	6	—
3. Synthesizes: relates to other commands on the system.	14	1
4. Repeats commands after study.	14	—
5. Looks through previous material.	7	—
6. Tries to figure out when you would use the command.	5	—
7. Relates to prior knowledge.	5	—
8. Underlines during study.	1	—
9. Generalizes: tries to formulate general rule to help remember command names.	—	—
10. Extends: makes hypotheses about the function of the command in some hypothetical state.	7	—
B. Problem solving.		
1. Eliminates other options.	5	1
2. Free associates.	1	1

Appendix B Continued

3. Tries to figure out based on other systems.	5	—
4. Recalls study materials verbatim.	3	1
5. Ignores facts that doesn't understand.	4	2
6. Tries to remember what did previously.	—	6
7. Guesses.	1	1
8. Uses sequence of commands to perform function of single command.	—	1
9. Assumes would work because we didn't give any other command that would.	1	—
IX. *Explicit Statements of Learning from Questions*		
A. Positive.	15	6
B. Negative.	3	7
X. *Benefit of Abstract Examples*		
A. Abstract example provides added information.	2	—
B. Concrete example confusing.	4	—

13
Conclusion: Outlines of a Field of Cooperative Systems

Scott P. Robertson

Department of Psychology
Rutgers University
New Brunswick, NJ

Wayne W. Zachary

CHI Systems Incorporated
Blue Bell, PA

In this book, we have tried to present an interdisciplinary set of chapters, all focusing on cooperative systems. We would like to come away from this exercise with a better theoretical understanding of cooperation and with suggestions for creating cooperative human–machine systems. In this conclusion, we first examine some of the common themes and issues that were addressed in the individual chapters, and then attempt to integrate them into a more coherent cognitive, computational model of cooperation. As a final point, we suggest a research agenda that arises from this model in specific and from the chapters in this book in general.

COMMON THEMES AND CONCERNS

The section introductions and afterwords have already identified some of the issues within each Section of the book, and discussed a few of these at length. In retrospect, we can see six major concerns which arose throughout the book. These are:

- an emphasis on the role of goals and plans in cooperative cognitive processes;

- the need to understand and develop expectations of other agents in a cooperative process;
- the pragmatic effects of time and temporal organization on cooperative processes;
- the changes that occur in knowledge states through cooperative interaction, and the relationship of these changes to the predictability of the agents' behavior in current and future situations;
- the inherent capabilities and limitations of agents and the effect of these limitations on cooperative processes, particularly in situations where humans are interacting with machines; and
- the need for special methodologies and tools for studying cooperative interaction and for designing systems that exhibit cooperation.

Some summary thoughts on each of these themes are given below.

GOALS AND PLANS

Authors in all of the chapters discussed the central role of goals and plans in organizing cooperative behavior. Cooperation by definition implies a goal, and goal-directed behavior implies planning and problem solving directed at finding or generating appropriate subgoals. Some chapters, like Graesser and Murray's, were concerned with the content and structure of intentional mechanisms that guide an agent's cooperative behavior. A related issue is the content and structure of a network of interacting agents, like those discussed by Malone, or Durfee et al. Other chapters, like Goodson and Schmidt's Kay and Black's, and Sebrecht's and Marsh's, stressed planning mechanisms for processing goals (usually a goal decomposition mechanism).

Two important and related issues in all of the chapters are the degree to which goals are shared and the degree to which distributed planning agents are autonomous. Many authors pointed out that knowledge about the rules of interaction and the structure of a problem-solving network must be shared by the agents. Durfee et al. find that autonomous nodes in a distributed system can work most effectively when they have knowledge of the abilities of other agents and about the communication paths in the network. Goodson and Schmidt describe "normative rules" that are communicated by two processes and that govern their interaction. At the extreme, Gibbs and Mueller suggest that knowledge about communicative conventions can override interpretation of the apparent intentions of other agents.

Both Malone and Durfee et al. point out that self interest can lead to cooperative behavior when mutual goals are involved. Malone shows that the market system in organizational behavior is based entirely on the assumption of mutual self interest, and that this type of organization leads to considerable autonomy of agents and maximal flexibility in organizational structure. Again, participants in a market system must understand the motives of other participants and be able to predict, or at least interpret, the behavior of others.

Knowledge about problem-solving strategies and communication protocols is always *shared* knowledge because it must be used the same way by all agents. Changes in the conventions of communication or strategies for problem solving must be propagated to all agents if they are to be successfully utilized in a shared effort. This type of shared knowledge is necessarily primary to any knowledge that is private to an agent because private knowledge is presumably the result of subprocesses that were initiated by the overall problem-solving mechanism.

UNDERSTANDING OTHER PARTICIPANTS

Different authors assume different degrees of understanding by each participant of the other participants' goals and knowledge states in cooperative situations. Also, the authors differ on the manner in which other participants' knowledge is modeled. Mack, Sebrechts, and Marsh, and Suchman all propose ways that one participant's knowledge state can be accommodated by another participant, but in all of these cases the modeling participant (instructional systems in all cases) comes with a built-in representation of the modeled participant. The stress in these cases is on developing the right models of humans to facilitate cooperative behavior, and the models are not mutable. On the other hand, Durfee et al. and Goodson and Schmidt stress that some participants in an interaction may need to know specific things about other participants at certain times, and therefore they focus on communication strategies for dynamically updating partial models.

Many proponents of intelligent tutoring systems suggest that one agent (the teacher) needs a full understanding of the knowledge states of the other (the learner) in order to diagnose problems and work cooperatively toward a solution. Newman explicitly differs with this position, suggesting that learners need to wind up with the same assumptions as teachers, but that teachers need not understand students' misconceptions. By structuring activities so that mismatches between the way that the teacher thinks and the way that students think are apparent, students are guided toward the target ways of thinking.

Suchman's observations, however, highlight how important it is that assumptions of the teacher about a system are expressed in the way that learners will eventually use them. Suchman's copier users had an understanding of their task which was different in some respects from the machine's interpretation. The machine made inferences about user goals based on local interpretations of actions. But Suchman suggests that goals were understood by the users in terms of their place in a sequence of planned actions. In some cases, this difference in interpretation led to copier instructions that were misleading to the users or just plain wrong in the context of their task.

Woods et al. also describe a system that has a different understanding of a task than some of its users. When the diagnostic system in their chapter interacted with nonexperts—people who presumably did not share many of the same assumptions as the system—the diagnostic process was slow and error-prone, when compared with an interaction with an expert. Woods et al.'s chapter makes two points relevant to shared knowledge. One is that a cooperative system works better when knowledge overlaps—a point made by the other contributors. The second is that when the participants in an interaction have no way to communicate about the interaction or problem-solving strategy, when they do not share metacommunicative strategies, their progress will be brittle. This makes Durfee et al.'s, Goodson and Schmidt's, and Malone's focus on communication networks and protocols a primary consideration in cooperative systems.

TEMPORAL EXTENT OF THE INTERACTION

The time period that a cooperative situation encompasses, or the complexity of the task that cooperating agents plan to accomplish, will have a significant impact on the role that the agents have to play. The situations described by Gibbs and Mueller consist of comprehension by one party of the utterance of a second party, and this occurs in a very brief period of time. Suchman's situation involves the length of time it takes for an individual to generate a copy of a document, a task requiring many more steps and decisions. Finally, the situations involving learning about a computer system or learning something in a classroom encompass much longer spans of time and even more complex decisions.

Time correlates with the complexity of the task when humans are involved. The complexity of the task is a factor in computational systems as well, although time may be a less significant consideration. The vehicle monitoring situations described by Durfee et al. and Goodson and Schmidt are extremely complex, although events may transpire quickly. Indeed, one can imagine the submarine tracking system of Goodson and Schmidt working the same way when tracking a missile, though in a shorter amount of time. In fact, the architectures and algorithms sought after by proponents of parallel systems are intended partly to reduce the time for complex computations.

So complexity and time constraints work together to place requirements on cooperative systems. One model of the comprehension of indirect requests, returning to Gibbs and Mueller's domain, involves the initial rejection of the literal meaning of an utterance (such as, "Do you have the time?") based on analysis of the speaker's state of knowledge and a "fall back" interpretation in terms of the indirect request (Clark & Lucy, 1975). This mechanism is fine if we assume a language comprehension system which (a) never adapts to novel uses of language by developing new comprehension rules, and (b) isn't going to encounter many such uses of language in a short period of time. From a design standpoint, by adopting the "conventional" approach to shared meaning in language, Gibbs and Mueller cut the computational resources required during comprehension once the convention is acquired. While it may make no difference in natural language, this strategy could have a significant consequence in situations involving complex tasks and requiring considerable cooperative problem solving.

Malone comments that "market" systems, with highly autonomous agents, are flexible in terms of their ability to reorganize structurally. He claims, however, that there is a tradeoff between flexibility and efficiency, with less flexible systems being extremely good at what they do. But Malone also points out that efficiency in this case is being thought of in the short term, in terms of accomplishing a specific goal. Long-term efficiency might involve pursuing many different, sometimes conflicting goals over a long period of time. As Woods et al. observe, the state of the problem-solving domain could easily change over long periods of time, requiring new plans for achieving the same goals. In this case, a flexible system is more desirable. Thus the extent of the cooperative enterprise over time and over situations is an important determinant of the cooperative strategy that will work best.

In general we would expect that the more complex the activity and the more severe the time constraints, the more complex the requirements for cooperative systems. When sequences of subgoals must be achieved by multiple agents, it will be necessary that each agent know about the activities of other agents in order to avoid redundancies and to schedule dependent tasks. Of course, if there is unlimited time for problem solving, relatively complex tasks could be achieved by simultaneous problem solving by noninteracting agents (for example, in situations where different algorithms are used and the results are polled for a decision). But the point is that a three-way tradeoff exists between time, task complexity, and the need for distributed, cooperative problem solving.

PREDICTABILITY AND MUTABILITY OF KNOWLEDGE STATES

Each chapter assumed different degrees of flexibility in the states of the cooperating agents and different amounts of variability in their problem-solving domains. In the vehicle monitoring situations discussed by Goodson and Schmidt and Durfee et al., the problem-solving domain was continually changing. In fact, characterization of the changes in the external world was the goal in these systems. Not coincidentally, these authors were some of the strongest proponents of highly interconnected, parallel-processing architectures in which each node had knowledge of communicative conventions and shared the capabilities of other nodes. The system discussed by Woods et al. also encountered many unexpected variations in the problem-solving environment, often because of user errors, but because it had no capability for communicating about its reasoning or its users' reasoning, it was difficult to use when unexpected variations arose. In all of these cases, it was advocated that allocation of responsibility be distributed or determined dynamically as resources changed through time.

Many of the chapters dealt with situations in which changes in the knowledge state of some or one of the participants was the major goal of a cooperative system (teaching and tutoring situations). Newman cleverly uses the mutability of the problem-solving domain and the teacher's reaction to it as a source of information about goals and ideas that students should acquire. All of the chapters on acquisition of computer concepts examined changes in users' cognitive representations with the development of expertise, and all suggested instructional environments appropriate to a certain phase or level. In all of the discussions of systems that were instructional or tutorial in nature, the locus of control for interaction was placed primarily with the expert, tutorial, or instructional system.

CAPABILITIES AND ROLES OF THE PARTICIPANTS

Goodson and Schmidt directly identified the capabilities of the components of cooperative systems as important considerations in the overall design of such systems. Specifically they noted that humans have working memory limitations, I/O channel capacity limitations, and variations in relevant knowledge. A cooperative system must be designed to accommodate and perhaps evaluate these capabilities. To different degrees, almost all of the other contributors addressed this issue as well.

Malone pointed out that in a market system, participants may differ in their abilities to provide products, services, or other commodities, or they may differ in their ability to participate in the communicative behavior of the organization. Other participants must take this into consideration when interacting with them and these concerns will affect the organizational structure that emerges. Similarly, in both Goodson and Schmidt's and Durfee et al.'s situations, resources at different nodes were utilized for different purposes, partially depending on their current capabilities. A node might be engaged in vehicle tracking and data transmission when a vehicle was in its region, for example, and therefore not be assigned other decision-making tasks. At a later time, when the same nodes are not engaged in vehicle tracking, their spare resources can be utilized.

Many of the errors observed in the learning situations described by Mack, Kay and Black, Graesser and Murray, and Sebrechts and Marsh were the result of failure of the computer systems or training materials to recognize their capabilities (especially the novices), specifically their state of knowledge.

The capability of a participant in a cooperative situation may change, either because it is sensitive to changes in the problem-solving environment or because it is inherently mutable. These capabilities must be taken into account during planning in a distributed system. Because changes can occur during problem solving, some mechanism must be available in highly variable systems solving complex problems for evaluating the relevant capabilities of participants at different times and reallocating responsibilities as necessary.

METHODOLOGICAL CONCERNS

Among the key problems that cognitive science faced as it emerged as a separate discipline were basic methodological uncertainties: What constituted legitimate data about cognitive processes? In what forms could such data be represented? How could they be collected? Were these data replicable, unbiased, etc.? These same problems re-emerge in the study of cognition from an interactive perspective. Almost all of the authors in the book have addressed some methodological issues, and many have had to create a method to meet their particular interest in studying cooperative systems.

Some of the authors were able to apply established techniques for data collection and analysis, particularly those concerned with conventional interactive tasks such as text editing. The design, internal knowledge, and capabilities of the computer in human–computer dyads are explicitly available, so most of the authors concerned with this class of cooperative system focused on identifying the knowledge and/or expectations of the human component. Kay and Black used a conventional technique (multidimensional scaling) to collect data on human knowledge about the computer. Both Mack and Sebrechts and March used verbal protocol analysis as a supplementary data source to assist in the analysis of essentially behavioral data. Mack recorded and analyzed human subject behavior in using text editors, and Sebrechts and Marsh used question–answering behavior to assess subject knowledge of specific computer systems. Graesser and Murray developed an extension to conventional verbal protocol analysis. Their method used the interactive situation as the basic unit of analysis, but rather than simply collecting thinking aloud data from the person, they intruded into the process with questions designed to focus the introspection process and elicit specific pieces of knowledge. Their method is similar to that developed separately by Clancey (1984, 1987) to support knowledge elicitation for intelligent tutoring systems.

The ease of analyzing interaction with conventional systems arises from the low degree of connectedness in the machine's actions. Although a person has a plan that integrates his or her actions over time, the conventional interactive program treats each human input as a separate event. When a computer action is related to a previous one, it is only because some previous action altered the data or problem conditions in some way (e.g., by changing a mode or a text editor). There is no *machine plan* that interacts with and modifies the *user plan*. When the computer system has more intelligence, this limitation goes away. In analyzing interactions between people and problem-solving machines, Woods et al. and Suchman had to develop new ways of recording and conceptualizing data about the interaction. Suchman developed a task analysis notation that included both the observed actions of the human–computer system and the availability of those actions to each of the components. She applied this notation to full videotapes of the person–machine interactions. Woods et al. worked from detailed recorded observations, as well as verbal protocol data, but developed a flow-chart-like notation for analyzing the sequence of information processing as it flowed between the two agents. Their charts show the abstracted steps in the process, expressed as information-processing states (e.g., choice, observation, entry, etc.), and indicate which agent is performing each step. In this way, the cognitive history of a dyadic problem-solving incident can be charted and traced. Interestingly, both Woods et al. and Suchman were concerned with notations that supported a *failure analysis* of the situations in which the human–computer team failed to achieve the situational goal.

In addition to these various methodological tools for the empirical observation and analysis of cooperation, other authors developed methods for the design of cooperative systems. Goodson and Schmidt focused their chapter on the presentation of a complete methodology for the design of *cooperative person–machine systems.* The steps in their design method can be summarized as:

1. formulate and decompose the problem-solving task
2. analyze human performance in each subtask,
3. allocate specific subtasks to man and machine, based on human performance limits,
4. design and implement a problem-solving process for the machine,
5. define and structure a relationship between the human and machine,
6. establish communicative norms based on the defined relationship, and
7. evaluate the system performance and repeat all steps if necessary.

They defined some tools for each of these steps, but focused primarily on the first, fourth, and fifth steps. Their idea of mapping out the temporal and informational dependencies among subtasks represents a well defined and practical method for problem decomposition. Although they cited specific examples of performance analysis, function allocation, and machine problem-solver design, their discussion is less general in these areas. (However, others, e.g., Zachary 1986, 1988, do deal with these issues in a more general manner.) Goodson and Schmidt also provided a novel and important methodological contribution in their rule-based scheme for defining and enforcing communicative norms based on desired person–machine relationships in support of the fourth and fifth steps in their process.

Durfee et al. did not so much define a design methodology as demonstrate one. Like Goodson and Schmidt, their method was both empirical and evolutionary. It began with a minimal system design and proceeded iteratively, at each cycle enhancing the current design and testing it against predefined system-level performance criteria, using the results to drive the next iteration. Unlike Goodson and Schmidt, Durfee et al. used a prescribed approach to task decomposition (Functionally Accurate and Cooperative, FAC). Still, the implicit design method used by Durfee et al. fits quite nicely into the framework defined by Goodson and Schmidt. Perhaps the most direct methodological contribution of Durfee et al.'s chapter is their definition of two related criteria by which a cooperative system may be designed and evaluated—coherence, a communication based measure, and efficiency, a time–computation based measure. These are criteria that will likely have relevance to a broad range of cooperative system designs.

Woods et al. dealt with methodological issues in a very general way, by providing principles for designing human–computer roles and relationships. Although dealing with conventional rather than parallel-distributed computers, Mack also provided design guidelines for developing "learnable" human–computer interfaces.

A COGNITIVE, COMPUTATIONAL FRAMEWORK FOR COOPERATION

So have we been able to develop a theory of cooperation from the varied chapters in this book? Probably not. But we can identify at least some elements of a definition of cooperation as these elements have emerged from the issues discussed above. This definition can form a framework for further discussion of cognition and computation in cooperative systems.

Cooperation, as a behavioral phenomenon, occurs in a specific context between two or more intelligent agents. It unfolds over time and through a process of communication. The number of agents involved can vary widely, from two to a large number. At or near the end points of this continuum, cooperation takes on different characteristics. Near the dyadic end, the context of the interaction, or situation, emerges as a primary factor in constraining the process. One way of operationalizing this notion of situation is to relate it to the two levels on which cooperation can be analyzed—the system level and the agent level. Viewed from the agent level, the local environment for cognition can be termed the *immediate context.* This immediate context provides the inputs and stimuli for the agent's cognitive processes. There is also a system level, however, and when the agent either shares in or is aware of the system-level goals, there is also a system-level context for the agent's cognitive process. This context includes the other agents and their involvement in the distributed process that the system represents, and the overall position of the system relative to the system-level goals. This system-level context, which subsumes the immediate context, can be termed the *local situation.* When cooperating in dyads or very small groups, the human agents individualize their communication and intentions in terms of their mutual relationship to the situation, and also expect the other agent(s) to do so. Empirically, this seems to give human dyadic interaction a particularly effective and cooperative quality. When computer agents are able to exhibit this capability, their effectiveness also improves.

Even in a dyad, the agents have a relationship with one another. This relationship consists (at a minimum) of a set of expectations, understandings or beliefs, and instrumental constructs (i.e., perceived or inferred capabilities). As the number of agents increases and cooperation moves toward the opposite end of the continuum (which Hewitt, 1985, has characterized as "open systems"), relationship emerges as the primary construct in constraining the cooperative process. In a dyad, there are only three relationships—each agent with the situation and the agents with each other. We note that the number of relationships among agents increases combinatorially (as the number of combinations of agents taken two at a time), while the number of relationships to the situation increases only linearly.

At some point, the sheer difference in numbers between the situational and interagent relationships qualitatively changes the behavior of cooperative systems. In large scale open systems, it is the set of relationships among the agents and not the situation that organizes and structures the interaction, and limits the kinds of behavior and cooperation that can occur.

Cooperation also involves some implicit or explicit means of coordinating the communications and instrumental actions involved. The means of coordination tend to be implicit in dyads or among very small sets of agents where they can be worked out *ad hoc* as the situation demands, or entrusted to well-established and deeply shared norms or standards. As more agents get involved, the need emerges for a more explicit form of coordination. In larger networks of agents, such as human organizations or local computer networks where the cooperation is based on explicit and well-defined goals, coordination mechanisms of convention or protocol dominate. Cases that are not incorporated into these conventions are handled by explicit negotiation. In very large and diffuse cooperative systems, such as speech communities or marketplaces, coordination becomes more implicit again. Individual relationships and individual or local interest emerge as the main mechanisms for coordinating activity in these very large-scale cooperating systems. These mechanisms tend to be more explicit as the group gets larger.

As a final point, cooperation is a recursive or at least a layered phenomenon. Specific cooperative contexts may themselves be only part of the mechanism for larger-scale forms of cooperation. For example, conversation is a cooperative act that may occur as part of and as the basis for classroom learning. Classroom learning, in turn, may be one component of a larger-scale process of education at a particular university. The university is a cooperative organization in which participants cooperate for differing reasons and to different degrees. It is itself part of a larger market system involving regional, national, or international economies on one hand, and scientific communities of ideas on the other. Higher extrapolations are also possible. Thus, while any single action may be analytically related to a specific cooperative context, it is likely to have many levels of meaning, particularly when the agent involved in a person.

This description of cooperation gives rise to some specific cognitive and computational characteristics of the agents involved. A minimalist model of cognition was discussed in the afterword to Section I. This minimalist model is adequate to explain (or design) noninteractive problem-solving processes, but needs some logical extensions to handle the aspects of cooperation defined above. The requirements suggested below are not offered as the only ones that need to be added to the minimalist cognitive model, but rather as the ones that emerged from the chapters here.

The first additional cognitive requirement for cooperation is a facility for plan recognition, i.e., an ability to infer the goals and plans of another agent based on communications or actions of that agent. The plan recognition facility has three separate components. The first component is the *keyhole* plan recognition ability—the simple capability to infer a plan based on observed actions of another agent. This capability does not assume any cooperative intent on the part of the agent being observed. The other two components do. The second component of a complete plan recognition capability is an implicative capability, which refers to an agent's ability to select specific actions or communications by which it intends another agent to infer its plan or goals. The third is the complement of this, an intended plan recognition ability that makes the intended recognitions of plans and goals in a sufficiently consistent way such that the ability can be assumed by the communicating agents. All three of these aspects of plan recognition come into play somewhere in the range of cooperative behavior.

A second set of cognitive abilities concerns representation of and reasoning about the immediate situation in which the cooperation is occurring. This is the capability by which situational intrusions on goals and plans are observed, understood, and pragmatically resolved through a plan adjustment process. People are able to perform this pragmatic plan adjustment extremely easily and often unconsciously; the ability dramatically contributes to the smoothness of human interaction, even in the face of unforeseen events. A related capability is a facility for internally representing social relationships (including expectations about other agents), and for reasoning about these relationships. This reasoning is involved in developing communicative plans, which specify what is to be communicated to whom and under what conditions. It is also involved in the process of adjustment of this communicative plan to situational factors.

A third cognitive requirement for cooperation is an ability to appropriate the actions of other agents purposively, and a companion ability to respond to action appropriation by learning. These abilities require an agent to recognize an opportunity in which appropriation can lead to learning, and a facility to communicate about the appropriated act. Similarly, appropriation requires an ability to examine and modify a knowledge structure in order to bring it into conformance with an appropriated action.

Several computational extensions to the minimalist model of a problem-solving device (as defined in the Afterword to Section II) are also required by the definition of cooperation given above. These include both representational requirements and processing requirements. The first set of requirements concerns the idea of the local situation. If cooperative behavior is based on situational factors, then a cooperative agent must compute and maintain some representation of the local situation. As suggested above, the situation involves the changing status of the complete problem-solving system (or at least as much of it as is available to a specific agent within it). The form of this situation representation is unclear. Durfee et al. have initially characterized it as including the hypotheses, actions, and/or plans recognized in or communicated by the other agents, and the relationship of the pattern of these hypotheses, actions, and plans to the system-wide goal. The agent is also required to be able to reason about this situation representation, and apply it to the local planning and plan-adjustment process.

A second set of computational requirements concern the different kinds of goals that can be encountered in a cooperative system. Shared goals are system-level goals, adopted by all component agents (typical in FAC systems). Common goals are agent-level goals that are local to an agent but that also may be common to many or all agents (typical in market place systems). Strictly local goals are those goals which may be individually held. They may be complementary between agents (typical in complementary dyadic relationships such as teacher–student).

Additional computational requirements arise from the other cognitive requirements for cooperation listed above. Plan recognition is known to be a computable process, and it is one that appears necessary for a fully cooperative agent. Representations of other agents, including expectations of their behavior and norms for interacting with them, are also required. It should be noted that relationships are not necessarily agent-unique, a fact that can be obscured by the dyadic context in which relationship was usually discussed in this book. The most useful relationships are in fact categorical (e.g., teacher, superior, helper, student, colleague, enemy), because they allow a communicative plan to be computed for an agent that has not been encountered previously. Because of this categorical relationship, the ability to instantiate relationship knowledge in specific agents is also required for a fully cooperative computational system.

Some final requirements arise when we are concerned with human–computer cooperation. People have a very powerful communication protocol through which cooperation can occur—natural language. Machine agents which must cooperate with people require a communication protocol that can support the range of functions performed by human language. These include instrumental use (e.g., as in speech acts), adaptation to the conventions of relationships and situations, and other functions as discussed by Gibbs and Mueller. This is not to say that a cooperative machine must possess natural language, only that it must possess a communication protocol that can be used to fulfill the same functions as human language. This protocol might be entirely nonlinguistic, e.g., a graphical language, as long as it has the required functionality.

As with the cognitive requirements, these computational requirements are not suggested as a final list of what is involved in cooperation, but rather as a first cut, based on the issues raised by the authors who contributed to this volume.

A PRELIMINARY RESEARCH AGENDA

In the introduction to this book, we hoped that a theory of cooperation would emerge from the interdisciplinary combination of studies contained here. This hope has been partially realized. A number of key issues have emerged, with substantial agreement on most of them. Still, some important differences remain, along with some open questions. We see these as elements on an agenda for future research on cognition, computation, and cooperation. This agenda, though, is admittedly filtered through our own biases and preferences, and is not meant as exhaustive in any sense.

The first item on this agenda is the difference between cooperative systems of different scales. Most of the authors here have focused on dyadic interaction, with substantial agreement on the problems and requirements inherent in it. There has also been substantial agreement on the smaller number of chapters that concerned multiagent systems. There remains some disconnection between these two levels of analysis, however. The emphasis in dyadic cooperation was placed on situational variables and pragmatics-level information processing, while in open systems the emphasis was placed on organizing relationships and effective problem decomposition. Additional research, possibly with machine-based systems and/or with given problems and teams of varying size, is needed to better understand the points and mechanisms by which the behavior of dyads shades into the behavior of open systems.

There is a general need to examine human interaction more intensively from a cognitive, computational perspective to build a more substantial base of data for analysis and modeling. The majority of cognitive analyses of person-to-person interaction have focused on speech and conversation, as represented here by the work of Gibbs and Mueller. While this work offers valuable insights, it seems strongly limited to the examination of dyadic or very small group situations. Some earlier work on the role of communication in team (i.e., multiperson) cooperation (Chapanis et al., 1972, 1977) demonstrated that pragmatic concerns overwhelmed other levels of analysis, even to the point that major elements of syntactic structure were ignored in communication. Other studies by the ethnomethodologists, cited by Suchman, have pointed out similar phenomena, but neither the work of Chapanis et al. nor of the ethnomethodologists is explicitly cognitive or computational. The authors in this book have developed some methodological tools for recording and analyzing behavioral data on cooperative interaction that could be used for this purpose, but at present, the base of cognitive data on human cooperation outside of everyday conversation is small. We hope future researchers expand it so that further elaborations, replications, and tests of the theories and models proposed here can be made.

There are several issues in computational research that arise from the concerns expressed in this book. These fall on a continuum from strictly machine-related issues of interest in artificial intelligence to human–computer issues of interest to human factors and computer–human interface design. Much of the work in artificial intelligence has built on an analogy to human cognitive psychology. Human cognitive mechanisms and processes provide a powerful model for the design of computer-based intelligence. But just as most of cognitive psychology has focused on human problem solving as an isolated (i.e., noninteractive) process, most of artificial intelligence has focused on building computational devices that solve problems by themselves (not interactively). As Suchman and Woods et al. pointed out, though, this is not a practical approach because the machine lacks the sensory abilities to acquire problem data without the intervention of a human being. This makes AI systems inherently interactive. Most of the computational studies in this book have shown that conventional problem-solving architectures lack a "something" that makes human problem solving naturally interactive. Durfee et al. systematically demonstrated some of the additional capabilities needed to make one class of computational architectures more cooperative with other agents of identical design, but this only scratches the surface. It would be useful to know what capabilities would allow heterogeneous problem-solving devices (i.e., with different fundamental architectures, such as production rule architectures and blackboard architectures) to cooperate effectively. A natural extension of this paradigm would be to include humans as agents with an even more different architecture, and identify what communicative and computational mechanisms are needed to support more effective human–computer interaction in heterogeneous AI systems.

FINAL COMMENTS

We hope that cognitive science is a cooperative enterprise. It is certainly the case that the issues addressed by cognitive scientists and the complexity of the project of cognitive science are significant enough to warrant a distributed problem-solving approach. The understanding of cooperative systems will require many passes like the one taken here, and an interdisciplinary approach seems not only appropriate, but essential, in pursuit of this problem. Since we are now engaged in designing cooperative computing systems, we are sure to see considerable progress on theories of interaction and cooperation in the future. If anything is to be learned from this particular study that will inform future studies, it is that shared goals and flexible communication strategies are essential for distributed problem solving.

REFERENCES

Chapanis, A., Ochsman, R., Parrish, R., & Weeks, G. (1972). Studies in interactive communication I: The effects of four communication modes on the behavior of teams during cooperative problem-solving. *Human Factors, 14*(6), 487–509.

Chapanis, A., Parrish, R., Ochsman, R., & Weeks, G. (1977). Studies in interactive communication II: The effects of four communication modes on the linguistic performance of teams during cooperative problem-solving. *Human Factors, 19*(2), 101–126.

Clancey, W. (1984). *Acquiring, representing, and evaluating a competence model of diagnosis* (HPP Memo 84-2). Stanford, CA: Stanford University.

Clancey, W. (1987). *Knowledge-based tutoring: The GUIDON Program.* Cambridge, MA: MIT Press.

Clark, H.H., & Lucy, P. (1975). Understanding what is meant from what is said: A study in conversationally conveyed requests. *Journal of Verbal Learning & Verbal Behavior, 14,* 56–72.

Clark, H.H. (1979). Responding to indirect speech acts. *Cognitive Psychology, 11,* 430–477.

Hewitt, C. (1985). The challenge of open systems. *Byte.*

Zachary, W. (1986). A cognitively-based functional taxonomy of decision support techniques. *Human Computer Interaction, 2*(1), 25–63.

Zachary, W. (1988). Decision support systems: Designing to extend the cognitive limits on user decision processes. In M. Hellander (Ed.), *Handbook of human–computer interaction.* New York: North Holland.

Author Index

Subject Index